T0211939

Lecture Notes in Computer Science   13706

Founding Editors

Gerhard Goos
Karlsruhe Institute of Technology, Karlsruhe, Germany

Juris Hartmanis
Cornell University, Ithaca, NY, USA

Editorial Board Members

Elisa Bertino
Purdue University, West Lafayette, IN, USA

Wen Gao
Peking University, Beijing, China

Bernhard Steffen
TU Dortmund University, Dortmund, Germany

Moti Yung
Columbia University, New York, NY, USA

More information about this series at https://link.springer.com/bookseries/558

Dimitrios Poulakis · George Rahonis (Eds.)

Algebraic Informatics

9th International Conference, CAI 2022
Virtual Event, October 27–29, 2022
Proceedings

Springer

Editors
Dimitrios Poulakis 🆔
Aristotle University of Thessaloniki
Thessaloniki, Greece

George Rahonis 🆔
Aristotle University of Thessaloniki
Thessaloniki, Greece

ISSN 0302-9743 ISSN 1611-3349 (electronic)
Lecture Notes in Computer Science
ISBN 978-3-031-19684-3 ISBN 978-3-031-19685-0 (eBook)
https://doi.org/10.1007/978-3-031-19685-0

© The Editor(s) (if applicable) and The Author(s), under exclusive license
to Springer Nature Switzerland AG 2022
This work is subject to copyright. All rights are reserved by the Publisher, whether the whole or part of the material is concerned, specifically the rights of translation, reprinting, reuse of illustrations, recitation, broadcasting, reproduction on microfilms or in any other physical way, and transmission or information storage and retrieval, electronic adaptation, computer software, or by similar or dissimilar methodology now known or hereafter developed.
The use of general descriptive names, registered names, trademarks, service marks, etc. in this publication does not imply, even in the absence of a specific statement, that such names are exempt from the relevant protective laws and regulations and therefore free for general use.
The publisher, the authors, and the editors are safe to assume that the advice and information in this book are believed to be true and accurate at the date of publication. Neither the publisher nor the authors or the editors give a warranty, expressed or implied, with respect to the material contained herein or for any errors or omissions that may have been made. The publisher remains neutral with regard to jurisdictional claims in published maps and institutional affiliations.

This Springer imprint is published by the registered company Springer Nature Switzerland AG
The registered company address is: Gewerbestrasse 11, 6330 Cham, Switzerland

Preface

This volume contains the papers presented at the 9th International Conference on Algebraic Informatics (CAI 2022). The conference was organized under the auspices of the Department of Mathematics of the Aristotle University of Thessaloniki, Greece, and took place online during October 27–29, 2022.

CAI is a biennial conference devoted to the intersection of theoretical computer science, algebra, and related areas. It was initiated in 2005, by Symeon Bozapalidis, in Thessaloniki, Greece. CAI 2007 and 2009 were hosted again in Thessaloniki, CAI 2011 in Linz, Austria, CAI 2013 in Porquerolles, France, CAI 2015 in Stuttgart, Germany, CAI 2017 in Kalamata, Greece, and CAI 2019 in Niš, Serbia. The organization of the conference in 2021 was shifted to 2022 due to the COVID-19 pandemic.

The contents of the volume consist of two abstracts, three full papers of invited speakers, and 12 contributed papers. The invited lectures were given by Manfred Droste, Yuri Matiyasevich, Panos Pardalos, Jean-Éric Pin, Robert Rolland, Paul Spirakis, and Moshe Vardi. In total 17 papers were submitted to CAI 2022 and 12 of them were carefully selected through a single-blind peer-review process. All papers (but one) were reviewed by at least three reviewers (respectively two) and on average 3.1. The papers contain original and unpublished research; the topics of them lie in automata theory, cryptography, coding theory, DNA computation, computer algebra, and theory of software architectures.

We are deeply grateful to the many people who supported CAI 2022 and helped to organize a successful event. First of all, we would like to thank the members of the Steering Committee who accepted our proposal for the organization of CAI 2022. We thank the colleagues in the Program Committee and further reviewers for their cooperation and excellent work in the review process. We thank Panagiotis Tzounakis for technical support with the website and the online system for conference's implementation.

The submission and the review process of the papers was done using the reliable EasyChair platform which made our job easier.

Last but not least, we are deeply grateful to Springer LNCS. Especially, to Ronan Nugent, Anna Kramer and Christine Reiss who helped us, to publish the proceedings of CAI 2022 in LNCS.

The financial support of the Special Account for Research Funds of the Aristotle University of Thessaloniki is acknowledged.

September 2022   Dimitrios Poulakis
                                                                   George Rahonis

Organization

CAI 2022 was organized under the auspices of the Department of Mathematics of the Aristotle University of Thessaloniki, Greece.

Program Committee Chairs

Dimitrios Poulakis	Aristotle University of Thessaloniki, Greece
George Rahonis	Aristotle University of Thessaloniki, Greece

Steering Committee

Symeon Bozapalidis	Aristotle University of Thessaloniki, Greece
Olivier Carton	Université de Paris, France
Manfred Droste	University of Leipzig, Germany
Werner Kuich	TU Wien, Austria
Dimitrios Poulakis	Aristotle University of Thessaloniki, Greece
Arto Salomaa	University of Turku, Finland

Program Committee

Ioannis Antoniou	Aristotle University of Thessaloniki, Greece
Yves Aubry	Université de Toulon and Aix-Marseille Université, France
Simon Bliudze	Inria Lille, France
Miroslav Ćirić	University of Niš, Serbia
Konstantinos Draziotis	Aristotle University of Thessaloniki, Greece
Frank Drewes	Umeå University, Sweden
Henning Fernau	Universität Trier, Germany
Zoltán Fülöp	University of Szeged, Hungary
István Gaál	University of Debrecen, Hungary
Dora Giammaresi	Università degli Studi di Roma "Tor Vergata", Italy
Mika Hirvensalo	University of Turku, Finland
Jarkko Kari	University of Turku, Finland
Lila Kari	University of Waterloo, Canada
Panagiotis Katsaros	Aristotle University of Thessaloniki, Greece
Stavros Konstantinidis	Saint Mary's University, Canada
Temur Kutsia	Johannes Kepler Universität Linz, Austria

Andreas Maletti	Universität Leipzig, Germany
Anastasia Mavridou	NASA Ames Research Center, USA
Sihem Mesnager	University of Paris VIII and Telecom Paris, France
Benjamin Monmege	Aix-Marseille Université, France
Abderahmane Nitaj	University of Caen Normandy, France
Kai Salomaa	Queen's University, Canada
Wolfgang Schreiner	Johannes Kepler Universität Linz, Austria
Pascal Véron	Université de Toulon, France
Heiko Vogler	TU Dresden, Germany
Mikhail Volkov	Ural Federal University, Russia
Michael Vrahatis	University of Patras, Greece.

Additional Reviewers

Eduard Baranov
Markus Bläser
Alexis Bonnecaze
Gustav Grabolle
Kalpana Mahalingam
Mircea Martin
Gerasimos Meletiou
Andy Oertel

Charles Olivier-Anclin
Kostantinos Parsopoulos
Eric Paul
Robert Rolland
Yannis Stamatiou
Charlotte Vermeylen
Martin Winter

Organizing Committee

Dimitrios Poulakis
George Rahonis
Panagiotis Tzounakis

Sponsoring Institution

Special Account for Research Funds of the Aristotle University of Thessaloniki

Abstracts of Invited Talks

Weighted Automata Over Monotonic Strong Bimonoids: Decidability and Undecidability of Finite Image

Manfred Droste

Institute of Computer Science, Leipzig University, Leipzig, Germany
droste@informatik.uni-leipzig.de

A weighted finite automaton $\mathcal{A}$ has finite image if the image of the weighted language associated with it is finite. First, we give a structural result characterizing when $\mathcal{A}$ has finite image. Then we characterize those past-finite monotonic strong bimonoids such that for each weighted finite automaton $\mathcal{A}$ it is decidable whether $\mathcal{A}$ has finite image. In particular, this is decidable over past-finite monotonic semirings.

Next, we give two undecidability results on the finite-image property of weighted finite automata over semirings, respectively strong bimonoids. We construct a computable idempotent commutative past-finite ordered semiring such that it is undecidable, for an arbitrary deterministic weighted finite automaton $\mathcal{A}$ over that semiring, whether $\mathcal{A}$ has finite image. Finally, we construct a computable commutative past-finite monotonic ordered strong bimonoid such that it is undecidable, for an arbitrary weighted finite automaton $\mathcal{A}$ over that strong bimonoid, whether $\mathcal{A}$ has finite image. This shows that the decidability results mentioned before cannot be extended to natural classes of ordered semirings and ordered strong bimonoids without further assumptions.

References

1. Droste, M., Fülöp, Z., Kószó, D., Vogler, H.: Finite-image property of weighted tree automata over past-finite monotonic strong bimonoids. Theor. Comput. Sci. **919**, 118–143 (2022). https://doi.org/10.1016/j.tcs.2022.03.036
2. Droste, M., Fülöp, Z., Kószó, D., Vogler, H.: Decidability boundaries for the finite-image property of weighted finite automata. Int. J. Found. Comput. Sci., to appear

Constraints, Graphs, Algebra, Logic, and Complexity

Moshe Y. Vardi

Department of Computer Science, Houston, Rice University,
TX, 77251–1892, USA
vardi@cs.rice.edu, http://www.cs.rice.edu/~vardi

Abstract. A large class of problems in AI and other areas of computer science can be viewed as constraint-satisfaction problems. This includes problems in database query optimization, machine vision, belief maintenance, scheduling, temporal reasoning, type reconstruction, graph theory, and satisfiability. All of these problems can be recast as questions regarding the existence of homomorphisms between two directed graphs. It is well-known that the constraint-satisfaction problem is NP-complete. This motivated an extensive research program into identify tractable cases of constraint satisfaction.

This research proceeds along two major lines. The first line of research focuses on non-uniform constraint satisfaction, where the target graph is fixed. The goal is to identify those traget graphs that give rise to a tractable constraint-satisfaction problem. The second line of research focuses on identifying large classes of source graphs for which constraint-satisfaction is tractable. We show in this talk how tools from graph theory, universal algebra, logic, and complexity theory, shed light on the tractability of constraint satisfaction.

References

1. Bulatov, A.A.: A dichotomy theorem for nonuniform CSPs. In: Proceedings of the 58th IEEE *Symposium* on Foundations of Computer Science, pp. 319–330. IEEE Computer Society (2017)
2. Feder, T.A., Vardi, M.Y.: The computational structure of monotone monadic SNP and constraint satisfaction: a study through Data log and group theory. SIAM J. Comput. 28, 57–104 (1998)
3. Kolaitis, P.G., Vardi, M.Y.: A logical approach to constraint satisfaction. In: Creignou, N., Kolaitis, P.G., Vollmer, H. (eds.) Complexity of Constraints. Lecture Notes in Computer Science, vol. 5250, pp. 125–155. Springer, Berlin, Heidelberg (2008). https://doi.org/10.1007/978-3-540-92800-3_6
4. Zhuk, D.: A proof of the CSP dichotomy conjecture. J. ACM, **67**(5), 30:1–30:78 (2020)

Contents

Invited Papers

Chaining Multiplications in Finite Fields with Chudnovsky-Type Algorithms and Tensor Rank of the k-Multiplication

Stéphane Ballet[1,2] and Robert Rolland[1,2(✉)]

[1] Aix Marseille Univ, CNRS, Centrale Marseille, I2M, Marseille, France
robert.rolland@acrypta.fr
[2] Case 907, 13288 Marseille Cedex 9, France

Abstract. We design a class of Chudnovsky-type algorithms multiplying k elements of a finite extension $\mathbb{F}_{q^n}$ of a finite field $\mathbb{F}_q$, where $k \geq 2$. We prove that these algorithms give a tensor decomposition of the k-multiplication for which the rank is in $O(n)$ uniformly in q. We give uniform upper bounds of the rank of k-multiplication in finite fields. They use interpolation on algebraic curves which transforms the problem in computing the Hadamard product of k vectors with components in $\mathbb{F}_q$. This generalization of the widely studied case of $k = 2$ is based on a modification of the Riemann-Roch spaces involved and the use of towers of function fields having a lot of places of high degree.

Keywords: Finite field · Tensor rank of the multiplication · Function field

1 Introduction

1.1 Context, Notation and Basic Results

Finite fields constitute an important area of mathematics. They arise in many applications, particularly in areas related to information theory. In particular, the complexity of multiplication in finite fields is a central problem. It is part of algebraic complexity theory. This paper is devoted to the study of the specific problem of the tensor rank of the k-multiplication in finite fields, i.e. the multiplication of k elements in a given extension of a finite field. Up to now, most of the work in this area has been devoted to the $k = 2$ case (cf. [4]), namely the so-called bilinear case. In this article, we are interested in the general case. More precisely, let q be a prime power, $\mathbb{F}_q$ the finite field with q elements and $\mathbb{F}_{q^n}$ the degree n extension of $\mathbb{F}_q$. Then, if k is an integer ≥ 2, the multiplication m_k of k elements in the finite field $\mathbb{F}_{q^n}$ is a k-multilinear map from $(\mathbb{F}_{q^n})^k$ into $\mathbb{F}_{q^n}$ over the field $\mathbb{F}_q$, thus it corresponds to a linear map M_k from the tensor power $(\mathbb{F}_{q^n})^{\otimes k}$ into $\mathbb{F}_{q^n}$. One can also represent M_k by a k-covariant and

© The Author(s), under exclusive license to Springer Nature Switzerland AG 2022
D. Poulakis and G. Rahonis (Eds.): CAI 2022, LNCS 13706, pp. 3–14, 2022.
https://doi.org/10.1007/978-3-031-19685-0_1

1-contravariant tensor $t_{M_k} \in (\mathbb{F}_{q^n}^*)^{\otimes k} \otimes \mathbb{F}_{q^n}$ where $\mathbb{F}_{q^n}^*$ denotes the algebraic dual of $\mathbb{F}_{q^n}$. Each decomposition

$$t_{M_k} = \sum_{i=1}^{s} \left(\otimes_{j=1}^{k} a_{i,j}^* \right) \otimes c_i \tag{1}$$

of the tensor t_{M_k}, where $a_{i,j}^* \in \mathbb{F}_{q^n}^*$ and $c_i \in \mathbb{F}_{q^n}$, brings forth a multiplication algorithm of k elements

$$\prod_{j=1}^{k} x_j = t_{M_k}(\otimes_{i=1}^{k} x_j) = \sum_{i=1}^{s} \left(\prod_{j=1}^{k} a_{i,j}^*(x_j) \right) c_i. \tag{2}$$

Definition 1. *A k-multilinear multiplication algorithm $\mathcal{U}_{q,n,k}$ in $\mathbb{F}_{q^n}$ is an expression*

$$\prod_{j=1}^{k} x_j = \sum_{i=1}^{s} \left(\prod_{j=1}^{k} a_{i,j}^*(x_j) \right) c_i.$$

where $a_{i,j}^* \in (\mathbb{F}_{q^n})^*$, and $c_i \in \mathbb{F}_{q^n}$.

The number s of summands in this expression is called the k-multilinear complexity of the algorithm $\mathcal{U}_{q,n,k}$ and is denoted by $\mu_M(\mathcal{U}_{q,n,k})$. If $k = 2$, it is called bilinear complexity of the algorithm.

Definition 2. *The minimal number of summands in a decomposition of the tensor T_{M_k} of the k-multilinear multiplication in $\mathbb{F}_{q^n}$ is called the k-multilinear complexity of the multiplication in $\mathbb{F}_{q^n}$ and is denoted by $\mu_{q,k}(n)$:*

$$\mu_{q,k}(n) = \min_{\mathcal{U}_{q,n,k}} \mu_M(\mathcal{U}_{q,n,k})$$

where $\mathcal{U}_{q,n,k}$ is running over all k-multilinear multiplication algorithms in $\mathbb{F}_{q^n}$ over $\mathbb{F}_q$. The complexity $\mu_{q,2}(n)$ will be denoted by $\mu_q(n)$, in accordance with the usual notation in the case of the product of two elements. It is called the bilinear complexity of the multiplication in $\mathbb{F}_{q^n}$.

It will be interesting to relate the k-multilinear complexity of the multiplication to the minimal number $\nu_{q,k}(n)$ of bilinear multiplications in $\mathbb{F}_q$ required to compute the product of k elements in the extension $\mathbb{F}_{q^n}$.

Lemma 1.

$$\nu_{q,k}(1) \le k - 1, \tag{3}$$

$$\nu_{q,k}(n) \le (k-1) \times \mu_q(n), \tag{4}$$

$$\mu_{q,k}(n) \le \mu_{q,k}(mn) \le \mu_{q,k}(m) \times \mu_{q^m,k}(n), \tag{5}$$

$$\nu_{q,k}(n) \le \nu_{q,k}(mn) \le \nu_{q,k}(m) \times \nu_{q^m,k}(n). \tag{6}$$

Proof. The two first inequalities are direct consequences of the definitions. The inequalities (5) and (6) follow from the embedding of the field $\mathbb{F}_{q^n}$ in the field $\mathbb{F}_{q^{mn}}$. □

When for any i the k linear forms $a_{i,j}^*$ are all the same linear form a_i^*, namely

$$t_{M_k} = \sum_{i=1}^{s} (a_i^*)^{\otimes k} \otimes c_i \qquad (7)$$

and

$$\prod_{j=1}^{k} x_j = t_{M_k}(\otimes_{i=1}^{k} x_j) = \sum_{i=1}^{s} \left(\prod_{j=1}^{k} a_i^*(x_j) \right) c_i,$$

the decomposition is called a symmetric decomposition of t_{M_k}. As a consequence of [7, Theorem 5], when $n > 1$ such a symmetric decomposition exists if and only if $k \leq q$. The symmetric k-multilinear complexity of the multiplication of k elements in $\mathbb{F}_{q^n}$ over $\mathbb{F}_q$, denoted by $\mu_{q,k}^{Sym}(n)$, is the minimum number of summands s in the decomposition (7). For $k = 2$ this complexity is denoted by $\mu_q^{Sym}(n)$. From the definitions we get $\mu_{q,k}(n) \leq \mu_{q,k}^{Sym}(n)$.

1.2 New Results and Organisation

From a generalization of Chudnovsky-type algorithms to the k-multiplication, obtained by Randriambololona and Rousseau in [10] (cf. also [11]), that we generalize to places of arbitrary degree, we obtain uniform upper bounds for the rank of the k-multiplication tensor in the finite fields (Theorem 3 in Sect. 4). In this aim, we apply this type of algorithms to an explicit tower of Garcia-Stichtenoth [9] and the corresponding descent tower described in Sect. 3.2. Note that Randriambololona and Rousseau only obtain an asymptotic upper bound in $O(n)$ by using Shimura curves used by Shparlinski, Tsfasman and Vladut in [12].

2 Theoretical Construction of a Multiplying Algorithm

2.1 Notations

Let $F/\mathbb{F}_q$ be an algebraic function field over the finite field $\mathbb{F}_q$ of genus g. We denote by $N_i(F/\mathbb{F}_q)$ the number of places of degree i of F over $\mathbb{F}_q$. If D is a divisor, $\mathcal{L}(D)$ denotes the Riemann-Roch space associated to D. Let Q be a place of $F/\mathbb{F}_q$. We denote by $\mathcal{O}_Q$ the valuation ring of the place Q and by F_Q the residue class field $\mathcal{O}_Q/Q$ of the place Q which is isomorphic to $\mathbb{F}_{q^{\deg(Q)}}$ where $\deg(Q)$ is the degree of the place Q. Let us recall that for any $g \in \mathcal{O}_Q$, $g(Q)$ denotes the class of g in $\mathcal{O}_Q/Q = F_Q$. Let us define the following Hadamard product in $\mathbb{F}_q^{N_1} \times \mathbb{F}_{q^2}^{N_2} \times \cdots \times \mathbb{F}_{q^d}^{N_d}$ where the N_i denote integers ≥ 0:

$$\bigodot_{i=1}^{k} (u^i_{1,1}, \cdots, u^i_{1,N_1}, \cdots, u^i_{d,1}, \cdots, u^i_{d,N_d}) =$$

$$\left(\prod_{i=1}^{k} u^i_{1,1}, \cdots, \prod_{i=1}^{k} u^i_{1,N_1}, \cdots, \prod_{i=1}^{k} u^i_{d,1}, \cdots, \prod_{i=1}^{k} u^i_{d,N_d} \right)$$

2.2 Algorithm

The following theorem generalizes the known results of the case $k = 2$.

Theorem 1 (Algorithm). *Let*

1. *q be a prime power and $k \geq 2$ be an integer,*
2. *$F/\mathbb{F}_q$ be an algebraic function field,*
3. *Q be a degree n place of $F/\mathbb{F}_q$,*
4. *$\mathcal{D}$ be a divisor of $F/\mathbb{F}_q$,*
5. *$\mathcal{P} = \{P_{1,1}, \cdots, P_{1,N_1}, \cdots, P_{d,1}, \cdots, P_{d,N_d}\}$ be a set of $N = \sum_{i=1}^{d} N_i$ places of arbitrary degree where $P_{i,j}$ denotes a place of degree i and N_i a number of places of degree i.*

We suppose that Q and all the places in $\mathcal{P}$ are not in the support of $\mathcal{D}$ and that:

1. *the map*

$$Ev_Q : \begin{cases} \mathcal{L}(\mathcal{D}) & \to & \mathbb{F}_{q^n} \simeq F_Q \\ f & \longmapsto & f(Q) \end{cases}$$

   is onto,
2. *the map*

$$Ev_{\mathcal{P}} : \begin{cases} \mathcal{L}(k\mathcal{D}) \longrightarrow \mathbb{F}_q^{N_1} \times \mathbb{F}_{q^2}^{N_2} \times \cdots \times \mathbb{F}_{q^d}^{N_d} \\ f \longmapsto (f(P_{1,1}), \cdots, f(P_{1,N_1}), \cdots, f(P_{d,1}), \cdots, f(P_{d,N_d})) \end{cases}$$

   is injective

Then, for any k elements $x_1, \dots, x_k$ in $\mathbb{F}_{q^n}$, we have

$$m_k(x_1, ..., x_k) = Ev_Q \left(Ev_{\mathcal{P}}^{-1} \left(\bigodot_{i=1}^{k} \left(Ev_{\mathcal{P}} \left(Ev_Q^{-1}(x_i) \right) \right) \right) \right),$$

and

$$\mu_{q,k}(n) \leq \sum_{i=1}^{d} N_i \mu_{q,k}(i), \tag{8}$$

$$\nu_{q,k}(n) \leq (k-1) \sum_{i=1}^{d} N_i \mu_q(i). \tag{9}$$

Proof. For any k elements $x_1, ..., x_k$ in $\mathbb{F}_{q^n}$, we have k elements $g_1, \cdots, g_k$ in $\mathcal{L}(\mathcal{D})$ such that $Ev_Q(g_i) = g_i(Q) = x_i$ where $g_i(Q)$ denotes the class of g_i in the residue class field F_Q.

Thus,

$$m_k(x_1, ..., x_k) = \prod_{i=1}^{k} x_i = \prod_{i=1}^{k} g_i(Q) = \left(\prod_{i=1}^{k} g_i \right)(Q). \tag{10}$$

Moreover, since the divisor $\mathcal{D}$ has a positive dimension, without loss of generality we can suppose that $\mathcal{D}$ is an effective divisor. Hence, $\mathcal{L}(\mathcal{D})^k \subset \mathcal{L}(k\mathcal{D})$ and $\prod_{i=1}^{k} g_i \in \mathcal{L}(k\mathcal{D})$. Now, we have to compute the function $h = \prod_{i=1}^{k} g_i$ and this is done by an interpolation process *via* the evaluation map $Ev_{\mathcal{P}}$. More precisely,

$$h = \prod_{i=1}^{k} g_i = Ev_{\mathcal{P}}^{-1} \left(\bigodot_{i=1}^{k} Ev_{\mathcal{P}}(g_i) \right),$$

namely

$$h = Ev_{\mathcal{P}}^{-1} \left(\bigodot_{i=1}^{k} \left(Ev_{\mathcal{P}} \left(Ev_Q^{-1}(x_i) \right) \right) \right).$$

Moreover, $\mathcal{L}(k\mathcal{D}) \subset \mathcal{O}_Q$ for any integer k, thus, we can apply the map Ev_Q over the element h which corresponds to the product of k elements $x_1, ..., x_k$ which gives the first inequality. Inequality (9) follows from Inequality (4) in Lemma 1 and we are done. $\qquad\square$

Theorem 2. *Let q be a prime power and let n be an integer > 1. Let $F/\mathbb{F}_q$ be an algebraic function field of genus g. Let $\mathcal{P}_i$ be a set of places of degree i in $F/\mathbb{F}_q$ and N_i the cardinality of $\mathcal{P}_i$. We denote $\mathcal{P} = \bigcup_{i=1}^{r} \mathcal{P}_i$. Let us suppose that there is a place of degree n and a non-special divisor of degree $g-1$. If there is an integer $r \geq 1$ such that:*

$$\sum_{i=1}^{r} iN_i > kn + kg - k \tag{11}$$

then

$$\mu_{q,k}(n) \leq \sum_{i=1}^{r} N_i \, \mu_{q,k}(i).$$

Let r_0, r_0' such that

$$\frac{\mu_{q,k}(r_0)}{r_0} = \sup_{1 \leq i \leq r} \frac{\mu_{q,k}(i)}{i} \quad and \quad \frac{\mu_q(r_0')}{r_0'} = \sup_{1 \leq i \leq r} \frac{\mu_q(i)}{i}.$$

Then,

$$\mu_{q,k}(n) \leq (kn + kg - k + r) \frac{\mu_{q,k}(r_0)}{r_0}. \tag{12}$$

$$\nu_{q,k}(n) \leq (k-1)(kn + kg - k + r) \frac{\mu_q(r_0')}{r_0'}. \tag{13}$$

Proof. Let $\mathcal{R}$ be a non-special divisor of degree $g-1$. Then we choose a divisor $\mathcal{D}_1$ such that $\mathcal{D}_1 = \mathcal{R}+Q$. Let $[\mathcal{D}_1]$ be the class of $\mathcal{D}_1$, then by [8], Lecture 14, Lemma 1, $[\mathcal{D}_1]$ contains a divisor $\mathcal{D}$ defined over $\mathbb{F}_q$ such that $ord_P\mathcal{D} = 0$ for all places $P \in \mathcal{P}$ and $ord_Q\mathcal{D} = 0$. Since $ord_Q\mathcal{D} = 0$, $\mathcal{L}(\mathcal{D})$ is contained in the valuation ring O_Q of Q. Hence Ev_Q is a restriction of the residue class mapping, and defines an $\mathbb{F}_q$-algebra homomorphism. The kernel of Ev_Q is $\mathcal{L}(\mathcal{D} - Q)$. But $\mathcal{D} - Q$ is non-special divisor of degree $g - 1$, then $l(\mathcal{D} - Q) = deg(\mathcal{D} - Q) - g + 1 = 0$ and Ev_Q is injective with $deg\ \mathcal{D} = n + g - 1$. Moreover, if $\mathcal{K}$ is a canonical divisor, we have $l(\mathcal{D}) = l(\mathcal{K} - \mathcal{D}) + n$ by the Riemann-Roch theorem. Hence, $l(\mathcal{D}) \geq n$ and as Ev_Q is injective, we obtain $l(\mathcal{D}) = n$. We conclude that Ev_Q is an isomorphism. The map $Ev_{\mathcal{P}}$ is well-defined because $\mathcal{L}(k\mathcal{D})$ is contained in the valuation ring of every place of $\mathcal{P}$. The kernel of $Ev_{\mathcal{P}}$ is $\mathcal{L}\left(k\mathcal{D} - \sum_{P \in \mathcal{P}} P\right)$ which is trivial because $\sum_{i=1}^{r} iN_i > kn + kg - k$. Therefore $Ev_{\mathcal{P}}$ is injective and $l(k\mathcal{D}) = kn + (k-1)g - k + 1$.

Let us remark that we can suppose that $kn + kg - k < \sum_{i=1}^{r} iN_i \leq kn + kg - k + r$. Then, by Inequality (8) in Theorem 1,

$$\mu_{q,k}(n) \leq \sum_{i=1}^{r} N_i \mu_{q,k}(i) = \sum_{i=1}^{r} iN_i \frac{\mu_{q,k}(i)}{i} \leq \frac{\mu_{q,k}(r_0)}{r_0} \sum_{i=1}^{r} iN_i,$$

which gives Inequality (12). Inequality (13) follows from Inequality (9) in Theorem 1 in the same way and the proof is complete.   □

Remark 1. A place of degree n and a non-special divisor of degree $g - 1$ exist if elementary numerical conditions are satisfied. More precisely, if

$$2g + 1 \leq q^{\frac{n-1}{2}}(q^{\frac{1}{2}} - 1), \tag{14}$$

then there exists a place of degree n by [13, Corollary]. Moreover, if $q \geq 4$ or $N_1 \geq g + 1$ then there exits a non-special divisor of degree $g - 1$ by [1] (cf. also [5]).

Remark 2. Note that it is an open problem to know if the bilinear complexity is increasing or not with respect to the extension degree, when $n \geq 2$.

3   On the Existence of These Algorithms

3.1   Strategy of Construction

In this section, we present our strategy of construction of these algorithms which is well adapted to an asymptotical study. More precisely, this strategy consists in fixing the definition field $\mathbb{F}_q$, the integer n and the parameter k of the algorithm 1. Then, we suppose the existence of family of algebraic function fields over $\mathbb{F}_q$ of the genus g growing to the infinity.

The main condition (11) of Theorem 2 supposes that we can find algebraic function fields having good properties. In particular, it is sufficient to have a family of function fields having sufficiently places with a certain degree r. In this aim, we focalize on sequences of algebraic functions fields with increasing genus attaining the Drinfeld-Vladut bound of order r (cf. [6, Definition 1.3]). Hence, let us find the minimal integer r for such a family, so that Condition (11) is satisfied.

Let $\mathcal{G}/\mathbb{F}_q = (G_i/\mathbb{F}_q)_i$ be a sequence of algebraic function fields G_i over $\mathbb{F}_q$ of genus g_i attaining the Drinfeld-Vladut Bound of order r, namely

$$\lim_{i\to+\infty} \frac{N_r(G_i)}{g_i} = \frac{1}{r}(q^{\frac{r}{2}} - 1).$$

The divisors D are such that $\deg(D) = n + g - 1$, thus $\deg kD = kn + kg - k$. To determine the minimal value of the integer r, we can suppose without less of generality that $rN_r > kn + kg - k$, which implies asymptotically that $q^{\frac{r}{2}} - 1 > k$, namely $r > 2\log_q(k+1)$.

3.2   Towers of Algebraic Function Fields

In this section, we present a sequence of algebraic function fields defined over $\mathbb{F}_q$ from the Garcia-Stichtenoth tower constructed in [9]. This tower is suitable with respect to the strategy defined in Sect. 3.1.

Let us consider a finite field $\mathbb{F}_{l^2}$ where l is a prime power such that $\mathbb{F}_{l^2}$ is an extension field of $\mathbb{F}_q$. We consider the Garcia-Stichtenoth's elementary abelian tower $\mathcal{F}$ over $\mathbb{F}_{l^2}$ constructed in [9] and defined by the sequence $\mathcal{F} = (F_0, F_1, \cdots, F_i, \cdots)$ where

$$F_0 := \mathbb{F}_{l^2}(x_0)$$

is the rational function field over $\mathbb{F}_{l^2}$, and for any $i \geq 0$, $F_{i+1} := F_i(x_{i+1})$ with x_{i+1} satisfying the following equation:

$$x_{i+1}^l + x_{i+1} = \frac{x_i^l}{x_i^{l-1} + 1}.$$

Let us denote by $g_i = g(F_i)$ the genus of F_i in $\mathcal{F}/\mathbb{F}_{l^2}$ and recall the following formulæ:

$$g_i = \begin{cases} (l^{\frac{i+1}{2}} - 1)^2 & \text{for odd } i, \\ (l^{\frac{i}{2}} - 1)(l^{\frac{i+2}{2}} - 1) & \text{for even } i. \end{cases} \tag{15}$$

Thus, as in [3], according to these formulæ, it is straightforward that the genus of any step of the tower satisfies:

$$(l^{\frac{i}{2}} - 1)(l^{\frac{i+1}{2}} - 1) < g_i < (l^{\frac{i+2}{2}} - 1)(l^{\frac{i+1}{2}} - 1). \tag{16}$$

Moreover, a tighter upper bound will be useful and can be obtained by expanding expressions in (15):

$$g_i \leq l^{i+1} - 2l^{\frac{i+1}{2}} + 1. \tag{17}$$

Then we can consider as in [2] the descent tower $\mathcal{G}/\mathbb{F}_q$ defined over $\mathbb{F}_q$ given by the sequence:

$$G_0 \subset G_1 \subset \cdots \subset G_i \subset \cdots$$

defined over the constant field $\mathbb{F}_q$ and related to the tower $\mathcal{F}$ by: $F_i = \mathbb{F}_{l^2} \otimes_{\mathbb{F}_q} G_i$ for all i.

Let us recall the known results concerning the number of places of degree one of the tower $\mathcal{F}/\mathbb{F}_{l^2}$, established in [9].

Proposition 1. *The number of places of degree one of $F_i/\mathbb{F}_{l^2}$ is:*

$$N_1(F_i/\mathbb{F}_{l^2}) = \begin{cases} l^i(l^2 - l) + 2l^2 & \text{if the caracteristic is even,} \\ l^i(l^2 - l) + 2l & \text{if the caracteristic is odd.} \end{cases}$$

Now, we are interested by supplementary properties concerning the descent tower $\mathcal{G}/\mathbb{F}_q$ in the context of Theorem 2, which we are going to apply to this tower. More precisely, we need that the steps of the tower $\mathcal{G}/\mathbb{F}_q$ verify the mandatory properties of the existence of a place of degree n and of sufficient number of places of certain degrees. In this aim, we introduce the notion of the action domain of an algebraic function field.

Definition 3. *Let us define the following quantities:*

1. $M_i = l^i(l^2 - l)$ *where* $l = q^{\frac{r}{2}}$,
2. $\Delta_{q,k,i} = M_i - kg_i + k$,
3. $\Theta_{q,k,i} = \{n \in \mathbb{N} \mid \Delta_{q,k,i} > kn\}$,
4. $R_{q,k,i} = \sup \Theta_{q,k,i}$,
5. *the set* $\phi_{q,i} = \{n \in \mathbb{N} \mid 2g_i + 1 \leq l^{\frac{n-1}{r}}(l^{\frac{1}{r}} - 1)\}$,
6. $\Gamma_{q,i} = \inf \phi_{q,i}$.
7. $I_{q,k,i} = \Theta_{q,k,i} \cap \phi_{q,i} = [\Gamma_{q,i}, R_{q,k,i}]$.

The set $I_{q,k,i}$ is called the action domain of $F_i/\mathbb{F}_q$.

Lemma 2. *Let q be a prime power and $k \geq 2$ be an integer and r the smallest even integer $> 2\log_q(k+1)$. Then $(\Delta_{q,k,i})_{i \in \mathbb{N}}$ is an increasing sequence such that $\lim_{i \to \infty} \Delta_{q,k,i} = +\infty$.*

Proof. By Inequality (17) we get:

$$\Delta_{q,k,i} = M_i - kg_i + k \geq q^{\frac{ri}{2}}\left(q^r - q^{\frac{r}{2}}\right) - k\left(q^{\frac{r(i+1)}{2}} - 2q^{r\frac{i+1}{4}}\right).$$

Hence:

$$\Delta_{q,k,i} \geq q^{\frac{ri}{2}}(q^r - q^{\frac{r}{2}}) - kq^{\frac{r(i+1)}{2}} = q^{\frac{r(i+1)}{2}}(q^{\frac{r}{2}} - 1 - k).$$

As $r > 2\log_q(k+1)$, we are done. $\square$

Proposition 2. *Let q be a prime power and $k \geq 2$ be an integer and r the smallest even integer $> 2\log_q(k+1)$. Then for any $i \geq 1$, the action domain $I_{q,k,i}$ of $F_i/\mathbb{F}_q$ is not empty.*

Proof. Let us compute bounds on the two values $\Gamma_{q,i}$ and $R_{q,k,i}$.

1. Bound on $\Gamma_{q,i}$. The set $\phi_{q,i}$ contains the set of integers n such that

$$2\left(l^{i+1} - 2l^{\frac{i+1}{2}} + 1\right) + 1 \leq l^{\frac{n-1}{r}}\left(l^{\frac{1}{r}} - 1\right). \tag{18}$$

If

$$n \geq r(i+1) + 1 + \log_{l^{\frac{1}{r}}}(2) - \log_{l^{\frac{1}{r}}}(l^{\frac{1}{r}} - 1),$$

Condition (18) is verified, then the integer n is in $\phi_{q,i}$. We conclude that

$$\Gamma_{q,i} \leq r(i+1) + 3.$$

2. Bound on $R_{q,k,i}$. The set $\Theta_{q,k,i}$ contains the set of integers n such that

$$q^{\frac{r(i+1)}{2}}(q^{\frac{r}{2}} - 1 - k) \geq kn. \tag{19}$$

As $r > 2\log_q(k+1)$, the integer $q^{\frac{r}{2}} - 1 - k$ is ≥ 1. Then the condition

$$n \leq \frac{1}{k} q^{\frac{r(i+1)}{2}}$$

implies Condition (19). As $q^{\frac{r}{2}} > k + 1$ the condition

$$n \leq (k+1)^i$$

implies $n \in \Theta_{q,k,i}$. Then

$$R_{q,k,i} \geq (k+1)^i.$$

3. Conclusion. For $i \geq 1$, the function $(k+1)^i - r(i+1) - 3$ is an increasing function of i. For $i = 1$ we have $[2r + 3, k + 1] \subset [\Gamma_{q,1}, R_{q,k,1}]$, and more generally $[r(i+1) + 3, (k+1)^i] \subset [\Gamma_{q,i}, R_{q,k,i}]$. □

Proposition 3. *Let q be a prime power and $k \geq 2$ be an integer and r the smallest even integer $> 2\log_q(k+1)$. Then:*

$$[2r + 3, +\infty[\subset \bigcup_{i \geq 1} I_{q,k,i}.$$

Proof. For any i, we have $[r(i+1) + 3, (k+1)^i] \subset I_{q,k,i}$ and $\bigcup_{i \geq 1} I_{q,k,i} = [2r + 3, +\infty[$ and the proof is complete. □

Hence, for any integer $n \geq 2r + 3$, there exists an action domain $I_{q,k,i}$ such that $n \in I_{q,k,i}$.

4 Uniform Upper Bounds

In this section, we study the complexity of the family of algorithms 1 constructed with a tower $\mathcal{G}/\mathbb{F}_q$ defined over $\mathbb{F}_q$ and defined in Sect. 3.2, such that this tower attains the Drinfeld-Vladut bound of order r where r is the smallest even integer $> 2\log_q(k+1)$. Moreover, for any integer i, we have $\sum_{j|r} jN_j(G_i) = N_1(F_i)$.

Theorem 3. *Let q be a prime power and $k \geq 2$ be an integer and r the smallest even integer $> 2\log_q(k+1)$. Let r_0, r_0' such that*

$$\frac{\mu_{q,k}(r_0)}{r_0} = \sup_{1 \leq i \leq r} \frac{\mu_{q,k}(i)}{i} \quad and \quad \frac{\mu_q(r_0')}{r_0'} = \sup_{1 \leq i \leq r} \frac{\mu_q(i)}{i}.$$

Then for any integer n, we have:

$$\mu_{q,k}(n) \leq \frac{k(kq^{\frac{r}{2}}+1)n - k + r}{r_0}\mu_{q,k}(r_0) \tag{20}$$

and consequently

$$\nu_{q,k}(n) \leq \frac{k(k-1)(kq^{\frac{r}{2}}+1)n - (k-1)(k-r)}{r_0'}\mu_q(r_0'). \tag{21}$$

Proof. We apply Theorem 2 to the tower descent tower $\mathcal{G}/\mathbb{F}_q$ by taking for any $n > 1$ the small step F_i satisfying the assumptions of Theorem 2. Let us set $M_i = l^i(l^2 - l)$ where $l = q^{\frac{r}{2}}$. For any integer $k \geq 2$ and any integer n, let i be the smallest integer such that $M_i > kn + kg_i - k$, then $kn < M_i - kg_i + k$ and $kn \geq M_{i-1} - kg_{i-1} + k$. So, we obtain

$$kn \geq q^{\frac{r(i-1)}{2}}\left(q^r - q^{\frac{r}{2}}\right) - k\left(q^{\frac{ri}{2}} - 2q^{\frac{ri}{4}}\right)$$

by Formula (17). Hence, $kn \geq q^{\frac{ri}{2}}\left(q^{\frac{r}{2}} - k - 1\right)$, and so:

$$q^{\frac{ri}{2}} \leq \frac{kn}{q^{\frac{r}{2}} - k - 1}.$$

But $q^{\frac{r}{2}} - k - 1 \geq 1$, hence:

$$i \leq \frac{2}{r}\log_q(kn). \tag{22}$$

Now, by Bound (12) in Theorem 2, we obtain:

$$\mu_{q,k}(n) \leq \frac{kn + kg_i - k + r}{r_0}\mu_{q,k}(r_0)$$

which gives by Formulae (17) and (22):

$$\mu_{q,k}(n) \leq \frac{kn + kq^{\frac{r(i+1)}{2}} - k + r}{r_0}\mu_{q,k}(r_0)$$

with

$$q^{\frac{r(i+1)}{2}} \leq q^{\log_q(kn)+\frac{r}{2}} = q^{\frac{r}{2}}kn,$$

which gives the first inequality. Inequality (21) follows from Inequality (13) in Theorem 2. □

Corollary 1. *Let q be a prime power and $k \geq 2$ be an integer and r the smallest even integer $> 2\log_q(k+1)$. Let us suppose that we are in the case where $k \geq r$. Let r_0, r_0' such that*

$$\frac{\mu_{q,k}(r_0)}{r_0} = \sup_{1 \leq i \leq r} \frac{\mu_{q,k}(i)}{i} \quad and \quad \frac{\mu_q(r_0')}{r_0'} = \sup_{1 \leq i \leq r} \frac{\mu_q(i)}{i}.$$

Then for any integer n, we have:

$$\mu_{q,k}(n) \leq \frac{k(kq^{\frac{r}{2}}+1)}{r_0}\mu_{q,k}(r_0)\, n \leq \frac{k(k(k+1)q+1)}{r_0}\mu_{q,k}(r_0)\, n \qquad (23)$$

$$\nu_{q,k}(n) \leq \frac{k(k-1)(kq^{\frac{r}{2}}+1)}{r_0'}\mu_q(r_0')\, n \leq \frac{k(k-1)(k(k+1)q+1)}{r_0'}\mu_q(r_0')\, n. \quad (24)$$

Proof. The first inequality of (23) follows immediately from Inequality (20) in Theorem 3. As r the smallest even integer $> 2\log_q(k+1)$, we have $r - 2 < 2\log_q(k+1)$ and so $q^{r/2-1} < k+1$. Hence, $q^{r/2} < q(k+1)$ which gives the first assertion. Assertion (24) is obtained with the same way. $\square$

Acknowledgement. We would like to thank the referees for their valuable comments.

References

1. Ballet, S., Le Brigand, D.: On the existence of non-special divisors of degree g and $g - 1$ in algebraic function fields over $\mathbb{F}_q$. J. Number Theory **116**, 293–310 (2006)
2. Ballet, S., Le Brigand, D., Rolland, R.: On an application of the definition field descent of a tower of function fields. In: Proceedings of the Conference Arithmetic, Geometry and Coding Theory (AGCT 2005), vol. 21, pp. 187–203. Société Mathématique de France, sér. Séminaires et Congrès (2009)
3. Ballet, S., Pieltant, J.: Tower of algebraic function fields with maximal Hasse-Witt invariant and tensor rank of multiplication in any extension of $\mathbb{F}_2$ and $\mathbb{F}_3$. J. Pure Appl. Algebra **222**(5), 1069–1086 (2018)
4. Ballet, S., Pieltant, J., Rambaud, M., Randriambololona, H., Rolland, R., Chaumine, J.: On the tensor rank of multiplication in finite extension of finite fields and related issues in algebraic geometry. Uspekhi Mat. Nauk **76**, 31–94 (2021)
5. Ballet, S., Ritzenthaler, C., Rolland, R.: On the existence of dimension zero divisors in algebraic function fields defined over $\mathbb{F}_q$. Acta Arith. **143**(4), 377–392 (2010)
6. Ballet, S., Rolland, R.: Families of curves over any finite field attaining the generalized Drinfeld-Vlăduţ bound. Publications Mathématiques de Besançon, Algèbre et Théorie des Nombres, pp. 5–18 (2011)
7. Bshouty, N.: Multilinear complexity is equivalent to optimal tester size. In: Electronic Colloquium on Computational Complexity (ECCC), Tr13(11) (2013)
8. Deuring, M.: Lectures on the Theory of Algebraic Functions of One Variable. Springer, Heidelberg (1973). https://doi.org/10.1007/BFb0060944
9. Garcia, A., Stichtenoth, H., Ruck, H.-G.: On tame towers over finite fields. Journal für die reine und angewandte Mathematik **557**, 53–80 (2003)

10. Randriambololona, H., Rousseau, É.: Trisymmetric multiplication formulae in finite fields. In: Bajard, J.C., Topuzoğlu, A. (eds.) WAIFI 2020. LNCS, vol. 12542, pp. 92–111. Springer, Cham (2021). https://doi.org/10.1007/978-3-030-68869-1_5

11. Rousseau, E.: ArithmÃtique efficace des extensions de corps finis. Ph.D. thesis, Institut Polytechnique de Paris (2021)

12. Shparlinski, I.E., Tsfasman, M.A., Vladut, S.G.: Curves with many points and multiplication in finite fileds. In: Stichtenoth, H., Tsfasman, M.A. (eds.) Coding Theory and Algebraic Geometry. LNM, vol. 1518, pp. 145–169. Springer, Heidelberg (1992). https://doi.org/10.1007/BFb0087999

13. Stichtenoth, H.: Algebraic Function Fields and Codes. Graduate Texts in Mathematics, vol. 254. Springer, Heidelberg (2008). https://doi.org/10.1007/978-3-540-76878-4

On Some Algebraic Ways to Calculate Zeros of the Riemann Zeta Function

Yuri Matiyasevich[(✉)] [iD]

Petersburg Department of Steklov Institute of Mathematics,
St. Petersburg 191023, Russia
yumat@pdmi.ras.ru
http://logic.pdmi.ras.ru/~yumat

Abstract. The Riemann zeta function is an important number-theoretical tool for studying prime numbers. The first part of the paper is a short survey of some known results about this function. The emphasis is given to the possibility to formulate the celebrated Riemann Hypothesis as a statement from class Π_1^0 in the arithmetical hierarchy.

In the second part of the paper the author demonstrates by numerical examples some non-evident ways for finding zeros of the zeta function. Calculations require the knowledge of the value of this function and of N its initial derivatives at one point and consist in solving N systems of linear equations with N unknowns.

These methods are not intended for practical calculations but are supposed to be useful for the study of the zeros.

Keywords: Riemann zeta function · Riemann hypothesis · Linear algebra

1 The Riemann Hypothesis

Prime numbers 2, 3, ... are one of the most important objects of investigations in Number Theory. An important and efficient tool for studying them is the celebrated *Riemann zeta function*. For a complex number s such that $\Re(s) > 1$ it can be defined by a *Dirichlet series*, namely,

$$\zeta(s) = \sum_{n=1}^{\infty} n^{-s}. \tag{1}$$

This function was studied already by L. Euler (for real values of s only). In particular, he gave another definition of this function:

$$\zeta(s) = \prod_{p \text{ is prime}} \frac{1}{1 - p^{-s}}. \tag{2}$$

In order to see the equivalence of the two definitions, (1) and (2), it suffice to notice that the factors in (2) are the sums of geometrical progressions,

© The Author(s), under exclusive license to Springer Nature Switzerland AG 2022
D. Poulakis and G. Rahonis (Eds.): CAI 2022, LNCS 13706, pp. 15–25, 2022.
https://doi.org/10.1007/978-3-031-19685-0_2

$$\frac{1}{1-p^{-s}} = 1 + p^{-s} + p^{-2s} + p^{-3s} + \ldots , \qquad (3)$$

substitute the right-hand side of (3) into (2) and expand the product. Essentially, the equality of the sum from (1) and the product from (2) is an analytical form of the *Fundamental Theorem of Arithmetic* which states that every natural number can be represented as a product of powers of prime numbers in a unique way. This very simple fact explains why the zeta function is so powerful tool for studying the primes number.

One question, which was of interest for mathematician for a long time, is as follows: *What is the number of primes below a given bound x?* Traditionally, this number is denoted as $\pi(x)$. Karl Gauss, examining available to him tables of primes, conjectured that

$$\pi(x) \approx \operatorname{Li}(x) = \int_2^x \frac{dy}{\ln(y)} = \frac{x}{\ln(x)} + o(x/\ln(x)) \qquad (4)$$

(Li is known as the *logarithmic integral function*). Informally speaking, one can say that in the vicinity of y the "probability" of an integer to be a prime is equal to $1/\ln(y)$.

In his seminal paper [38] B. Riemann gave an explicit formula for $\pi(x)$ in which $\operatorname{Li}(x)$ *formally* is the main term. However, it is not easy to estimate the contribution of the remaining terms in this formula because they involve summations over the zeros of the zeta function.

Already Euler indicated that $\zeta(s)$ vanishes whenever s is a negative even integer, and these numbers are called the *trivial zeros* of the zeta function. Riemann extended the definition of the zeta function to the whole complex plane (except for the point $s = 1$) and proved that all other, the *non-trivial zeros* of this function are not real and reside inside the *critical strip*

$$0 \leq \Re(s) \leq 1. \qquad (5)$$

J. Hadamard [17] and Ch. de la Vallée Poussin [40] narrowed this strip to

$$0 < \Re(s) < 1. \qquad (6)$$

This improvement implied that $\operatorname{Li}(x)$ is indeed the main term, that is

$$\pi(x) = \operatorname{Li}(x) + o(x/\ln(x)). \qquad (7)$$

This equality is known as the *Prime Number Theorem*.

The celebrated *Riemann Hypothesis* predicts that all the non-trivial zeta zeros lie on the *critical line*

$$\Re(s) = \frac{1}{2}. \qquad (8)$$

This implies that

$$\pi(x) = \operatorname{Li}(x) + O(x^{1/2} \ln(x)); \qquad (9)$$

moreover, the hypothesis itself follows from (9).

In Number Theory there is an whole host of results so far proved only under the assumption of the validity of the Riemann Hypothesis. Surprisingly, there are such conditional results in Computer Science as well.

In 2002 M. Agrawal, N. Kayal, and N. Saxena [2] discovered a long-desired polynomial algorithm for recognizing the primality of a given integer p. Their original proof gave an upper bound $O(\ln^{12+\epsilon}(p))$ for the required time; by 2005 this marvelous result was improved to $O(\ln^{6+\epsilon}(p))$ by H. W. Lenstra and C. Pomerance (see [22]). However, already in 1976 G. L. Miller [32] found an algorithm with time complexity $O(\ln^{4+\epsilon}(p))$ *under the assumption of the validity of (generalized) Riemann Hypothesis*

2   The Riemann Hypothesis in the Arithmetical Hierarchy

A. Turing was fond of Number Theory. In particular, he investigated the Riemann zeta function. To calculate its values he invented an analogous machine. In 1939 Turing got a grant for its physical implementation and started the construction. The process was broken by the WW2 and never resumed.

After the war Turing was the first who used a digital computer for verifying the Riemann Hypothesis. To this end he substantially improved the known technique. The so called "Turing's method" is in use up to now for verifying the hypothesis for the initial zeros of the zeta function.

More details about Turing's contribution to Number Theory can be found in [3–5,13]. However, number-theorists seem to be unaware of yet another study of Turing related to the Riemann Hypothesis. It was done in his logical dissertation [39]. The prehistory was as follows.

K. Gödel developed a very powerful technique of *arithmetization* which allows us to represent different mathematical statements by formulas expressing properties of natural numbers. However, it is not straightforward to see from the definition (1) of the zeta function that the Riemann Hypothesis can be expressed by such a formula. Nevertheless, the hypothesis, as many other great problems, has many equivalent formulations (see, for example, [5,6]) and some of them are more amenable to arithmetization. For example, to achieve this goal one can use (9) and replace $\mathrm{Li}(x)$ by a finite sum approximating this integral.

Different reformulation of the Riemann Hypothesis would lead to formulas of different intricacy. A complexity of an arithmetical formula can be measured by its position in the *arithmetical hierarchy* which is defined by the number of alternations of the universal and existential quantifiers.

In [39] Turing introduced the notion of *number-theoretical theorems*. He wrote:

> By a number-theoretic theorem we shall mean a theorem of the form "$\theta(x)$ vanishes for infinitely many natural numbers x", where $\theta(x)$ is a primitive recursive function. ... An alternative form for number-theoretic theorems is "for each natural number x there exists a natural number y such that $\phi(x, y)$ vanishes", where $\phi(x, y)$ is primitive recursive.

Respectively, a problem is called number-theoretical if its solution could be given in the form of a number-theoretical theorem. It is easy to see that the set of such problems is exactly the class Π_2^0 of the arithmetical hierarchy.

As one of the examples of number-theoretical problems Turing proved that the Riemann Hypothesis can be reformulated as Π_2^0 statement.

This result of Turing was improved to Π_1^0 by G. Kreisel [20] in 1958. However, neither Turing's no Kreisel's reformulations of the Riemann Hypothesis immediately attracted attention of specialists in Number Theory. The situation changed in 1970 when the author made the last step in the proof of what is nowadays referred to as *DPRM-theorem*[1]. This theorem establishes that every formula from Π_1^0 with parameters $a_1, \ldots, a_m$ is equivalent to a formula of the special form

$$\forall x_1 \ldots x_n [P(a_1, \ldots, a_m, x_1 \ldots x_n) \neq 0] \tag{10}$$

where P is a polynomial with integer coefficients.

Together with the above mentioned result of Kreisel, DPRM-theorem has the following corollary: *one can construct a particular polynomial $R(x_1 \ldots x_n)$ with integer coefficients such that the Riemann Hypothesis is equivalent to the statement that the Diophantine equation*

$$R(x_1 \ldots x_n) = 0 \tag{11}$$

has no solutions.

A method for an actual construction of such an equation (11) was described in [15, Section 2]. Later a simplified version was presented in [25, Section 6.4]; more technical details were supplied in [7,35–37]. More recently the author [26] found yet another reformulation of the Riemann Hypothesis as Π_1^0 statement which is especially suitable for constructing an equivalent Diophantine equation.

B. Fodden [16] described Diophantine reformulations of extended Riemann Hypothesis.

DPRM-theorem was worked out as a tool to establish the undecidability of Hilbert's 10th problem. It is one of the 23 mathematical problems posed by D. Hilbert in 1900 in [18]. In this problem he asked for an algorithm for recognizing whether a given arbitrary Diophantine equation has a solution.

The Riemann Hypothesis is a part of Hilbert's 8th problem. Now, equation (11) shows that this hypothesis is a very special case of the 10th problem; such a relationship (found via the Computability Theory) between the 8th and 10th Hilbert's problems seems to have never been anticipated by specialists in Number Theory.

In the last decades some researchers were interested in yet another reformulation of the Riemann Hypothesis. C. S. Calude, E. Calude, and M. J. Dinneen ([11], for further development see [8–10]) suggested that the complexity of a Π_1^0 statement can be defined as some measure of the simplest machine (or program) that never halts if and only if the statement is true. Among other famous mathematical problems they estimated (from above) the complexity of the Riemann

---

[1] After M. Davis, H. Putnam, J. Robinson and Yu. Matiyasevich; for detailed proofs see, for example, [14,19,23,25,30].

```
from math import gcd
d=m=p=0
f0=f1=f3=n=q=1
while p**2*(m-f0)<f3:
    d=2*n*d-(-1)**n*f1
    n=n+1
    g=gcd(n,q)
    q=n*q//g
    if g==1: p=p+1
    m=0; q2=q
    while q2>1:
        q2=q2//2; m=m+d
    f1=2*f0
    f0=2*n*f0
    f3=(2*n+3)*f3
```

Fig. 1. Python 3 program that never halts if and only the Riemann Hypothesis is true

Hypothesis. Of course, such a bound heavily depends on our current level of knowledge.

The numerical value of such a complexity measure depends also on the formalism used for describing computations. In [10–12] a version of *register machines* was used for estimating the complexity of mathematical problems. Such models of computational devices were proposed in 1961 by J. Lambek [21], by Z. A. Melzak [31], and by M. L. Minsky [34] (see also [33]). Types of admissible instructions of register machines can vary. In [11], the Riemann Hypothesis was presented by a register machine with 290 rather powerful instructions (this was improved to 178 instructions in [12]).

More recently, A. Yedidia and S. Aaronson [41] constructed a classical Turing machine with two-letter tape alphabet which, having started with the empty tape, will never halt if and only if the Riemann Hypothesis is true. Their machine had 5372 state; this was improved later to 744 states (see [1]).

The author [24] found yet another reformulations of the Riemann Hypothesis as a Π_1^0 statement and constructed a register machine which never halts if and only if the hypothesis is true; this machine has 29 registers and 130 primitive instructions. Another incarnations of this reformulation of the hypothesis is presented on Fig. 1 as a Python 3 program.

3   Calculation of the Zeta Zeros by Solving Algebraic Equations

There are many efficient methods to calculate (with an arbitrary accuracy) the values of the zeta functions and its derivatives for a given value of the argument.

In principle, these methods allow one to calculate (with an arbitrary accuracy) a zero ρ of the transcendental equation

$$\zeta(s) = 0 \tag{12}$$

starting from its approximative value a:

$$a \approx \rho, \quad \zeta(\rho) = 0. \tag{13}$$

A standard way to do it is to use Newton's iterations:

$$a_0 = a, \tag{14}$$

$$a_{n+1} = a_n - \frac{\zeta(a_n)}{\zeta'(a_n)}. \tag{15}$$

However, it is difficult to analyze such an iterating process and prove something about its limiting value ρ.

Another possibility is to calculate $N + 1$ first derivatives of the zeta function at point $s = a$ and consider an initial fragment of the Taylor expansion,

$$\zeta(s) = P_{a,N}(s) + O\big((s - a)^{N+1}\big), \tag{16}$$

where $P_{a,N}(s)$ is polynomial of degree N; let

$$P_{a,N}(s) = \sum_{n=0}^{N} p_{a,N,n} s^n. \tag{17}$$

If a is sufficiently close to ρ, then

$$P_{a,N}(\rho) \approx \zeta(\rho) = 0 \tag{18}$$

and (approximate) solving the transcendental equation (12) can be replaced by solving the algebraic equation

$$P_{a,N}(s) = 0. \tag{19}$$

Unfortunately, we cannot analytically indicate which of the N solutions of this equation is close to a zero of the zeta function.

Here we present numerical examples of non-obvious ways to calculate a zeta zero by solving systems of linear equations. These ways seem to be specific for the zeta function and other functions defined by Dirichlet series. So far there was no theoretical justification of this technique.

Let us introduce new unknowns

$$s_1, \ldots, s_N \tag{20}$$

and replace s^n in (17) by s_n – let

$$P_{a,N}(s_1, \ldots, s_N) = p_{a,N,0} + \sum_{n=1}^{N} p_{a,N,n} s_n. \tag{21}$$

The intended values of the unknowns (20) are

$$s_1 = \rho, \ \ldots, \ s_N = \rho^n, \tag{22}$$

and we rewrite equation (19) as

$$P_{a,N}(s_1, \ldots, s_N) = 0. \tag{23}$$

This single equation is not sufficient for calculating values of N unknowns (20). But where could we find additional equations?

Our answer is not evident. Number ρ, being a solution of equation (12), is also a solution of equations

$$m^{-s}\zeta(s) = 0, \qquad m = 1, \ldots N. \tag{24}$$

Starting from these equations, we can get parameterized versions of (16), (17), (21), and (19):

$$m^{-s}\zeta(s) = P_{a,N,m}(s) + O\big((s-a)^{N+1}\big), \tag{25}$$

$$P_{a,N,m}(s) = \sum_{n=0}^{N} p_{a,N,m,n} s^n, \tag{26}$$

$$P_{a,N,m}(s_1, \ldots, s_N) = p_{a,N,m,0} + \sum_{n=1}^{N} p_{a,N,m,n} s_n, \tag{27}$$

$$P_{a,N,m}(s_1, \ldots, s_N) = 0, \qquad m = 1, \ldots N. \tag{28}$$

Of course, equations (24) all are equivalent one to another, but equations (28) are not, and solving this linear system we can get very accurate approximation to (powers of) zeros of the zeta function. Below are two numerical examples.

For $a = 0.4 + 14\mathrm{i}$ and $N = 5$

$$s_1 = 0.499999828490\ldots + 14.134725265432\ldots\mathrm{i}, \tag{29}$$

the distance between s_1 and the value of the first nontrivial zeta zero,

$$\rho_1 = 0.5 + 14.134725141734\ldots\mathrm{i}, \tag{30}$$

is less than 3×10^{-7}.

For $a = 2 + 14\mathrm{i}$ and $N = 100$

$$|s_1 - \rho_1| < 2 \times 10^{-68}. \tag{31}$$

More numerical examples can be found in [28].

4 Calculation of Summands from Dirichlet Series for the Zeta Function at Its Zeros

The same idea – usage of multiple equations (24) – can be applied for direct calculation of the individual summands, n^{-s}, from (1) when s is equal to a zero of the zeta function.

To this end we replace polynomials (17) in (16) by finite Dirichlet series: let

$$m^{-s}\zeta(s) = D_{a,N,m}(s) + O\big((s-a)^N\big) \tag{32}$$

where

$$D_{a,N,m}(s) = \sum_{n=1}^{N} d_{a,N,m,n}\, n^{-s} \tag{33}$$

(finding the coefficients $d_{a,N,m,n}$ consists in solving linear systems induced by (32)). Then we consider corresponding finite *Dirichlet series with independent exponents*:

$$D_{a,N,m}(s_1,\ldots,s_N) = \sum_{n=1}^{N} d_{a,N,m,n}\, n^{-s_n} \tag{34}$$

(this notion was introduced by the author in [27]) and linear forms

$$Q_{a,N,m}(q_1,\ldots,q_N) = \sum_{n=1}^{N} d_{a,N,m,n}\, q_n. \tag{35}$$

Now we solve the system consisting of equations

$$Q_{a,N,m}(q_1,\ldots,q_N) = 0, \qquad m = 1,\ldots N-1 \tag{36}$$

and *normalization condition*

$$q_1 = 1; \tag{37}$$

we expect to get a solution in which

$$q_n \approx n^{-\rho}. \tag{38}$$

Table 1 presents a numerical example—the result of such calculations for $N = 5$ and $a = 0.4 + 14i$. For the same value of a and $N = 200$ the accuracy is much higher: for $n = 2,\ldots,200$

$$\big|q_n/n^{-\rho_1} - 1\big| < 4 \times 10^{-325}. \tag{39}$$

More numerical data can be found in [29].

If a is between two consecutive non-trivial zeta zeros, say, ρ_k and ρ_{k+1}, then the values of q_n approximate neither $n^{-\rho_k}$ nor $n^{-\rho_{k+1}}$. However, using $q_1,\ldots,q_N$ one can calculate the coefficients of quadratic equations with zeros close to $n^{-\rho_k}$ and $n^{-\rho_{k+1}}$ for $n < N^{1/3}$; for details and other numerical examples see [29].

Table 1. The solution of system (36)–(37) and its relative accuracy for $N = 5$ and $a = 0.4 + 14i$.

n	q_n	$\left| q_n/n^{-\rho_1} - 1 \right|$
2	$-0.658570722632\ldots + 0.257458025275\ldots i$	$4.8927\ldots \cdot 10^{-8}$
3	$-0.568086335195\ldots - 0.103010905955\ldots i$	$4.6606\ldots \cdot 10^{-8}$
4	$0.367430765659\ldots - 0.339108615925\ldots i$	$5.8663\ldots \cdot 10^{-8}$
5	$-0.324829272639\ldots + 0.307385716024\ldots i$	$9.3121\ldots \cdot 10^{-8}$

References

1. Aaronson, S.: The blog. http://www.scottaaronson.com/blog/?p=2741. Accessed 25 Aug 2022
2. Agrawal, M., Kayal, N., Saxena, N.: PRIMES is in P. Ann. Math. (2) **160**(2), 781–793 (2004). https://doi.org/10.4007/annals.2004.160.781
3. Booker, A.R.: Artin's conjecture, Turing's method, and the Riemann hypothesis. Exp. Math. **15**(4), 385–407 (2006). https://doi.org/10.1080/10586458.2006.10128976
4. Booker, A.R.: Turing and the Riemann hypothesis. Notices Am. Math. Soc. **53**(10), 1208–1211 (2006)
5. Broughan, K.: Equivalents of the Riemann Hypothesis. Volume 1: Arithmetic Equivalents. Cambridge University Press, Cambridge (2017). https://doi.org/10.1017/9781108178228
6. Broughan, K.: Equivalents of the Riemann Hypothesis. Volume 2: Analytic Equivalents. Cambridge University Press, Cambridge (2017). https://doi.org/10.1017/9781108178266
7. Caceres, J.M.H.: The Riemann hypothesis and Diophantine equations. Master's thesis Mathematics, Mathematical Institute, University of Bonn (2018)
8. Calude, C.S., Calude, E.: The complexity of mathematical problems: an overview of results and open problems. Int. J. Unconv. Comput. **9**(3–4), 327–343 (2013)
9. Calude, C.S., Calude, E.: Evaluating the complexity of mathematical problems. I. Complex Syst. **18**(3), 267–285 (2009)
10. Calude, C.S., Calude, E.: Evaluating the complexity of mathematical problems. II. Complex Syst. **18**(4), 387–401 (2010)
11. Calude, C.S., Calude, E., Dinneen, M.J.: A new measure of the difficulty of problems. J. Mult.-Val. Log. Soft Comput. **12**(3–4), 285–307 (2006)
12. Calude, E.: The complexity of Riemann's hypothesis. J. Mult.-Val. Log. Soft Comput. **18**(3–4), 257–265 (2012)
13. Cooper, S.B., van Leeuwen, J. (eds.): Alan Turing - His Work and Impact. Elsevier Science, Amsterdam (2013)
14. Davis, M.: Hilbert's tenth problem is unsolvable. Am. Math. Mon. **80**, 233–269 (1973). https://doi.org/10.2307/2318447
15. Davis, M., Matijasevič, Y., Robinson, J.: Hilbert's tenth problem: diophantine equations: positive aspects of a negative solution. Proc. Symp. Pure Math. **28**, 323–378 (1976). https://doi.org/10.1090/pspum/028.2
16. Fodden, B.: Diophantine equations and the generalized Riemann hypothesis. J. Number Theory **131**(9), 1672–1690 (2011). https://doi.org/10.1016/j.jnt.2011.01.017

17. Hadamard, J.: Sur la distribution des zéros de la fonction $\zeta(s)$ et ses conséquences arithmétiques. Bulletin de la Société Mathématique de France **24**, 199–220 (1896). https://doi.org/10.24033/bsmf.545
18. Hilbert, D.: Mathematische Probleme. Vortrag, gehalten auf dem internationalen Mathematiker Kongress zu Paris 1900. Nachr. K. Ges. Wiss., Göttingen, Math.-Phys. Kl, pp. 253–297 (1900). Reprinted in Gesammelte Abhandlungen, Springer, Berlin 3 (1935); Chelsea, New York (1965). English translation: Bull. Amer. Math. Soc. **8**, 437–479 (1901–1902); reprinted. In: Browder (ed.) Mathematical Developments arising from Hilbert Problems, Proceedings of Symposia in Pure Mathematics 28, American Mathematical Society, pp. 1–34 (1976)
19. Jones, J.P., Matiyasevich, Y.V.: Register machine proof of the theorem on exponential Diophantine representation of enumerable sets. J. Symb. Log. **49**, 818–829 (1984). https://doi.org/10.2307/2274135
20. Kreisel, G.: Mathematical significance of consistency proofs. J. Symb. Log. **23**(2), 155–182 (1958)
21. Lambek, J.: How to program an infinite abacus. Can. Math. Bull. **4**, 295–302 (1961). https://doi.org/10.4153/CMB-1961-032-6
22. Lenstra Jr, H.W., Pomerance, C.: Primality testing with Gaussian periods. http://www.math.dartmouth.edu/~carlp/aks041411.pdf. Accessed 25 Aug 2022
23. Manin, Y.I., Panchishkin, A.A.: Introduction to number theory (in Russian). Itogi Nauki Tekh., Ser. Sovrem. Probl. Mat., Fundam. Napravleniya **49**, 348 p. (1990). Translated as: Introduction to modern number theory. Fundamental problems, ideas and theories. Second edition. Encyclopaedia of Mathematical Sciences, 49. Springer-Verlag, Berlin, 2005. xvi+514 pp. ISBN: 978-3-540-20364-3; 3-540-20364-8
24. Matiyasevich, Y.: The Riemann hypothesis in computer science. Theor. Comput. Sci. **807**, 257–265 (2020). https://doi.org/10.1016/j.tcs.2019.07.028
25. Matiyasevich, Y.V.: Hilbert's Tenth Problem (in Russian). Fizmatlit (1993). http://logic.pdmi.ras.ru/~yumat/H10Pbook. English translation: MIT Press, Cambridge (Massachusetts) London (1993). http://mitpress.mit.edu/9780262132954/. French translation: Masson, Paris Milan Barselone (1995). Greek translation: EURYALOS editions, Athens, 2022
26. Matiyasevich, Y.V.: The Riemann Hypothesis as the parity of binomial coefficients (in Russian). Chebyshevskii Sb. **19**, 46–60 (2018). https://doi.org/10.22405/2226-8383-2018-19-3-46-60
27. Matiyasevich, Y.V.: Hunting zeros of Dirichlet series by linear algebra. I. POMI Preprints (01), 18 p. (2020). https://doi.org/10.13140/RG.2.2.29328.43528
28. Matiyasevich, Y.V.: Hunting zeros of Dirichlet series by linear algebra. II. POMI Preprints (01), 18 p. (2022). https://doi.org/10.13140/RG.2.2.20434.22720
29. Matiyasevich, Y.V.: Hunting zeros of Dirichlet series by linear algebra III (in Russian). POMI Preprints (03), 31 p. (2022). https://doi.org/10.13140/RG.2.2.28325.99044. Extended English abstract http://logic.pdmi.ras.ru/~yumat/publications/papers/139_paper/eng_abstract_ext.pdf. Accessed 25 Aug 2022
30. Matiyasevich, Y.: Hilbert's tenth problem: what was done and what is to be done. In: Hilbert's Tenth Problem: Relations with Arithmetic and Algebraic Geometry. Proceedings of the workshop, Ghent University, Belgium, 2–5 November 1999, pp. 1–47. American Mathematical Society, Providence (2000)
31. Melzak, Z.A.: An informal arithmetical approach to computability and computation. Can. Math. Bull. **4**, 279–293 (1961). https://doi.org/10.4153/CMB-1961-031-9
32. Miller, G.L.: Riemann's hypothesis and tests for primality. J. Comput. Syst. Sci. **13**, 300–317 (1976). https://doi.org/10.1016/S0022-0000(76)80043-8

33. Minsky, M.L.: Computation: finite and infinite machines. Prentice-Hall Series in Automatic Computation, vol. VII, 317 p. Prentice-Hall, Inc., Englewood Cliffs (1967)
34. Minsky, M.L.: Recursive unsolvability of Post's problem of "Tag" and other topics in theory of Turing machines. Ann. Math. **2**(74), 437–455 (1961). https://doi.org/10.2307/1970290
35. Moroz, B.Z.: The Riemann hypothesis and Diophantine equations (in Russian). St. Petersburg Mathematical Society Preprints (03) (2018). http://www.mathsoc.spb.ru/preprint/2018/index.html#03. Accessed 25 Aug 2022
36. Murty, M.R., Fodden, B.: Hilbert's Tenth Problem. An Introduction to Logic, Number Theory, and Computability. Student Mathematical Library, vol. 88. American Mathematical Society (AMS), Providence (2019). https://doi.org/10.1090/stml/088
37. Nayebi, A.: On the Riemann hypothesis and Hilbert's tenth problem. Unpublished Manuscript, February 2012. http://web.stanford.edu/~anayebi/projects/RH_Diophantine.pdf. Accessed 25 Aug 2022
38. Riemann, B.: Über die Anzhal der Primzahlen unter einer gegebenen Grösse. Monatsberichter der Berliner Akademie (1859), included into: Riemann, B. Gesammelte Werke. Teubner, Leipzig, 1892; reprinted by Dover Books, New York (1953). http://www.claymath.org/publications/riemanns-1859-manuscript. English translation. http://www.maths.tcd.ie/pub/HistMath/People/Riemann/Zeta/EZeta.pdf
39. Turing, A.M.: Systems of logic based on ordinals. Proc. Lond. Math. Soc. **2**(45), 161–228 (1939). https://doi.org/10.1112/plms/s2-45.1.161
40. de la Vallée Poussin, C.J.: Recherches analytiques de la théorie des nombres premiers. Annales de la Société Scientifique de Bruxelles **20B**, 183–256 (1896)
41. Yedidia, A., Aaronson, S.: A relatively small Turing machine whose behavior is independent of set theory. Complex Syst. **25**, 297–327 (2016). https://doi.org/10.25088/ComplexSystems.25.4.297

Shuffle Product of Regular Languages: Results and Open Problems

Jean-Éric Pin[(✉)]

IRIF, CNRS and Université Paris-Cité, Case 7014, 75205 Paris Cedex 13, France
`Jean-Eric.Pin@irif.fr`

Abstract. This survey paper presents known results and open problems on the shuffle product applied to regular languages. We first focus on varieties and positive varieties closed under shuffle. Next we turn to the class of intermixed languages, the smallest class of languages containing the letters and closed under Boolean operations, product and shuffle. Finally, we study Schnoebelen's sequential and parallel decompositions of languages and discuss various open questions around this notion.

Keywords: Shuffle product · Regular languages · Open problems

1 Introduction

The shuffle product is a standard tool for modeling concurrency which has long been studied in formal language theory. A nice survey on this topic was proposed by Restivo [12] in 2015. Restivo's article is divided into two parts: the first part is devoted to language theory and the second part to combinatorics on words. Our current survey paper focuses on the shuffle product applied to regular languages and can therefore be seen as an extension of the first part of [12].

The main part of the article, Sect. 3, is devoted to the study of various classes of regular languages closed under shuffle. We first examine the varieties closed under shuffle, for which a complete description is known. We then turn to positive varieties of languages closed under shuffle, for which the situation is more contrasted and several questions are still open. Next we turn to the class of intermixed languages, the smallest class of languages containing the letters and closed under Boolean operations, product and shuffle. Here again, only partial results are known.

In the last part of the paper, we study Schnoebelen's sequential and parallel decompositions of languages and discuss various open questions around this notion.

2 Shuffle and Recognition

2.1 Shuffle Product

The *shuffle* $u \shuffle v$ of two words u and v is the set of words obtained by shuffling u and v. Formally, it is the set of words of the form $u_1 v_1 \cdots u_n v_n$, where the u_i's

© The Author(s), under exclusive license to Springer Nature Switzerland AG 2022
D. Poulakis and G. Rahonis (Eds.): CAI 2022, LNCS 13706, pp. 26–39, 2022.
https://doi.org/10.1007/978-3-031-19685-0_3

and v_i's are possibly empty words such that $u_1 \cdots u_n = u$ and $v_1 \cdots v_n = v$. For instance,

$$ab \sqcup ba = \{abba, baab, baba, abab\}$$

This definition extends by linearity to languages: the *shuffle* of two languages L and K is the language

$$L \sqcup K = \bigcup_{u \in L, v \in K} u \sqcup v$$

The shuffle product is a commutative and associative operation on languages and it distributes over union.

2.2 Monoids and Ordered Monoids

The *algebraic approach* to the study of regular languages is based on the use of monoids to recognise languages. There are two versions, one using monoids and the other one using ordered monoids. The ordered version is more suitable for classes of languages not closed under complementation. Let us briefly recall the relevant definitions.

An *ordered monoid* is a monoid M equipped with a partial order $\leqslant$ compatible with the product on M: for all $x, y, z \in M$, if $x \leqslant y$ then $zx \leqslant zy$ and $xz \leqslant yz$. Note that the equality relation makes any monoid an ordered monoid.

Let P be a subset of M. It is a *lower set* if, for all $s, t \in P$, the conditions $s \in P$ and $t \leqslant s$ imply $t \in P$. It is an *upper set* if $s \in P$ and $s \leqslant t$ imply $t \in P$. Finally, *the lower set generated by P* is the set

$$\downarrow P = \{t \in M \mid \text{ there exists } s \in P \text{ such that } t \leqslant s\}.$$

Given two ordered monoids M and N, a *morphism of ordered monoids* $\varphi \colon M \to N$ is an order-preserving monoid morphism from M to N. In particular, if $(M, \leqslant)$ is an ordered monoid, the identity on M is a morphism of ordered monoids from $(M, =)$ to $(M, \leqslant)$.

A monoid M *recognizes* a language L of A^* if there exist a morphism $\varphi \colon A^* \to M$ and a subset U of M such that $L = \varphi^{-1}(U)$. In the ordered version, the definition is the same, but U is required to be an upper set of the ordered monoid.

Let L be a language of A^*. The *syntactic preorder* of L is the preorder $\preccurlyeq_L$ defined on A^* by $u \preccurlyeq_L v$ if and only if, for every $x, y \in A^*$,

$$xuy \in L \Rightarrow xvy \in L. \tag{1}$$

The *syntactic congruence* of L is the associated equivalence relation $\sim_L$, defined by $u \sim_L v$ if and only if $u \preccurlyeq_L v$ and $v \preccurlyeq_L u$.

The *syntactic monoid* of L is the quotient monoid $\mathrm{Synt}(L)$ of A^* by $\sim_L$ and the natural homomorphism $\eta : A^* \to \mathrm{Synt}(L)$ is called the *syntactic morphism* of L. The syntactic preorder $\preccurlyeq_L$ induces a partial order $\leqslant_L$ on $\mathrm{Synt}(L)$. The resulting ordered monoid is called the *syntactic ordered monoid*[1] of L. Recall that a language is regular if and only if its syntactic monoid is finite.

2.3   Power Monoids and Lower Set Monoids

Let M be a monoid. Then the set $\mathcal{P}(M)$ of subsets of M is a monoid, called the *power monoid* of M, under the multiplication of subsets defined by

$$XY = \{xy \mid x \in X \text{ and } y \in Y\}$$

The ordered counterpart of this construction works as follows. Let $(M, \leqslant)$ be an ordered monoid and let $\mathcal{P}^{\downarrow}(M)$ be the set of all lower sets of M. The *product of two lower sets* X and Y is the lower set

$$XY = \{z \in M \mid \text{there exist } x \in X \text{ and } y \in Y \text{ such that } z \leqslant xy\}.$$

This operation makes $\mathcal{P}^{\downarrow}(M)$ a monoid. Furthermore, set inclusion is compatible with this product and thus $(\mathcal{P}^{\downarrow}(M), \subseteq)$ is an ordered monoid, called the *lower set monoid* of M.

The connection between the shuffle and the operator $\mathcal{P}$ was given in [10].

Proposition 1. *Let L_1 and L_2 be two languages and let M_1 and M_2 be monoids recognizing L_1 and L_2 respectively. Then $L_1 \shuffle L_2$ is recognized by the monoid $\mathcal{P}(M_1 \times M_2)$.*

Since a language is regular if and only if it is recognised by a finite monoid, Proposition 1 gives an algebraic proof of the well-known fact that regular languages are closed under shuffle. The following ordered version was given in [4].[2]

Proposition 2. *Let L_1 and L_2 be two languages and let M_1 and M_2 be ordered monoids recognizing L_1 and L_2 respectively. Then $L_1 \shuffle L_2$ is recognized by the ordered monoid $\mathcal{P}^{\downarrow}(M_1 \times M_2)$.*

3   Classes of Languages Closed Under Shuffle

It is a natural question to look for classes of regular languages closed under shuffle. In this section, we focus successively on varieties of languages and on positive varieties of languages. The last subsection is devoted to the class of intermixed languages and its subclasses.

---

[1] The syntactic ordered monoid of a language was first introduced by Schützenberger in 1956, but he apparently only made use of the syntactic monoid later on. I rediscovered this notion in 1995 [11], but unfortunately used the opposite order for several years, in particular in [3–6], before I switched back to the original order.

[2] As explained in the first footnote, the opposite of the syntactic order was used in this paper, and consequently, upper set monoids were used in place of lower set monoids.

3.1   Varieties of Languages Closed Under Shuffle

Following the success of the variety approach to classify regular languages (see [17] for a recent survey), Perrot proposed in 1977 to find the varieties of languages closed under shuffle. Let us first recall the definitions.

A *class of languages* is a correspondence $\mathcal{V}$ which associates with each alphabet A a set $\mathcal{V}(A^*)$ of languages of A^*. A class of regular languages is a *variety of languages* if it is closed under Boolean operations (that is, finite union, finite intersection, and complementation), left and right quotients and inverses of morphisms between free monoids.

A *variety of finite monoids* is a class of finite monoids closed under taking submonoids, quotients and finite direct products. If $\mathbf{V}$ is a variety of finite monoids, let $\mathcal{V}(A^*)$ denote the set of regular languages of A^* recognized by a monoid of $\mathbf{V}$. The correspondence $\mathbf{V} \mapsto \mathcal{V}$ associates with each variety of finite monoids a variety of languages. Conversely, to each variety of languages $\mathcal{V}$, we associate the variety of finite monoids generated by the syntactic monoids of the languages of $\mathcal{V}$. Eilenberg's variety theorem states that these two correspondences define mutually inverse bijective correspondences between varieties of finite monoids and varieties of languages. For instance, Schützenberger [15] proved that star-free languages correspond to aperiodic monoids.

To start with, let us describe the smallest nontrivial variety of languages closed under shuffle. Let $[u]$ denote the *commutative closure* of a word u, which is the set of words commutatively equivalent to u. For instance, $[aab] = \{aab, aba, baa\}$. A language L is *commutative* if, for every word $u \in L$, $[u]$ is contained in L. Equivalently, a language is *commutative* if its syntactic monoid is commutative. A characterisation of the star-free commutative languages related to the shuffle product is given in [2, Proposition 1.2].

Proposition 3. *A language of A^* is star-free commutative if and only if it is a finite union of languages of the form $[u] \shuffle B^*$ where u is a word and B is a subset of A.*

This leads to the following result of Perrot [10].

Proposition 4. *The star-free commutative languages form a variety of languages, which is the smallest nontrivial variety of languages closed under shuffle. It is also the smallest class of languages closed under Boolean operations and under shuffle by a letter.*

Perrot actually characterised all commutative varieties of languages closed under shuffle. They correspond, via Eilenberg's correspondence, to the varieties of finite commutative monoids whose groups belong to a given variety of finite commutative groups. Perrot also conjectured that the only non-commutative variety of languages closed under shuffle was the variety of all regular languages. This conjecture remained open for twenty years, but was finally solved by Ésik and Simon [8].

Theorem 1. *The unique non-commutative variety of languages closed under shuffle is the variety of all regular languages.*

The story of the proof of Theorem 1 is worth telling. A *renaming* is a *length preserving morphism* $\varphi: A^* \to B^*$. This means that $|\varphi(u)| = |u|$ for each word u of A^*, or, equivalently, that each letter of A is mapped by φ to a letter of B. The characterisation of varieties of languages closed under renaming was known for a long time. For a variety $\mathbf{V}$ of finite monoids, let $\mathbf{PV}$ denote the variety of finite monoids generated by the monoids of the form $\mathcal{P}(M)$, where $M \in \mathbf{V}$. A variety of finite monoids $\mathbf{V}$ is a *fixed point* of the operator $\mathbf{P}$ if $\mathbf{PV} = \mathbf{V}$.

Reuteunauer [13] and Straubing [16] independently proved the following result:

Proposition 5. *A variety of languages is closed under renaming if and only if the corresponding variety of finite monoids is a fixed point of the operator $\mathbf{P}$.*

It was also known that the unique non-commutative variety of languages satisfying this condition is the variety of all regular languages. Ésik and Simon managed to find an ingenious way to link renamings and the shuffle operation.

Proposition 6. *Let $\varphi: A^* \to B^*$ be a surjective renaming and let $C = A \cup \{c\}$, where c is a new letter. Then there exist monoid morphisms $\pi: C^* \to A^*$, $\gamma: C^* \to \{a, b\}^*$ and $\eta: B^* \to C^*$ such that*

$$\varphi(L) = \eta^{-1}\Big(\big(\pi^{-1}(L) \cap \gamma^{-1}((ab)^*)\big) \sqcup\!\sqcup A^*\Big)$$

It follows that if a variety of languages containing the language $(ab)^*$ is closed under shuffle, then it is also closed under renaming, a key argument in the proof of Theorem 1.

3.2   Positive Varieties of Languages Closed Under Shuffle

A variation of Eilenberg's variety theorem was proposed by the author in [11]. It gives two mutually inverse bijective correspondences between varieties of finite ordered monoids and positive varieties of languages.

The definition of varieties of finite ordered monoids is similar to that of varieties of finite monoids: they are classes class of finite ordered monoids closed under taking ordered submonoids, quotients and finite direct products.

A class of regular languages is a *positive variety of languages* if it is closed under finite union, finite intersection, left and right quotients and inverses of morphisms between free monoids. The difference with language varieties is that positive varieties are not necessarily closed under complementation.

It is now natural to try to describe all positive varieties closed under shuffle. As in the case of varieties, this study is related to the closure under renaming. For a variety $\mathbf{V}$ of finite ordered monoids, let $\mathbf{P}^{\downarrow}\mathbf{V}$ denote the variety of finite monoids generated by the ordered monoids of the form $\mathcal{P}^{\downarrow}(M)$, where $M \in \mathbf{V}$. A variety of finite ordered monoids $\mathbf{V}$ is a *fixed point* of the operator $\mathbf{P}^{\downarrow}$ if $\mathbf{P}^{\downarrow}\mathbf{V} = \mathbf{V}$. The following result was proved in [3]:

Proposition 7. *A positive variety of languages is closed under renaming if and only if the corresponding variety of finite ordered monoids is a fixed point of the operator* $\mathbf{P}^{\downarrow}$.

Coming back to the shuffle, Proposition 2 leads to the following corollary.

Corollary 1. *If a variety of finite ordered monoids is a fixed point of the operator* $\mathbf{P}^{\downarrow}$, *then the corresponding positive variety of languages is closed under shuffle.*

Note that, contrary to Proposition 7, Corollary 1 only gives a sufficient condition for a positive variety of languages to be closed under shuffle.

The results of Sect. 3.1 show that the unique maximal proper variety of languages closed under shuffle is the variety of commutative languages. The counterpart of this result for positive varieties was given in [4].

Theorem 2. *There is a largest proper positive variety of languages closed under shuffle.*

Let $\mathcal{W}$ denote this positive variety and let $\mathbf{W}$ be its corresponding variety of finite ordered monoids. Their properties are summarized in the next statements, also proved in [4,6].

Theorem 3. *The positive variety $\mathcal{W}$ is the largest positive variety of languages such that, for $A = \{a, b\}$, the language $(ab)^*$ does not belong to $\mathcal{W}(A^*)$.*

A key result is that $\mathbf{W}$, and hence $\mathcal{W}$, are decidable. A bit of semigroup theory is needed to make this statement precise.

Two elements s and t of a monoid are *mutually inverse* if $sts = s$ and $tst = t$. An *ideal* of a monoid M is a subset I of M such that $MIM \subseteq I$. It is *minimal* if, for every ideal J of M, the condition $J \subseteq I$ implies $J = \emptyset$ or $J = I$. Every finite monoid admits a unique minimal ideal. Finally, let us recall that every element s of a finite monoid has a unique idempotent power, traditionally denoted by s^{ω}. In the following, we will also use the notation $x^{\omega+1}$ as a shortcut for xx^{ω}.

Theorem 4. *A finite ordered monoid M belongs to $\mathbf{W}$ if and only if, for any pair (s, t) of mutually inverse elements of M, and any element z of the minimal ideal of the submonoid of M generated by s and t, $(stzst)^{\omega} \leqslant st$.*

Thus a regular language belongs to $\mathcal{W}$ if and only if its ordered syntactic monoid satisfies the decidable condition stated in Theorem 4. An equivalent characterisation in term of the minimal automaton of the language is given in [6].

The next theorem shows that the positive variety $\mathcal{W}$ is very robust.

Theorem 5. *The positive variety $\mathcal{W}$ is closed under the following operations: finite union, finite intersection, left and right quotients, product, shuffle, renaming and inverses of morphisms. It is not closed under complementation.*

The positive variety $\mathcal{W}$ can be defined alternatively as the largest proper positive variety of languages satisfying (1) (respectively (2) or (3)):

(1) not containing the language $(ab)^*$;
(2) closed under shuffle;
(3) closed under renaming;

Despite its numerous closure properties, no constructive description of $\mathcal{W}$, similar to the definition of star-free or regular languages, is known. For instance, the least positive variety of languages satisfying conditions (1)–(3) is the positive variety of polynomials of group languages, which is strictly contained in $\mathcal{W}$.

Problem 1. Find a constructive description of $\mathcal{W}$, possibly by introducing more powerful operators on languages of $\mathcal{W}$.

Let us come back to the problem of finding all positive varieties of languages closed under shuffle. The first question is to know in which case the converse of Corollary 1 holds. More precisely,

Problem 2. For which positive varieties of languages closed under shuffle is the corresponding variety of finite ordered monoids a fixed point of the operator $\mathbf{P}^{\downarrow}$?

We know this is the case for $\mathcal{W}$, but the general case is unknown. That said, an in-depth study of the fixed points of the operator $\mathbf{P}^{\downarrow}$ can be found in [1]. This paper actually covers the more general case of lower set semigroups and studies the fixed points of the operator $\mathbf{P}^{\downarrow}$ on varieties of finite ordered semigroups. An important property is the following:

Proposition 8. *Every intersection and every directed union of fixed points of* $\mathbf{P}^{\downarrow}$ *is also a fixed point for* $\mathbf{P}^{\downarrow}$.

The article [1] gives six independent basic types of such fixed points, from which many more may be constructed using intersection. Moreover, it is conjectured that all fixed points of $\mathbf{P}^{\downarrow}$ can be obtained in this way. The presentation of these basic types would be too technical for this survey article, but one of them is the variety $\mathbf{W}$.

3.3   Intermixed Languages

In the early 2000s, Restivo proposed as a challenge to characterise the smallest class of languages containing the letters and closed under Boolean operations, product and shuffle. Let us call *intermixed* the languages of this class.

Problem 3 (Restivo). Is it decidable to know whether a given regular language is intermixed?

This problem is still widely open, and only partial results are known. To start with, the smallest class of languages containing the letters and closed under Boolean operations and product is by definition the class of star-free languages. It is not immediate to see that star-free languages are not closed under shuffle, but an example was given in [10]: the languages $(abb)^*$ and a^* are star-free, but their shuffle product is not star-free. This led Castiglione and Restivo [7] to propose the following question:

Problem 4. Determine conditions under which the shuffle of two star-free languages is star-free.

The following result, which improves on the results of [7], can be seen as a reasonable answer to this problem.

Theorem 6. *The shuffle of two star-free languages of the positive variety* $\mathcal{W}$ *is star-free.*

Proof. This is a consequence of the fact that the intersection of $\mathbf{W}$ and the variety of finite aperiodic monoids is a fixed point of the operator $\mathbf{P}^{\downarrow}$, a particular instance of [1, Theorem 7.4].

Here are the known closure properties of intermixed languages obtained in [2]. A morphism $\varphi \colon A^* \to B^*$ is said to be *length-decreasing* if $|\varphi(u)| \leqslant |u|$ for every word u of A^*.

Proposition 9. *The class of intermixed languages is closed under left and right quotients, Boolean operations, product and shuffle. It is also closed under inverses of length-decreasing morphisms, but it is not closed under inverses of morphisms.*

We now give an algebraic property of the syntactic morphism of intermixed languages, which is the main result of [2].

Theorem 7. *Let* $\eta : A^* \to M$ *be the syntactic morphism of a regular language of* A^* *and let* $x, y \in \eta(A) \cup \{1\}$. *If* L *is intermixed, then* $x^{\omega+1} = x^{\omega}$ *and* $(x^{\omega}y^{\omega})^{\omega+1} = (x^{\omega}y^{\omega})^{\omega}$.

Theorem 7 shows that intermixed languages form a proper subclass of the class of regular languages, since the language $(aa)^*$ does not satisfy the first identity. Unfortunately, we do not know whether our two identities suffice to characterise the intermixed languages and hence the decidability of this class remains open.

The reader will find in [2] several partial results on subclasses of the class of intermixed languages, but only one of these subclasses is known to be decidable. It is actually a rather small class in which the use of the shuffle is restricted to shuffling a language with a letter.

It is shown in [2] that the smallest class with these properties is the class of commutative star-free languages. Let us set aside this case by considering classes containing at least one noncommutative language. In fact, for technical reasons which are partly justified by [2, Proposition 4.1], our classes will always contain the languages of the form $\{ab\}$ where a and b are two distinct letters of the alphabet.

In summary, we consider the smallest class of languages $\mathcal{C}$ containing the languages of the form $\{ab\}$, where a, b are distinct letters, and which is closed under Boolean operations and under shuffle by a letter. The following results are obtained in [2].

Proposition 10. *A language L belongs to $\mathcal{C}$ if and only if there exists a star-free commutative language C such that the symmetric difference $L \bigtriangleup C$ is finite.*

In view of this result, it is natural to call *almost star-free commutative* the languages of the class $\mathcal{C}$. These languages admit the following algebraic characterisation.

Theorem 8. *Let $\eta : A^* \to M$ be the syntactic morphism of a regular language of A^* and let $x, y, z \in \eta(A^+)$. Then L is almost star-free commutative if and only if $x^\omega = x^{\omega+1}$, $x^\omega y = yx^\omega$ and $x^\omega yz = x^\omega zy$.*

Corollary 2. *It is decidable whether a given regular language is almost star-free commutative.*

4   Sequential and Parallel Decompositions

We now switch to a different topic, which is still related to the shuffle product. Sequential and parallel decompositions of languages were introduced by Schnoebelen [14] for some model-cheking applications. A reminder of the notions of rational and recognizable subsets of a monoid is in order to define these decompositions properly.

Let M be a monoid. A subset P of M is *recognizable* if there exist a finite monoid F and a monoid morphism $\varphi : M \to F$ such that $P = \varphi^{-1}(\varphi(P))$. It is well known that the class $\mathrm{Rec}(M)$ of recognizable subsets of M is closed under finite union, finite intersection and complement.

The class $\mathrm{Rat}(M)$ of *rational* subsets of M is the smallest set $\mathcal{R}$ of subsets of M satisfying the following properties:

(1) For each $m \in M$, $\{m\} \in \mathcal{R}$
(2) The empty set belongs to $\mathcal{R}$, and if X, Y are in $\mathcal{R}$, then $X \cup Y$ and XY are also in $\mathcal{R}$.
(3) If $X \in \mathcal{R}$, the submonoid X^* generated by X is also in $\mathcal{R}$.

Let τ and σ be the transductions from A^* into $A^* \times A^*$ defined as follows:

$$\tau(w) = \{(u, v) \in A^* \times A^* \mid w = uv\}$$
$$\sigma(w) = \{(u, v) \in A^* \times A^* \mid w \in u \shuffle v\}$$

Observe that σ is a monoid morphism from A^* into the monoid $\mathcal{P}(A^* \times A^*)$, that is, $\sigma(x_1 x_2) = \sigma(x_1)\sigma(x_2)$ for all $x_1, x_2 \in A^*$.

4.1   Definitions and Examples

We are now ready to give the definitions of the two types of decomposition. Let $\mathcal{S}$ be a set of languages. A language K admits a *sequential decomposition* over $\mathcal{S}$ if $\tau(K)$ is a finite union of sets of the form $L \times R$, where $L, R \in \mathcal{S}$.

A language K admits a *parallel decomposition* over $\mathcal{S}$ if $\sigma(K)$ is a finite union of sets of the form $L \times R$, where $L, R \in \mathcal{S}$.

A *sequential (resp. parallel) system* is a finite set $\mathcal{S}$ of languages such that each member of $\mathcal{S}$ admits a sequential (resp. parallel) decomposition over $\mathcal{S}$. A language is *sequentially decomposable* if it belongs to some sequential system. It is *decomposable* if it belongs to a system which is both sequential and parallel. Thus, for each decomposable language L, one can find a sequential and parallel system $\mathcal{S}(L)$ containing L.

Example 1. Let $K = \{abc\}$. Then

$$\tau(K) = (\{1\} \times \{abc\}) \cup (\{a\} \times \{bc\}) \cup (\{ab\} \times \{c\}) \cup (\{abc\} \times \{1\})$$

and

$$\sigma(K) = (\{1\} \times \{abc\}) \cup (\{a\} \times \{bc\})$$
$$\cup (\{b\} \times \{ac\}) \cup (\{c\} \times \{ab\}) \cup (\{ab\} \times \{c\})$$
$$\cup (\{bc\} \times \{a\}) \cup (\{ac\} \times \{b\}) \cup (\{abc\} \times \{1\})$$

One can verify that the set $\mathcal{S} = \{\{1\}, \{a\}, \{b\}, \{c\}, \{ab\}, \{ac\}, \{bc\}, \{abc\}\}$ is a sequential and parallel system, and hence K is a decomposable language.

Here is a more complex example. Recall that a word $u = a_1 a_2 \cdots a_n$ (where $a_1, \ldots, a_n$ are letters) is a *subword* of a word v if v can be factored as $v = v_0 a_1 v_1 \cdots a_n v_n$. For instance, ab is a subword of $cacbc$. Given two words u and v, let $\binom{v}{u}$ denote the number of distinct ways to write u as a subword of v.

Example 2. Let L be the set of words of $\{a, b\}^*$ having ab as a subword an odd number of times. Its minimal automaton is represented below:

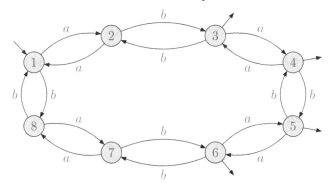

The transition monoid of L is the dihedral group D_4, a non-abelian group of order 8. For $i, j, k \in \{0, 1\}$ and $c \in A$, let

$$M_k^{i,j} = \left\{ x \in A^* \mid |x|_a \equiv i \bmod 2, \ |x|_b \equiv j \bmod 2 \text{ and } \binom{x}{ab} \equiv k \bmod 2 \right\}$$

$$M^{i,j} = \left\{ x \in A^* \mid |x|_a \equiv i \bmod 2, \ |x|_b \equiv j \bmod 2 \right\}$$

$$M_c^{i,j} = M^{i,j} \cap A^* c A^*$$

Let $\mathcal{F}$ be the set of finite union of languages of the form $M_k^{i,j}$, $M_c^{i,j}$ or $\{1\}$. A non-trivial verification [5] shows that $\mathcal{F}$ is a sequential and parallel system for L. Thus L is a decomposable language.

4.2   Closure Properties

The following result is stated in [5], but partly relies on results from [14], where (3) implies (1) is credited to Arnold and Carton.

Theorem 9. *Let K be a language. The following conditions are equivalent:*

(1) K is regular,
(2) $\tau(K)$ is recognizable,
(3) K is sequentially decomposable.

Consequently, if K is decomposable, then K is regular and $\sigma(K)$ is recognizable.

As observed by Schnoebelen, it follows that the language $(ab)^*$ is not decomposable, since the set $\sigma((ab)^*)$ is not recognizable.

The following theorem summarises the closure properties of decomposable languages obtained in [14] and [5].

Theorem 10. *The class of decomposable languages is closed under finite union, product, shuffle and left and right quotients. It is not closed under intersection, complementation and star. It is also closed under inverses of length preserving morphisms, but not under inverses of morphisms.*

The negative parts of this theorem are obtained from the following counterexamples: the languages $(ab)^+ \cup (ab)^*bA^*$ and $(ab)^+ \cup (ab)^*aaA^*$ are decomposable but their intersection $(ab)^+$ is not. Furthermore, the language $L = (aab)^* \cup A^*b(aa)^*abA^*$ is decomposable, but if $\varphi \colon A^* \to A^*$ is the morphism defined by $\varphi(a) = aa$ and $\varphi(b) = b$, then $\varphi^{-1}(L) = (ab)^*$ is not decomposable.

4.3   Schnoebelen's Problem

Schnoebelen [14] asked for a description of the class of decomposable languages, which implicitly leads to the following problem:

Problem 5. Is it decidable to know whether a regular language is decomposable?

As a first step, Schnoebelen [14] proved the following result.

Proposition 11. *Every commutative regular language is decomposable.*

Denote by Pol($\mathcal{C}om$) the *polynomial closure* of the class of commutative regular languages, that is, the finite unions of products of commutative regular languages. Since, by Theorem 10, decomposable languages are closed under finite union and product, Proposition 11 can be improved as follows:

Theorem 11 (Schnoebelen). *Every language of* Pol($\mathcal{C}om$) *is decomposable.*

Schnoebelen originally conjectured that a language is decomposable if and only if it belongs to Pol($\mathcal{C}om$). However, this conjecture has been refuted in [5], where it is shown that the decomposable language of Example 2 is not in Pol($\mathcal{C}om$).

Describing the class of decomposable languages seems to be a difficult question and Problem 5 is still widely open. One could hope for an algebraic approach, but decomposable languages do not form a positive variety of languages for two reasons. First, they are not closed under inverses of morphisms. This is a minor issue, since they are closed under inverses of renamings, and one could still hope to use Straubing's positive *lp*-varieties instead (see [17] for more details). However, they are also not closed under intersection, and hence we may have to rely on the conjunctive varieties defined by Klíma and Polák [9].

Even if decomposable languages are not closed under intersection, a weaker closure property still holds.

Proposition 12 (Arnold). *The intersection of a decomposable language with a commutative regular language is decomposable.*

This result can be used to give a non-trivial example of indecomposable language.

Proposition 13. *Let $A = \{a, b, c\}$. The language $(ab)^*cA^*$ is not decomposable.*

Proof. Let $L = (ab)^*cA^*$. If L is decomposable, the language

$$Lc^{-1} = (ab)^* \cup (ab)^*cA^*$$

is decomposable by Theorem 10. The intersection of this language with the commutative regular language $\{a, b\}^*$ is equal to $(ab)^*$, and thus by Proposition 12, $(ab)^*$ should also be decomposable. But we have seen this is not the case and hence L is not decomposable.

Let us conclude this section with a conjecture. A *group language* is a regular language recognized by a finite group. Let Pol($\mathcal{G}$) be the *polynomial closure* of the class of group languages, that is, the finite unions of languages of the form $L_0a_1L_1 \cdots a_nL_n$, where each L_i is a group language and the a_i's are letters. The class Pol($\mathcal{G}$) is a well studied positive variety, with a simple characterisation: a regular language belongs to Pol($\mathcal{G}$) if and only if, in its ordered syntactic monoid, the relation $1 \leqslant e$ holds for all idempotents e. We propose the following conjecture as a generalisation of Example 2:

Conjecture 1. Every language of Pol($\mathcal{G}$) is decomposable.

Since decomposable languages are closed under finite union and product, it would suffice to prove that every group language is decomposable. The following result could potentially help solve the conjecture.

Proposition 14. *Let G be a finite group, let $\pi : A^* \to G$ be a surjective morphism and let $L = \pi^{-1}(1)$.*

(1) If the language L is decomposable, then every language recognized by π is decomposable.

(2) The following formula holds

$$\sigma(L) = \bigcup_{\substack{r,s \leqslant |G|^4 \\ (a_1 \cdots a_r \sqcup\!\sqcup b_1 \cdots b_s) \cap L \neq \emptyset}} (La_1 La_2 L \cdots La_r L) \times (Lb_1 Lb_2 L \cdots Lb_s L)$$

The bound $|G|^4$ is probably not optimal. If it could be improved to $|G|$, this may lead to a parallel system containing L.

5   Conclusion

The problems presented in this article give evidence that there is still a lot to be done in the study of the shuffle product, even for regular languages. We urge the reader to try to solve them!

References

1. Almeida, J., Cano, A., Klíma, O., Pin, J.É.: Fixed points of the lower set operator. Internat. J. Algebra Comput. **25**(1–2), 259–292 (2015)
2. Berstel, J., Boasson, L., Carton, O., Pin, J.É., Restivo, A.: The expressive power of the shuffle product. Inf. Comput. **208**, 1258–1272 (2010)
3. Cano, A., Pin, J.É.: Upper set monoids and length preserving morphisms. J. Pure Appl. Algebra **216**, 1178–1183 (2012)
4. Cano Gómez, A., Pin, J.É.: Shuffle on positive varieties of languages. Theoret. Comput. Sci. **312**, 433–461 (2004)
5. Gómez, A.C., Pin, J.É.: On a conjecture of schnoebelen. In: Ésik, Z., Fülöp, Z. (eds.) DLT 2003. LNCS, vol. 2710, pp. 35–54. Springer, Heidelberg (2003). https://doi.org/10.1007/3-540-45007-6_4
6. Gómez, A.C., Pin, J.É.: A robust class of regular languages. In: Ochmański, E., Tyszkiewicz, J. (eds.) MFCS 2008. LNCS, vol. 5162, pp. 36–51. Springer, Heidelberg (2008). https://doi.org/10.1007/978-3-540-85238-4_3
7. Castiglione, G., Restivo, A.: On the shuffle of star-free languages. Fund. Inform. **116**(1–4), 35–44 (2012)
8. Ésik, Z., Simon, I.: Modeling literal morphisms by shuffle. Semigroup Forum **56**(2), 225–227 (1998)
9. Klíma, O., Polák, L.: On varieties of meet automata. Theoret. Comput. Sci. **407**(1–3), 278–289 (2008)
10. Perrot, J.F.: Variétés de langages et opérations. Theoret. Comput. Sci. **7**, 197–210 (1978)
11. Pin, J.É.: A variety theorem without complementation. Russian Math. (Izvestija vuzov. Matematika) **39**, 80–90 (1995)

12. Restivo, A.: The shuffle product: new research directions. In: Dediu, A.-H., Formenti, E., Martín-Vide, C., Truthe, B. (eds.) LATA 2015. LNCS, vol. 8977, pp. 70–81. Springer, Cham (2015). https://doi.org/10.1007/978-3-319-15579-1_5
13. Reutenauer, C.: Sur les variétés de langages et de monoïdes. In: Theoretical Computer Science (Fourth GI Conference Aachen), vol. 67, pp. 260–265 (1979)
14. Schnoebelen, P.: Decomposable regular languages and the shuffle operator. EATCS Bull. **67**, 283–289 (1999)
15. Schützenberger, M.P.: On finite monoids having only trivial subgroups. Inf. Control **8**, 190–194 (1965)
16. Straubing, H.: Recognizable sets and power sets of finite semigroups. Semigroup Forum **18**, 331–340 (1979)
17. Straubing, H., Weil, P.: Varieties. In: Pin, J.É. (ed.) Handbook of Automata Theory. Volume I. Theoretical Foundations, pp. 569–614. European Mathematical Society (EMS), Berlin (2021)

Contributed Papers

Ordering the Boolean Cube Vectors by Their Weights and with Minimal Change

Valentin Bakoev[✉]

"St. Cyril and St. Methodius" University of Veliko Tarnovo,
Veliko Tarnovo, Bulgaria
v.bakoev@ts.uni-vt.bg

Abstract. The topic of generating all subsets of a given set occupies an important place in the books on combinatorial algorithms. The considered algorithms, like many other generating algorithms, are of two main types: for generating in lexicographic order or in Gray code order. Both use binary representation of integers (i.e., binary vectors) as characteristic vectors of the subsets.

Ordering the vectors of the Boolean cube according to their weights is applied in solving some problems—for example, in computing the algebraic degree of Boolean functions, which is an important cryptographic parameter. Among the numerous orderings of Boolean cubic vectors by their (Hamming) weights, two are most important. The weight ordering, where the second criterion for sorting the vectors of equal weights, is the so-called weight-lexicographic order. It is considered in detail in [2]. Considered here is the second weight ordering, where the vectors of equal weights are arranged so that every two consecutive vectors differ in exactly two coordinates, i.e., they are ordered by minimal change. The properties of this ordering are derived. Based on these, an algorithm was developed that generates the vectors of the Boolean cube in this ordering. It uses a binary representation of integers and only performs additions of integers instead of operations on binary vectors. Its time and space complexities are of a linear type with respect to the number of vectors generated. The algorithm was used in the creation of sequence A351939 in OEIS [8].

Keywords: n-dimensional Boolean cube · Binary vector · Weight ordering relation · Minimal change · Generating all subsets · Gray code · Revolving door algorithm

1 Introduction

The topic of generating all subsets of a given set—say A of size n—occupies an important place in a lot of the known books on combinatorial algorithms, for example, [3–7] and others. The algorithms considered there use binary vectors

ⓒ The Author(s), under exclusive license to Springer Nature Switzerland AG 2022
D. Poulakis and G. Rahonis (Eds.): CAI 2022, LNCS 13706, pp. 43–54, 2022.
https://doi.org/10.1007/978-3-031-19685-0_4

(bit strings, binary words) of length n, i.e., binary representation of integers in n bits, as characteristic vectors of the subsets. These algorithms, like many other combinatorial algorithms, are of two main types, depending on the order in which these vectors are generated: in lexicographic order or in Gray code order. But there are problems where different arrangements of the binary vectors must be studied and used. For example, such a problem is "Given a Boolean function f of n variables by its truth table vector. Find (if exists) a vector $\alpha \in \{0,1\}^n$ of maximal (or minimal) weight, such that $f(\alpha) = 1$.", which is discussed in [2]. This problem is closely related to the problem of computing the algebraic degree of Boolean functions which is an important cryptographic parameter. To solve this problem efficiently (as shown in [1]), we studied the orderings of the vectors of the n-dimensional Boolean cube $\{0,1\}^n$ according to their weights. Among the huge number of weight orderings (see A051459 in the OEIS [8]), we chose to define and investigate the so-called Weight-Lexicographic Ordering (WLO), where the vectors of $\{0,1\}^n$ are sorted first by their (Hamming) weights, and then lexicographically. Two algorithms have been proposed that generate the vectors of $\{0,1\}^n$ in WLO. The second only performs integer additions to obtain a sequence of integers whose n-bit binary representations are in WLO. The sequence A294648 in OEIS [8] is obtained by this algorithm.

This article is a continuation of [2]. We use a similar approach to obtain another weight ordering of the vectors of $\{0,1\}^n$, where they are ordered first by their weights and second—every two consecutive vectors of the same weight differ in exactly two coordinates. In other words, the vectors of the same weight are sorted in the same way as by the well-known **revolving door algorithm** used in generating combinations with minimal change [3–7], etc. The algorithm proposed here performs only integer additions and the binary representation of these integers forms the ordering under consideration. It provides an alternative way to sort the vectors to be tested when computing the algebraic degree of Boolean functions. The algorithm can be considered as a non-recursive extension of the revolving door algorithm because it generates **all subsets** of a given set in the order discussed. It was used to create the sequence A351939 in OEIS [8].

The rest of this article is organized as follows. The main concepts and preliminary results are presented in Sect. 2. Section 3 begins with Theorem 1 on the serial numbers and weights of the Boolean cube vectors when arranged in Gray code. Then a new weight ordering called *ordering the Boolean cube vectors by their weights and with minimal change* is defined and its properties are derived in Theorem 2. In Sect. 4, an algorithm that generates the Boolean cube vectors in this ordering is proposed and discussed. Section 5 contains some concluding remarks.

2   Basic Notions and Preliminary Results

Let $\mathbb{N}$ be the set of natural numbers. We assume that $0 \in \mathbb{N}$ and so $\mathbb{N}^+ = \mathbb{N}\backslash\{0\}$ means the set of positive natural numbers.

The set of all n-dimensional binary vectors is known as *n-dimensional Boolean cube* (hypercube) and is defined as $\{0,1\}^n = \{(x_1, x_2, \ldots, x_n) | x_i \in$

$\{0,1\}$, for $i = 1,2,\ldots,n\}$. The number of all binary vectors is $|\{0,1\}^n| = |\{0,1\}|^n = 2^n$.

If $\alpha = (a_1, a_2, \ldots, a_n) \in \{0,1\}^n$ is an arbitrary vector, the natural number $\#\alpha = \sum_{i=1}^{n} a_i.2^{n-i}$ is called a *serial number* of the vector α. Thus $\#\alpha$ means the natural number having n-digit binary representation $a_1 a_2 \ldots a_n$. A *weight* (or *Hamming weight*) of α is the natural number $wt(\alpha)$, equal to the number of non-zero coordinates of α and so $wt(\alpha) = \sum_{i=1}^{n} a_i$. For example, if $\alpha = (0,1,1,0,1,1,0,1) \in \{0,1\}^8$, then $\#\alpha = 2^6 + 2^5 + 2^3 + 2^2 + 2^0 = 109$, and $wt(\alpha) = \sum_{i=1}^{n} a_i = 5$.

Let $\alpha = (a_1, a_2, \ldots, a_n)$ and $\beta = (b_1, b_2, \ldots, b_n)$ be arbitrary vectors of $\{0,1\}^n$. A *Hamming distance* between α and β is the natural number $d(\alpha, \beta)$ equal to the number of coordinates in which α and β differ. If $d(\alpha, \beta) = 1$, then α and β are called *adjacent*, or more precisely *adjacent in ith coordinate*, if they differ in this coordinate only.

For arbitrary vectors $\alpha = (a_1, a_2, \ldots, a_n)$ and $\beta = (b_1, b_2, \ldots, b_n) \in \{0,1\}^n$ we say that α *lexicographically precedes* β and write $\alpha \leq \beta$ when $\alpha = \beta$ or if $\exists i, 0 \leq i < n$, such that $a_1 = b_1, a_2 = b_2, \ldots, a_i = b_i$, but $a_{i+1} < b_{i+1}$. The relation $R_\leq \subseteq \{0,1\}^n \times \{0,1\}^n$, defined as $(\alpha, \beta) \in R_\leq$ when $\alpha \leq \beta$ is called a *lexicographic ordering* relation. It is easy to check that $R_\leq$ is a *total ordering* relation in $\{0,1\}^n$.

For an arbitrary $k \in \mathbb{N}, 0 \leq k \leq n$, the subset of all n-dimensional binary vectors of weight k is called a *kth layer* of $\{0,1\}^n$ and denoted by $L_{n,k} = \{\alpha| \alpha \in \{0,1\}^n : wt(\alpha) = k\}$. It is clear that $|L_{n,k}| = \binom{n}{k}$, for $k = 0,1,\ldots,n$. All these numbers (binomial coefficients) form the nth row of Pascal's triangle and then $\sum_{k=0}^{n} \binom{n}{k} = 2^n = |\{0,1\}^n|$. Obviously, the family of all layers $L_n = \{L_{n,0}, L_{n,1}, \ldots, L_{n,n}\}$ is a *partition* of the n-dimensional Boolean cube into layers. The *sequence of layers* $L_{n,0}, L_{n,1}, \ldots, L_{n,n}$ defines an order of the vectors of $\{0,1\}^n$ in accordance with their weights as follows: if $\alpha, \beta \in \{0,1\}^n$ and $wt(\alpha) < wt(\beta)$, then α precedes β in the sequence of layers, and when $wt(\alpha) = wt(\beta) = k$, then $\alpha, \beta \in L_{n,k}$ and there is no precedence between them. The corresponding relation $R_{<_{wt}}$ is defined as follows: for arbitrary $\alpha, \beta \in \{0,1\}^n$, $(\alpha, \beta) \in R_{<_{wt}}$ if $wt(\alpha) < wt(\beta)$. We set $(\alpha, \alpha) \in R_{<_{wt}}$, because we want $R_{<_{wt}}$ to be reflexive. If $(\alpha, \beta) \in R_{<_{wt}}$ we say that "α *precedes by weight* β" and write also $\alpha <_{wt} \beta$. It is easy to verify that $R_{<_{wt}}$ is a *partial ordering* in $\{0,1\}^n$ and we refer to it as a *Weight Ordering* (WO). Since the vectors of $L_{n,k}$ can be arranged in $\binom{n}{k}!$ ways, for $k = 0,1,\ldots,n$, there are $\prod_{k=0}^{n} \binom{n}{k}!$ WOs of the vectors of $\{0,1\}^n$ [2]. This number is the same as "Number of orderings of the subsets of a set with n elements that are compatible with the subsets' sizes; i.e., if A, B are two subsets with $A <= B$ then $Card(A) <= Card(B)$"[1]—see A051459 in the OEIS [8]. Among these numerous orderings, the most important are those in which the lexicographic order or the order obtained by minimal change is chosen as a second criterion for ordering all vectors of equal weights (i.e., those in the same layers). The so-called Weight-Lexicographic Ordering (WLO) is discussed in detail in [2]. We will continue with the next option—generating the sequence

---

[1] Here, $A <= B$ means $A \subseteq B$ and $Card(A) <= Card(B)$ means $|A| <= |B|$.

of layers so that any two consecutive vectors in the same layer differ in exactly two coordinates. For an arbitrary layer, this arrangement means that each pair of consecutive vectors is obtained by *minimal change*.

The vectors of $\{0,1\}^n$ are *ordered in Gray code* if any two adjacent vectors differ exactly in one coordinate only. When the first and the last vectors differ exactly in one coordinate too, the corresponding Gray code is called a *cyclic* one. Instead of the usual recursive definition, we will use the following inductive definition of the *binary reflected cyclic Gray code*. But hereafter we use the shorter term *Gray code*, which stands for binary reflected cyclic Gray code.

Definition 1. 1) The vectors (0) and (1) of $\{0,1\}^1$ are in Gray code.
2) Let $\alpha_0, \alpha_1, \ldots, \alpha_{2^{n-1}-1}$ be the sequence of all vectors of $\{0,1\}^{n-1}$ in Gray code.
3) We perform two basic steps to obtain the vectors of $\{0,1\}^n$ in Gray code. First, we take the sequence of vectors $\alpha_0, \alpha_1, \ldots, \alpha_{2^{n-1}-1}$ of $\{0,1\}^{n-1}$ in Gray code and add a zero at the beginning of each of its vector. Second, we take the same sequence in reverse order, i.e., $\alpha_{2^{n-1}-1}, \ldots, \alpha_1, \alpha_0$—it is obtained by *reflection* of the original sequence. Then we add one at the beginning of each vector of the reflected sequence. The resulting sequence

$$(0, \alpha_0), (0, \alpha_1), \ldots, (0, \alpha_{2^{n-1}-2}), (0, \alpha_{2^{n-1}-1}),$$
$$(1, \alpha_{2^{n-1}-1}), (0, \alpha_{2^{n-1}-2}), \ldots, (1, \alpha_1), (1, \alpha_0)$$

contains all vectors of $\{0,1\}^n$ ordered in *Gray code*.

Definition 2 (Addition a number to a sequence). Let $n, m \in \mathbb{N}^+$ and $s = s_1, s_2, \ldots, s_n$ be a sequence of integers. We define the operation *addition of the natural number m to the sequence a* as: $s + m = s_1 + m, s_2 + m, \ldots, s_n + m$.

We denote by $\mathbf{0}_n = (0, 0, \ldots, 0)$ the *zero vector* of n-coordinates, and by $\mathbf{1}_n = (1, 1, \ldots, 1)$ the *all-ones vector* of n-coordinates.

3   Ordering the Vectors of the Boolean Cube by Their Weights and with Minimal Change

Theorem 1. *Let the vectors of $\{0,1\}^n$ be obtained according to Definition 1. Then:*

1) The serial numbers of the vectors form the sequence of natural numbers $s(n)$, recursively defined as

$$s(n) = \begin{cases} 0, 1, & \text{if } n = 1, \\ s(n-1), s^r(n-1) + 2^{n-1}, & \text{if } n > 1. \end{cases}$$

Thus, $s(n)$ is a concatenation of two sequences: $s(n-1)$ and $s^r(n-1) + 2^{n-1}$, where $s^r(n-1) + 2^{n-1}$ means the same sequence $s(n-1)$, but taken in reverse order and the number 2^{n-1} is added to each of its terms.

2) *The weights of the vectors form the sequence of natural numbers $w(n)$, recursively defined as*

$$w(n) = \begin{cases} 0,1, & \text{if } n = 1, \\ w(n-1), w^r(n-1) + 1, & \text{if } n > 1. \end{cases}$$

Analogously, $w(n)$ is a concatenation of two sequences: $w(n-1)$ and $w^r(n-1) + 1$, where $w^r(n-1) + 1$ means the sequence $w(n-1)$, taken in reverse order and each of its terms is incremented by one.

Theorem 1 determines the bijection between the vectors of $\{0,1\}^n$ in Gray code, their serial numbers and their weights. The statements of Theorem 1 can be easily proved by induction on n and therefore the proof is omitted. Instead, the statements are illustrated in Fig. 1.

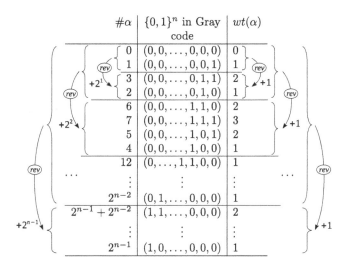

Fig. 1. Illustration of the statements of Theorem 1: the sequences $s(n)$ and $w(n)$ in the left and right columns, respectively.

Let us consider another weight ordering of the vectors of $\{0,1\}^n$, which is different from WLO. We will represent this WO by the sequence of serial numbers of the vectors in the layers, as in [2]. Let $L_{n,k} = \{\alpha_1, \alpha_1, \ldots, \alpha_m\}$ be an arbitrary layer of $\{0,1\}^n$—hence $m = \binom{n}{k}$. We denote by $g_{n,k} = \#\alpha_1, \#\alpha_2, \ldots, \#\alpha_m$ the *sequence of serial numbers*, corresponding to the vectors of $L_{n,k}$, for $0 \leq k \leq n$. Then $G_n = g_{n,0}, g_{n,1}, \ldots, g_{n,k}, \ldots, g_{n,n}$ means the sequence of natural numbers representing the vectors of $\{0,1\}^n$ in this WO. Since $g_{n,k}$ contains $\binom{n}{k}$ terms, for $k = 0, 1, \ldots, n$, hence G_n has $\sum_{k=0}^{n} \binom{n}{k} = 2^n$ terms. Following Definition 1 and Theorem 1, we define G_n inductively as follows:

Definition 3. 1) The WO sequence of $\{0,1\}^1$ is $G_1 = g_{1,0}, g_{1,1} = 0, 1$.

2) Let $G_{n-1} = g_{n-1,0}, g_{n-1,1}, \ldots, g_{n-1,n-1}$ be the WO sequence of $\{0,1\}^{n-1}$.

3) The WO sequence of n-dimensional Boolean cube $G_n = g_{n,0}, g_{n,1}, \ldots, g_{n,n}$ is defined as:

- $g_{n,0} = 0$ (it represents the WO of the layer $L_{n,0} = \{\mathbf{0}_n\}$);
- $g_{n,n} = 2^n - 1$ (it represents the WO of the layer $L_{n,n} = \{\mathbf{1}_n\}$);
- $g_{n,k} = g_{n-1,k}, g_{n-1,k-1}^r + 2^{n-1}$, for $k = 1, 2, \ldots, n - 1$. Thus $g_{n,k}$ is a **concatenation** of two sequences: first the sequence $g_{n-1,k}$ is taken (or copied) and then the sequence $g_{n-1,k-1}^r + 2^{n-1}$ is added. Here $g_{n-1,k-1}^r$ means the terms of $g_{n-1,k-1}$ in **reverse order**. The sequence $g_{n,k}$ represents the WO of the layer $L_{n,k}$.

For arbitrary $n \in \mathbb{N}^+$, the sequence of integers $g_{n,k}$ can be defined recursively as:

$$
g_{n,k} = \begin{cases} 0, & \text{if } k = 0, \\ 2^n - 1, & \text{if } k = n \\ g_{n-1,k}, g_{n-1,k-1}^r + 2^{n-1}, & \text{if } 0 < k < n, \end{cases}
$$

where $g_{n-1,k-1}^r + 2^{n-1}$ means the terms of $g_{n-1,k-1}$ in reverse order and each term is increased by 2^{n-1}. This formula resembles formulas (25) on [3, p. 362], (5.4) on [7, p. 127], [4, p. 48], and others.

For example, the sequences obtained in accordance with Definition 3, for $n = 1, 2$ and 3, are shown in Table 1, as well as in Figs. 2 and 3.

Table 1. $\{0,1\}^n$ in WO, obtained by Definition 3, for $n = 1, 2, 3$

n	$\{0,1\}^n$ in WO	$g_{n,k}$ and G_n
1	$L_{1,0}:$ (0),	$g_{1,0} = 0,$
	$L_{1,1}:$ (1)	$g_{1,1} = 1$
2	$L_{2,0}:$ (0,0),	$g_{2,0} = 0,$
	$L_{2,1}:$ (0,1),(1,0),	$g_{2,1} = 1, 2,$
	$L_{2,2}:$ (1,1)	$g_{2,1} = 3$
3	$L_{3,0}:$ (0,0,0),	$g_{3,0} = 0,$
	$L_{3,1}:$ (0,0,1),(0,1,0),(1,0,0),	$g_{3,1} = 1, 2, 4,$
	$L_{3,2}:$ (0,1,1),(1,1,0),(1,0,1),	$g_{3,2} = 3, 6, 5,$
	$L_{3,3}:$ (1,1,1)	$g_{3,3} = 7$

Remark 1. As we noted, $g_{n,k}$ (resp. $L_{n,k}$) contains $\binom{n}{k}$ terms (resp. vectors), for $k = 0, 1, \ldots, n$. Pascal's equality $\binom{n}{k} = \binom{n-1}{k-1} + \binom{n-1}{k}$ corresponds to Definition 3, but $g_{n,k}$ is a concatenation (instead of sum) of $g_{n-1,k}$ and $g_{n-1,k-1}^r$, and the number of terms of $g_{n,k}$ is a sum of the number of terms in $g_{n-1,k-1}$ and $g_{n-1,k}$. As we said, $G_n = g_{n,0}, g_{n,1}, \ldots, g_{n,n}$ contains a total of $\sum_{k=0}^{n} \binom{n}{k} = 2^n$ terms, and thus the sequences $G_1, G_2, \ldots, G_n, \ldots$ correspond to the rows numbered $1, 2, \ldots, n$ of Pascal's triangle.

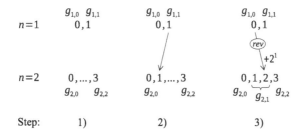

Fig. 2. How the sequence g_2 is derived from g_1

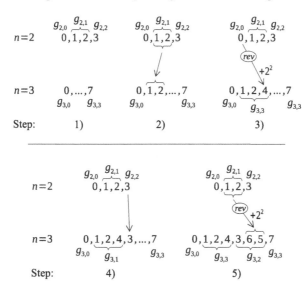

Fig. 3. How the sequence g_3 is derived from g_2

Definition 4. The sequence of vectors of equal weights $\alpha_1, \alpha_2, \ldots, \alpha_m$ are in a *minimal change ordering*, if every two consecutive vectors differ in exactly two coordinates, i.e., α_{i+1} is obtained from α_i by inverting exactly two of its coordinates, for $i = 1, 2, \ldots, m-1$. If α_1 and α_m differ in exactly two coordinates too, the sequence is in a *cyclic minimal change ordering*.

Theorem 2. *Let $G_n = g_{n,0}, g_{n,1}, \ldots, g_{n,n}$ be the WO sequence of $\{0,1\}^n$ obtained according to Definition 3 for a given integer $n \in \mathbb{N}^+$. Then the vectors corresponding to:*

(1) the first terms of $g_{n,k}$ and $g_{n,k+1}$ are adjacent vectors;
(2) the last terms of $g_{n,k}$ and $g_{n,k+1}$ are also adjacent vectors, for $k = 0, 1, \ldots, n-1$.

In addition, (3) the vectors corresponding to the terms of $g_{n,k}$ are in a cyclic minimal change ordering, for $k = 1, 2, \ldots, n-1$.

Then G_n defines ordering the Boolean cube vectors first by their weights and second with minimal change.

Proof. We will prove the statements of the theorem by induction on n, following Definition 3.

1) *Basic step*: Obviously, the theorem is true for $n = 1$ and $n = 2$, as shown in Table 1 and Fig. 2.
2) *Inductive hypothesis*: We suppose that the theorem is true for fixed $n > 1$.
3) *Inductive step*: We consider the sequence G_{n+1}.
   - If $k = 0$, then $g_{n+1,0} = 0$, which corresponds to the layer $L_{n+1,0} = \{\mathbf{0}_{n+1}\}$. The vector $\mathbf{0}_{n+1}$ is adjacent to each vector of $L_{n+1,1}$, since all vectors of $L_{n+1,1}$ have a weight $= 1$. Hence the theorem is true when $k = 0$.
   - If $k = n + 1$, then $g_{n+1,n+1} = 2^{n+1} - 1$, which corresponds to the layer $L_{n+1,n+1} = \{\mathbf{1}_{n+1}\}$. The vector $\mathbf{1}_{n+1}$ is adjacent to each vector of $L_{n+1,n}$, since all vectors of this layer have a weight $= n$. So, the theorem is true when $k = n + 1$.
   - Let k be an arbitrary integer, $0 < k < n$. Then:

$$g_{n+1,k} = g_{n,k}, \ g_{n,k-1}^r + 2^n \ \text{ and}$$
$$g_{n+1,k+1} = g_{n,k+1}, \ g_{n,k}^r + 2^n.$$

We consider the sequences $g_{n,k-1}$, $g_{n,k}$ and $g_{n,k+1}$, and denote their first and last terms as follows:

$$g_{n,k-1} : \#\alpha_1, \ldots, \#\alpha_p$$
$$g_{n,k} : \#\beta_1, \ldots, \#\beta_q$$
$$g_{n,k+1} : \#\gamma_1, \ldots, \#\gamma_r$$

According to the inductive hypothesis we have:

(1) $\{\alpha_1, \beta_1\}$, and $\{\beta_1, \gamma_1\}$ are pairs of adjacent vectors;
(2) $\{\alpha_p, \beta_q\}$ and $\{\beta_q, \gamma_r\}$ are pairs of adjacent vectors;
(3) $g_{n,k-1}$, $g_{n,k}$ and $g_{n,k+1}$ are such that the corresponding sequences $\alpha_1, \alpha_2, \ldots, \alpha_p$; $\beta_1, \beta_2 \ldots, \beta_q$ and $\gamma_1, \gamma_2, \ldots, \gamma_r$ are in a cyclic minimal change ordering.

According to Definition 3 we have:

$$g_{n+1,k} = \#\beta_1, \ldots, \#\beta_q, \ \#\alpha_p + 2^n, \ldots, \#\alpha_1 + 2^n \ \text{ and}$$
$$g_{n+1,k+1} = \#\gamma_1, \ldots, \#\gamma_r, \ \#\beta_q + 2^n, \ldots, \#\beta_1 + 2^n.$$

Therefore:

(1) The vectors $(0, \beta_1)$ and $(0, \gamma_1)$, corresponding to the first terms of $g_{n+1,k}$ and $g_{n+1,k+1}$ are adjacent.

(2) The vectors $(1, \alpha_1)$ and $(1, \beta_1)$, corresponding to the last terms of $g_{n+1,k}$ and $g_{n+1,k+1}$ are also adjacent.

(3) We consider $g_{n+1,k}$ and the corresponding sequence $(0, \beta_1), \ldots, (0, \beta_q)$, $(1, \alpha_p), \ldots, (1, \alpha_1)$ of the layer $L_{n+1,k}$. According to the inductive hypothesis:

  – since β_q and α_p are adjacent vectors, hence $(0, \beta_q)$ and $(1, \alpha_p)$ differ in exactly two coordinates;
  – since β_1 and α_1 are adjacent vectors, hence $(0, \beta_1)$ and $(1, \alpha_1)$ differ in exactly two coordinates too.

From these conclusions and the inductive hypothesis (3), it follows that $g_{n+1,k}$ is such that the sequence of the corresponding vectors $(0, \beta_1), \ldots, (0, \beta_q), (1, \alpha_p)$, $\ldots, (1, \alpha_1)$ is in a cyclic minimal change ordering. Analogously, the same is true for $g_{n+1,k+1}$ and the corresponding sequence $(0, \gamma_1), \ldots, (0, \gamma_r), (1, \beta_q), \ldots$, $(1, \beta_1)$.

Hence the statements (1), (2) and (3) also hold for every k, $0 < k < n$. Therefore G_{n+1} defines an ordering of the Boolean cube vectors first by their weights and second with minimal change.

Therefore the theorem is true for every $n \in \mathbb{N}^+$. $\qquad\qquad\square$

For brevity, we call the ordering of the vectors of $\{0,1\}^n$ according to the sequence G_n *ordering by weights and with minimum change*. By the notion of a characteristic vector, this ordering is transformed into *ordering by cardinalities and with minimal change* of all subsets of a given n-element set. Hence a theorem about the subsets, corresponding to Theorem 2 can be formulated and proved. Statements similar to those in Theorem 2, but only for $g_{n,k}$ have been proved in [3,4,6,7], and others.

4   Algorithm for Generating the Vectors of the Boolean Cube Ordered by Weights and with Minimal Change

Based on Definition 3, we developed an algorithm that generates the sequence G_n for a given integer $n \in \mathbb{N}^+$. Since the definition is iterative, the algorithm generates all sequences G_m, for $m = 1, 2, \ldots, n-1$, before generating G_n. We call this algorithm *Algorithm A*. It performs **three main steps**. In the **first**, it computes the binomial coefficients $\binom{m}{k}$, i.e., the lengths of the subsequences $g_{m,k}$, for $m = 1, 2, \ldots, n$ and for $k = 0, 1, \ldots, m$. So it fills the rows numbered by $0, 1, \ldots, n$ of Pascal's triangle by storing these numbers in a two-dimensional array denoted by P_t in the C/C++ code below. Using these numbers, the algorithm performs the **second step**: computing the places (indices) where each subsequence $g_{m,k}$ of G_m starts, for $m = 1, 2, \ldots, n$ and for $k = 0, 1, \ldots, m$. The results are stored in a two-dimensional array called ss_beg. **Third step**: following Definition 3, the algorithm computes the terms of the sequences G_m, for $m = 1, 2, \ldots, n$ and fills them in a two-dimensional array called g. Here is the code of a function that implements the third step.

Listing 1.1. Implementation of the third step of Algorithm A—computing the sequence G_n for a given integer n

```
typedef unsigned long long ull;
void fill_g_seqs (int n) {
  g[1][0] = 0; // initialization
  g[1][1] = 1; // of G_1
  ull m = 2; // to be added to subsequences
  for (int row = 2; row <= n; row++) {
    g[row][0] = 0;  // zero term
    ull k = m + m - 1;
    g_seqs[r][k] = k; // last term
    k = 1; // number of the serial term
    for (int col = 1; col < row; col++) {
      // Preparing to copy
      ull seq_len = P_t[row-1][col];
      ull seq_beg = ss_beg[row-1][col];
      // I: copying a subsequence
      for (ull j = 0; j < seq_len; j++)
        g[row][k++] = g[row-1][seq_beg+j];
      // Preparing for II
      seq_len = P_t[row-1][col-1];
      seq_beg = ss_beg[row-1][col-1];
      // II: reverse and add m to a subseq.
      for (ull j= seq_len -1; j >= 0; j--)
        g[row][k++]= g[row-1][seq_beg+j] + m;
    }
    m += m; //or m *= 2; - for the next row
  }
}
```

Definition 3 and Theorem 2 imply the **correctness** of Algorithm A. When $n \le 64$, then the unsigned integers are represented in a single 64-bit (or less-bit) computer word. Then calculating the **time complexity** is easy. In first and second steps, the algorithm fills $1 + 2 + \cdots + n + 1 = (n+2)(n+1)/2 = \Theta(n^2)$ cells in the arrays P_t and ss_beg. It only performs additions whose number is less than the number of filled cells, but again $\Theta(n^2)$. In the third step, the algorithm computes $2^1 + 2^2 + \cdots + 2^n = 2^{n+1} - 1$ terms (and fills as many cells) corresponding to the number of terms in $G_1, G_2, \ldots, G_n$. As the attached code shows, the number of additions and subtractions in the third step is proportional to 2^{n+1}, most of which are for computing indices. Since the time complexity of this step dominates (over those of the previous steps), the time complexity of Algorithm A is $\Theta(2^{n+1})$. Regarding the **space complexity**, we note that: in step 1, a two-dimensional array of size $n \times (n+1)$ is used (for Pascal's triangle), and in step 2, another array of the same type is used. In step 3, a two-dimensional array of size $n \times 2^n$ is used, and therefore the total space complexity is $\Theta(n.2^n)$. But it can be reduced to $\Theta(2^n)$ by using two one-dimensional dynamic arrays of size 2^n in the third step. If G_i is computed and its terms are filled in the first array, then the algorithm can compute and fill the terms of G_{i+1} in the second array. The first and second arrays then swap their source and destination roles for the next step. So, the time complexity and space complexity are of an exponential type with respect to the size of the input n. But both are of a linear type with respect to the size of the output—the number of combinatorial objects (i.e., integers, vectors, subsets) generated by Algorithm A.

Some results, obtained by Algorithm A are shown in Table 2 and Table 3. The algorithm and Definition 3 were used in the creation of the sequence A351939 in OEIS [8].

Table 2. Results obtained by Algorithm A, for $n = 1, 2, \ldots, 5$.

n	G_n
1	0, 1
2	0, 1, 2, 3
3	0, 1, 2, 4, 3, 6, 5, 7
4	0, 1, 2, 4, 8, 3, 6, 5, 12, 10, 9, 7, 13, 14, 11, 15
5	0, 1, 2, 4, 8, 16, 3, 6, 5, 12, 10, 9, 24, 20, 18, 17, 7, 13, 14, 11, 25, 26, 28, 21, 22, 19, 15, 27, 30, 29, 23, 31

Table 3. G_3 and corresponding subsets of $A = \{a, b, c\}$ ordered by cardinalities and with minimal change

G_3	α	$S \subseteq A$
0	(0,0,0)	$\emptyset$
1	(0,0,1)	$\{c\}$
2	(0,1,0)	$\{b\}$
4	(1,0,0)	$\{a\}$
3	(0,1,1)	$\{b, c\}$
6	(1,1,0)	$\{a, b\}$
5	(1,0,1)	$\{a, c\}$
7	(1,1,1)	$\{a, b, c\}$

5   Conclusions

The proposed algorithm A combines two main operations in a most natural way: first, it sorts the vectors of the Boolean cube by their weights, and second, it sorts the vectors of the same weights as the revolving door algorithm. Algorithm A is of the same type as the algorithm that generates the vectors of the Boolean cube in WLO, discussed in [2]. Both algorithms can be used to generate all subsets of a given set in the corresponding ordering. Both are iterative algorithms, use binary representation of integers and perform only additions of integers instead of operations on binary vectors when $n \leq 64$. Both work similarly and have the same type of complexity, which is proportional to the number of integers (vectors, subsets) generated. The difference between these orderings is substantial, but the difference between the algorithms is very small, as is the formal difference between the corresponding orderings—if we take the same sequence without changing its original order (i.e., if we ignore the reflection) in point 3 of Definition 1, we get a definition of a lexicographic ordering of the vectors of the Boolean cube.

We hope that the approach proposed in creating Algorithm A can be used in other combinatorial algorithms.

References

1. Bakoev, V.: A method for fast computing the algebraic degree of boolean functions. In: Proceedings of the 21st International Conference on Computer Systems and Technologies (CompSysTech 2020), Ruse, Bulgaria, 19–20 June 2020, pp. 141–147 (2020). https://doi.org/10.1145/3407982.3408005
2. Bakoev, V.: Some problems and algorithms related to the weight order relation on the n-dimensional boolean cube. Discrete Math. Algorithms Appl. **13**(3), 2150021 (23 p) (2021). https://doi.org/10.1142/S179383092150021X
3. Knuth, D.: The Art of Computer Programming, Volume 4A: Combinatorial Algorithms, Part 1. Addison-Wesley, New Jersey (2011)
4. Kreher, D., Stinson, D.: Combinatorial Algorithms: Generation, Enumeration and Search. CRC Press, Boca Raton (1999)
5. Nijenhuis, A., Wilf, H.: Combinatorial Algorithms for Computers and Calculators, 2nd edn. Academic Press, Cambridge (1978)
6. Reingold, E., Nievergelt, J., Deo, N.: Combinatorial Algorithms, Theory and practice. Prentice-Hall, New Jersey (1977)
7. Ruskey, F.: Combinatorial generation: working version (1j-CSC 425/520) (2003). http://page.math.tu-berlin.de/~felsner/SemWS17-18/Ruskey-Comb-Gen.pdf
8. Sloane, N.J., et al.: The on-line encyclopedia of integer sequences. https://oeis.org

Designated-Verifier Linkable Ring Signatures with Unconditional Anonymity

Danai Balla[ID], Pourandokht Behrouz[✉][ID], Panagiotis Grontas[ID], Aris Pagourtzis[ID], Marianna Spyrakou[ID], and Giannis Vrettos[ID]

School of Electrical and Computer Engineering, National Technical University of Athens, 9, Iroon Polytechniou St, 157 80 Athens, Greece
{pbehrouz,mspyrakou}@mail.ntua.gr, pgrontas@corelab.ntua.gr,
pagour@cs.ntua.gr

Abstract. In this work we propose Designated-Verifier Linkable Ring Signatures with unconditional anonymity, a cryptographic primitive that protects the privacy of signers in two different ways. First, it allows signers to hide inside a ring (an *anonymity set* of signers), which they can create by collecting a set of public keys all of which must be used for verification. Second, it allows a designated entity to simulate signatures, thus making it difficult for an adversary to deduce a signer's identity from the content of the exchanged messages. To the best of our knowledge, our scheme is the first to combine linkable ring signatures with designated-verifier signatures, achieving unconditional anonymity without compromising any other security property of these schemes.

Keywords: Ring signatures · Unconditional anonymity · Designated verifier · Non-transferability · Linkability

1 Introduction

Ring signatures (RS) [14] provide anonymity to the signer of a message by allowing them to hide within a crowd of peers. While an RS is created by a single private key, it is verified using a *set* of public keys. This set is called a *ring* and can be created in an ad-hoc manner. In RS constructions, anonymity can be unconditional, which means that the signer can be revealed with probability no better than that of a random guess. However, anonymity can be a double-edged sword; while it is necessary for some applications, it may be a thwarting factor in others. For instance, in e-voting, it allows voters to freely express their preferences. If left uncontrolled, though, it can be an enabler for an attacker, e.g. by facilitating double-voting.

Linkable Ring Signatures (LRS) [10] limit anonymity while preserving its essence; signatures of the same signer can be grouped together (i.e., be *linked*)

This work was supported in part by the Algorand Centres of Excellence programme managed by Algorand Foundation. Any opinions, findings, and conclusions or recommendations expressed in this material are those of the author(s) and do not necessarily reflect the views of Algorand Foundation.

© The Author(s), under exclusive license to Springer Nature Switzerland AG 2022
D. Poulakis and G. Rahonis (Eds.): CAI 2022, LNCS 13706, pp. 55–68, 2022.
https://doi.org/10.1007/978-3-031-19685-0_5

without giving away the signer's identity. To achieve this, LRS schemes embed a *linking tag* or *pseudoidentity* in the signature. This tag, however, downgrades LRS anonymity from unconditional to computational. This allows a strong (e.g., a quantum) adversary to learn the identity of the signer.

The protection offered by RS is also vulnerable to a side-channel attack against anonymity, by exploiting the messages exchanged. Indeed, messages may contain information that 'leaks' the signer's identity. To thwart such an attack, an additional entity, able to simulate signatures, can be used. Designated-Verifier signatures (DVS) [5] facilitate this approach, by including the verifier's public key during signing. As a result, a signature effectively states that it originates either from the original signer or from the designated verifier (DV). Therefore, the DV is the only entity that is able to verify the authenticity of a signature that has not originated from itself. In contrast, the public will verify a signature successfully, no matter whether it has been created by the DV (a simulation) or by the original signer.

Designated-Verifier Linkable Ring Signatures (DVLRS) [2], a recent proposal, combine LRS and DVS. Its aim is to increase the privacy of the ring members by also allowing a DV to add 'noise' through simulated signatures. Therefore, it is more difficult to identify a ring member by the content of their messages, because one cannot tell if they are original or simulations. However, DVLRS provide only computational anonymity, a feature inherited from LRS.

Our Contribution. In this work we resolve an open problem stated in [2] by enhancing DVLRS with unconditional anonymity. We call our new primitive UDVLRS (Designated-Verifier Linkable Ring Signatures with Unconditional anonymity). In addition, our construction yields a shorter signature than that of [2]. We formally define the security model of UDVLRS, propose an instantiation, and prove its security. Table 1 lists the security properties of UDVLRS and presents a comparison between the new scheme and DVLRS in terms of computational assumptions on which the security properties of the two schemes are based.

Table 1. Security properties of UDVLRS and DLVRS and corresponding computational assumptions. MDLR is an assumption that we define in this paper; we prove that it is equivalent to DLOG.

Property	DVLRS	UDVLRS
Unforgeability	DLOG	MDLR/DLOG
Anonymity	DDH	Unconditional
Non-transferability	Unconditional	Unconditional
Linkability	DLOG	MDLR/DLOG
Non-slanderability	DLOG	DLOG

Related Work. Liu et al. [10] provide the first LRS construction and define its security properties: *unforgeability, linkability* and *anonymity*. They prove that the former two properties hold if the Discrete Logarithm Problem (DLOG) is hard, while the latter rests on the Decisional Diffie Hellman (DDH) assumption. These security properties were refined in [4,11,12]. Liu et al. [9] proposed the first unconditionally anonymous LRS scheme. Based upon their construction we introduce a more realistic model for anonymity, by allowing the adversary to access previously known signatures and corrupted public keys. Their work also defined a new security property, *non-slanderability*, which aimed to prevent an adversary from arbitrarily linking signatures to signers that did not create them. DVLRS combined this property with linkability, thus making the security model of the primitive simpler. We find however that this stronger definition of linkability is not compatible with unconditional anonymity, as will be discussed in Subsect. 3.7. As a result, we revert to the definition of [9].

The pseudoidentity parameter of [10] is a function of the public keys of the ring and the secret signing key. As a result, linking applies only to signatures originating from the same ring. In [9] the pseudoidentity is generalized, since it is a function of the secret signing key and a commonly shared string called event (ev). As a result, signatures from different rings can also be linked.

The essential security property for DVS, besides unforgeability, is *non-transferability*. It was formalized in [15] and states that no party except the DV can be convinced that a signature is not a simulation. DVS can be either publicly or privately verifiable. In the latter case they are called *strong* designated-verifier signatures and require the secret key of the DV for verification.

RS and DVS exhibit a similarity with regard to anonymity, as discussed in [14]. In particular, in both cases, the public cannot tell who created a signature. Therefore, a ring of two members provides in effect designated-verifier signatures. However, as stated in [2], linkability breaks this connection, because the pseudoidentity can be used to prove that a particular signature is not a simulation. As a result, there have been some attempts in the literature to combine LRS and DVS, but none of them employs the semantics of DVLRS and UDVLRS. In [6,16] only private verifiability is provided, while in [6,7] linkability is not considered. DVLRS was the first scheme to combine linkability with a designated verifier and public verifiability, thus achieving noise via simulated signatures. UDVLRS extend DVLRS with unconditional anonymity, without sacrifying public verifiability and perfect non-transferability. Unforgeability and linkability remain conditional on the hardness of DLOG. Lastly, UDVLRS generalize linkability, by allowing the linking of signatures among different rings with different designated verifiers.

2 Preliminaries

2.1 Notation

We denote by λ the security parameter. L is a ring consisting of n public keys. In particular, L is a subset of the set of all possible public keys $\mathcal{U}$; the size of $\mathcal{U}$

is denoted by $\mu(\lambda)$. As usual, $[n]$ denotes the set $\{1, \ldots, n\}$. Equality is denoted by '=', assignment by '←', definition by '$\triangleq$', while '←$_s$' denotes a selection of an element from a set uniformly at random. Our security definitions are in the form of games between a challenger $\mathcal{C}$ and an adversary $\mathcal{A}$. Their input is always the security parameter and their output is the truth value of the condition that will make $\mathcal{A}$ win the game; for notation brevity, our games return the condition itself. The absence of an output is denoted by '⊥'. Values of no interest in a particular context are denoted by '·'. We refer to the cryptographic parameters of our constructions as params. They are required in all our algorithms, but we do not include them for the sake of conciseness. We denote a public key as pk and a secret key as sk. A pseudoidentity is denoted by pid and the set of all possible pids by $\mathcal{PID}$. Linking is based on a common string ev which originates from a set $\mathcal{EID}$. We will use D to denote the designated verifier, and π to denote a signer belonging to the ring. Most of the algorithms that constitute our primitive take as common input the values ev, L, m, pk_D to denote respectively an event, a ring, a message, and the public key of the designated verifier. We collectively refer to these values as the signature parameters.

2.2 Security Assumptions

The security of some facets of our construction rests on a variation of the Discrete Logarithm Problem (DLOG), which is easier to use with ring signatures. It was first defined in [9] and it was used as a problem computationally equivalent to DLOG. A similar version was introduced in [1] as the Discrete Logarithm Relation. We will be using [9]'s version, which we reintroduce as the Modified Discrete Logarithm Relation (MDLR):

Definition 1. *Modified Discrete Logarithm Relation (MDLR). Let $\mathbb{G}$ be a cyclic group of prime order q generated by g and $Y_1, Y_2, \ldots, Y_n ←_s \mathbb{G}, Y_1 \neq 1_{\mathbb{G}}$. A solution to the MDLR problem is a tuple $(\phi_1, \phi_2, \ldots, \phi_n) \in \mathbb{Z}_q^n$ such that $Y_1 \cdot Y_2^{\phi_2} \cdots Y_n^{\phi_n} = g^{\phi_1}$ and $\sum_{i=1}^n \phi_i \neq 0 \pmod{q}$.*

Note that for $n = 1$, MDLR is the standard DLOG problem.

Proposition 1. DLOG *is computationally equivalent to MDLR.*

Proof. Assuming a DLOG oracle and a MDLR instance $Y_1, Y_2, \ldots, Y_n$ compute $x_1, x_2, \ldots, x_n$ such that $Y_i = g^{x_i}, i \in [n]$. Select $\{\phi_i \in \mathbb{Z}_q\}_{i=2}^n$ and set $\phi_1 ← x_1 + \sum_{i=2}^n \phi_i x_i \mod q$. In (the negligible) case that $\sum_{i=1}^n \phi_i = 0 \pmod{q}$ repeat the process. Clearly, $g^{\phi_1} = g^{x_1 + \sum_{i=2}^n \phi_i x_i} = Y_1 \cdot Y_2^{\phi_2} \cdots Y_n^{\phi_n}$.

Assuming a MDLR oracle and a DLOG instance $Y = g^x$, select $x_2, \ldots, x_n$ and compute $\{Y_i ← g^{x_i}\}_{i=2}^n$. Query the MDLR oracle with $Y, Y_2, \ldots, Y_n$ and receive $\phi_1, \phi_2, \ldots, \phi_n \in \mathbb{Z}_q$ such that $Y \cdot Y_2^{\phi_2} \cdots Y_n^{\phi_n} = g^{\phi_1}$ and $\sum_{i=1}^n \phi_i \neq 0 \pmod{q}$. Thus $Y = g^{\phi_1 - \sum_{i=2}^n \phi_i x_i}$ and the DLOG of Y is $x = \phi_1 - \sum_{i=2}^n \phi_i x_i \pmod{q}$. □

3   UDVLRS Definition and Security Model

3.1   UDVLRS Definition

Our definition of UDVLRS is a combination of the respective ones in [2,9]:

Definition 2. *A UDVLRS scheme is a tuple of PPT algorithms* Π = (Setup, KGen, Sign, Extract, Sim, Vrfy, Link) *where:*

- params ← Setup(λ). *The* Setup *algorithm generates the cryptographic groups for UDVLRS operations as well as (implicitly) the message, signature, pseudoidentity and event spaces* ($\mathcal{MSG}, \mathcal{SG}, \mathcal{PID}, \mathcal{EID}$ *respectively*).
- (sk, pk) ← KGen(). *The key generation algorithm.*
- σ ← Sign(ev, L, m, pk$_D$, sk$_\pi$). Sign *is the algorithm that is used to sign a message* m *by some* $\pi \in [n]$ *for event* ev, *ring* L, *and designated verifier* D.
- pid ← Extract(σ). Extract *is an algorithm that can obtain the pseudoidentity* pid *from a signature. It is publicly executable.*
- σ ← Sim(ev, L, m, pk$_D$, sk$_D$, pid). Sim *is the signature simulation algorithm that allows the designated verifier* D *to produce indistinguishable signatures for pseudoidentity* pid.
- $\{0,1\}$ ← Vrfy(ev, L, m, pk$_D$, σ). Vrfy *is the public verification algorithm which outputs 1 if the signature is valid or 0 if it is not.*
- $\{0,1\}$ ← Link(σ_1, ev$_1$, σ_2, ev$_2$). Link *is the public linking algorithm which outputs 1 if* σ_1 *and* σ_2 *originate from the same signer or if they are simulated to look like they originate from the same signer.*

Note that the Sim algorithm requires a pid for a linkable simulation. Thus, it must have seen a signature before. This does not restrict the applications of our primitive, since an initialization phase could be designed where each signer would publish a signed registration message, without sensitive data. Alternatively, signatures with a random pid could be used before or during the registration phase. These would not be linked to any signer.

3.2   Correctness

The completeness of UDVLRS rests on the following two properties:

Verification Correctness. A signature or simulation, for a specific event, ring, message and designated verifier is valid if and only if it was honestly generated (i.e., Vrfy(ev, L, m, pk$_D$, σ) = 1 $\Leftrightarrow$ σ = Sign(ev, L, m, pk$_D$, sk$_\pi$), pk$_\pi$ $\in L$ or σ = Sim(ev, L, m, pk$_D$, sk$_D$, pid), (sk$_D$, pk$_D$) = KGen(), pid $\in \mathcal{PID}$).

Linking Correctness. Two valid signatures σ_1, σ_2 are said to be linked (in terms, Link(σ_1, ev$_1$, σ_2, ev$_2$) = 1) if and only if ev$_1$ = ev$_2$ and one of the following conditions hold:

i σ_1 = Sign(ev$_1$, L_1, m$_1$, pk$_{D_1}$, sk$_\pi$), pk$_\pi$ $\in L_1$ and σ_2 = Sign(ev$_2$, L_2, m$_2$, pk$_{D_2}$, sk$_\pi$), pk$_\pi$ $\in L_2$. Both signatures are honestly generated by the same signer π.

ii $\sigma_1 = \mathsf{Sign}(\mathsf{ev}_1, L_1, \mathsf{m}_1, \mathsf{pk}_{D_1}, \ \mathsf{sk}_\pi)$, $\sigma_2 = \mathsf{Sim}(\mathsf{ev}_2, L_2, \mathsf{m}_2, \mathsf{pk}_{D_2}, \mathsf{sk}_{D_2}$, $\mathsf{Extract}(\sigma_1))$, $\mathsf{pk}_\pi \in L_1$, $(\mathsf{sk}_{D_2}, \mathsf{pk}_{D_2}) \leftarrow \mathsf{KGen}()$. σ_1 is honestly generated by signer π and σ_2 by designated verifier D_2 using the pseudoidentity extracted from σ_1.

iii $\sigma_1 = \mathsf{Sim}(\mathsf{ev}_1, L_1, \mathsf{m}_1, \mathsf{pk}_{D_1}, \mathsf{sk}_{D_1}, \mathsf{pid})$, $(\mathsf{sk}_{D_1}, \mathsf{pk}_{D_1}) = \mathsf{KGen}()$, $\mathsf{pid} \in \mathcal{PID}$ and $\sigma_2 = \mathsf{Sim}(\mathsf{ev}_2, L_2, \mathsf{m}_2, \mathsf{pk}_{D_2}, \mathsf{sk}_{D_2}, \mathsf{pid})$, $(\mathsf{sk}_{D_2}, \mathsf{pk}_{D_2}) = \mathsf{KGen}()$. Both σ_1, σ_2 are simulated by designated verifiers D_1, D_2 using the same pid.

3.3 Adversarial Capabilities

To model the security UDVLRS, we consider a strong adaptive adversary that has the ability to add new users, corrupt any set of users, and request signatures and simulations from any user and designated verifier. To formally model the capabilities of $\mathcal{A}$ we use the following oracles similar to [2,8,9]:

- $\mathsf{pk} \leftarrow \mathcal{JO}()$. The *Joining Oracle*, adds a public key to the list of public keys $\mathcal{U}$, and returns it.
- $\mathsf{sk} \leftarrow \mathcal{CO}(\mathsf{pk})$. The *Corruption Oracle*, receives a public key pk that is an output of $\mathcal{JO}$ and returns the corresponding secret key sk. It models the ability of $\mathcal{A}$ to control some members of $\mathcal{U}$.
- $\sigma \leftarrow \mathcal{SO}(\mathsf{ev}, L, \mathsf{m}, \mathsf{pk}_D, \mathsf{pk}_\pi)$. The *Signing Oracle* receives the signature parameters and public key $\mathsf{pk}_\pi \in L$ and outputs a signature σ such that $\sigma \leftarrow \mathsf{Sign}(\mathsf{ev}, L, \mathsf{m}, \mathsf{pk}_D, \mathsf{sk}_\pi)$ and $(\mathsf{pk}_\pi, \mathsf{sk}_\pi) \leftarrow \mathsf{KGen}()$.
- $\sigma \leftarrow \mathcal{MO}(\mathsf{ev}, L, \mathsf{m}, \mathsf{pk}_D, \mathsf{pid})$. The *Simulation Oracle* receives the signature parameters and a pseudoidentity pid and outputs a signature σ such that $\sigma \leftarrow \mathsf{Sim}(\mathsf{ev}, L, \mathsf{m}, \mathsf{pk}_D, \mathsf{sk}_D, \mathsf{pid})$ and $(\mathsf{pk}_D, \mathsf{sk}_D) \leftarrow \mathsf{KGen}()$.

We model hash functions as random oracles [3] denoted as $\mathcal{RO}$. For simplicity, we abuse notation, and also denote the set of an oracle's answers by its name.

3.4 Unforgeability

Unforgeability requires that only a ring member or the designated verifier can produce signatures or simulations that verify successfully. To formally define it for a UDVLRS scheme Π, we consider the experiment $\mathsf{Exp}^{\mathsf{unf}}_{\mathcal{A},\Pi}$ in Game 1.1. The adversary queries all the oracles according to any adaptive strategy. Then $\mathcal{A}$ chooses the signature parameters and creates a forged signature σ^*. The adversary succeeds if the signature verifies, and none of the keys of L, nor pk_D, have been queried to $\mathcal{CO}$ and if the signature is not a query output of $\mathcal{SO}$ or $\mathcal{MO}$.

Definition 3. *Unforgeability. A UDVLRS scheme Π is unforgeable if for any* PPT *adversary $\mathcal{A}$ the following holds:*

$$\mathsf{Adv}^{\mathsf{unf}}_{\mathcal{A}}(\lambda) \triangleq \Pr\left[\mathsf{Exp}^{\mathsf{unf}}_{\mathcal{A},\Pi}(\lambda) = 1\right] \leq \mathsf{negl}(\lambda)$$

Game 1.1: Unforgeability experiment $\mathsf{Exp}_{\mathcal{A},\Pi}^{\mathrm{unf}}$

---

params $\leftarrow \Pi.\mathsf{Setup}(1^\lambda)$

$\mathcal{U} \leftarrow \big\{(\mathsf{pk}_i, \mathsf{sk}_i) \leftarrow \Pi.\mathsf{KGen}()\big\}_{i=1}^{\mu(\lambda)}$

$(\sigma^*, \mathsf{ev}, L = \{\mathsf{pk}_i\}_{i=1}^n, \mathtt{m}, \mathsf{pk}_D) \leftarrow \mathcal{A}^{\mathcal{RO},\mathcal{JO},\mathcal{CO},\mathcal{SO},\mathcal{MO}}(\mathcal{U})$

return

$\mathsf{Vrfy}(\mathsf{ev}, \sigma, L, \mathtt{m}, \mathsf{pk}_D) = 1$ AND $\forall i \in \mathcal{CO}$, $\mathsf{pk}_i \notin L$ AND $D \notin \mathcal{CO}$ AND

$\sigma^* \notin \mathcal{SO}$ AND $\sigma^* \notin \mathcal{MO}$

3.5   Anonymity

Unconditional anonymity means that for any (unbounded) adversary $\mathcal{A}$ it should be impossible to find the public key of the signer of a specific signature for a ring L with probability greater than that of random sampling.

More formally, consider the interaction between an unbounded adversary $\mathcal{A}$ and a challenger $\mathcal{C}$ in Game 1.2. The adversary queries the $\mathcal{JO}, \mathcal{CO}, \mathcal{SO}, \mathcal{MO}$ according to any adaptive strategy, sets pk_D, and forms a ring L with any subset of n private keys. $\mathcal{MO}$ does not give any advantage to $\mathcal{A}$ that cannot be obtained with the calls to $\mathcal{SO}$ since the signatures are not constructed with any key from the ring, but are simulations. Assume that from the calls of the $\mathcal{CO}$, $\mathcal{A}$ has obtained m_1 private keys. Using its unlimited computational power and the pseudoidentity, $\mathcal{A}$ might also obtain m_2 private keys using signatures from $\mathcal{SO}$, where the signer is known. The public keys obtained in this way may be contained in L, but $n > m_1 + m_2 + 1$ should hold. $\mathcal{A}$ gives $\mathcal{C}$ an event ev, a message $\mathtt{m}$, the set of public keys L, and the designated verifier public key pk_D. $\mathcal{C}$ picks $\pi \leftarrow_\$ [n]$ and constructs a challenge signature $\sigma_c = \mathsf{Sign}(\mathsf{ev}, L, \mathtt{m}, \mathsf{pk}_D, \mathsf{sk}_\pi)$ and gives it to $\mathcal{A}$, who must guess π. The adversary wins the game if it correctly guesses the signer, and its private key has not been obtained through the $\mathcal{CO}$ or $\mathcal{SO}$.

Definition 4. *Anonymity. A UDVLRS scheme Π is unconditionally anonymous if for any unbounded adversary $\mathcal{A}$ the following holds:*

$$\mathsf{Adv}_{\mathcal{A}}^{\mathrm{anon}}(\lambda) \triangleq \Pr[\mathsf{Exp}_{\mathcal{A},\Pi}^{\mathrm{anon}}(\lambda) = 1] - \frac{1}{n - m_1 - m_2} = 0$$

Note that in schemes like [2,10], where there is *one-to-one* correspondence between private and public key, unconditional anonymity is unattainable, since an unbounded adversary might recover the key from the linking tag by 'reversing' the function that connects them. Our construction avoids this pitfall.

Game 1.2: Anonymity experiment $\mathsf{Exp}^{\text{anon}}_{\mathcal{A},\Pi}$

---

params $\leftarrow \Pi.\mathsf{Setup}(1^\lambda)$

$\mathcal{U} \leftarrow \left\{(\mathsf{pk}_i, \mathsf{sk}_i) \leftarrow \Pi.\mathsf{KGen}()\right\}_{i=1}^{\mu(\lambda)}$

$(\mathsf{ev}, L = \{\mathsf{pk}_i\}_{i=1}^n, \mathtt{m}, \mathsf{pk}_D) \leftarrow \mathcal{A}^{\mathcal{JO},\mathcal{CO},\mathcal{SO},\mathcal{MO}}(\mathcal{U})$

$\pi \leftarrow_\$ [n]$

$\sigma_c \leftarrow \Pi.\mathsf{Sign}(\mathsf{ev}, L, \mathtt{m}, \mathsf{pk}_D, \mathsf{sk}_\pi)$

$\xi \leftarrow \mathcal{A}^{\mathcal{CO},\mathcal{SO},\mathcal{MO}}(L, \mathtt{m}, \sigma_c)$

return $\xi = \pi$ AND $\pi \notin \mathcal{CO}$ AND π cannot be obtained from $\sigma \in \mathcal{SO}$

3.6 Non-transferability

Non-Transferability means that the public cannot tell if a valid signature originates from the Sign or the Sim algorithm. In essence, this property ensures that a simulation is only distinguishable by the designated verifier and the owner of the pseudoidentity used to create it. The formal definition is given in Game 1.3.

The adversary is considered to be computationally unbounded, and it is given access to $\mathcal{CO}, \mathcal{SO}, \mathcal{MO}$ for the same reasons as in Subsect. 3.5. It may query them with any adaptive strategy at any point. The adversary $\mathcal{A}$ chooses the signature parameters. The challenger then produces the signature σ and the simulation σ' with the same pseudoidentity pid and one of the two is given randomly to $\mathcal{A}$, who must now guess whether it received a signature or a simulation.

Definition 5. *Non-transferability. A UDVLRS scheme Π is perfectly non-transferable if for any unbounded adversary $\mathcal{A}$ the following holds:*

$$\mathsf{Adv}^{\text{trans}}_{\mathcal{A}}(\lambda) \triangleq \Pr\left[\mathsf{Exp}^{\text{trans}}_{\mathcal{A},\Pi}(\lambda) = 1\right] - \frac{1}{2} = 0$$

Game 1.3: Non-transferability experiment $\mathsf{Exp}^{\text{trans}}_{\mathcal{A},\Pi}$

---

params $\leftarrow \Pi.\mathsf{Setup}(1^\lambda)$

$\mathcal{U} \leftarrow \left\{(\mathsf{pk}_i, \mathsf{sk}_i) \leftarrow \Pi.\mathsf{KGen}()\right\}_{i=1}^{\mu(\lambda)}$

$(\mathsf{ev}, L = \{\mathsf{pk}_i\}_{i=1}^n, \mathtt{m}, \mathsf{pk}_D, \mathsf{pk}_\pi) \leftarrow \mathcal{A}^{\mathcal{JO},\mathcal{CO},\mathcal{SO},\mathcal{MO}}(\mathcal{U})$

$\sigma_0 \leftarrow \Pi.\mathsf{Sign}(\mathsf{ev}, L, \mathtt{m}, \mathsf{pk}_D, \mathsf{sk}_\pi)$

$\mathsf{pid} \leftarrow \Pi.\mathsf{Extract}(\sigma_0)$

$\sigma_1 \leftarrow \Pi.\mathsf{Sim}(\mathsf{ev}, L, \mathtt{m}, \mathsf{pk}_D, \mathsf{sk}_D, \mathsf{pid})$

$b \leftarrow_\$ \{0,1\}$

$b' \leftarrow \mathcal{A}^{\mathcal{CO},\mathcal{SO},\mathcal{MO}}(L, \mathtt{m}, \sigma_b)$

return $b = b'$

3.7 Linkability

Linkability means that all signatures of the same signer should be linked, while all other security properties are preserved. Simulated signatures generated by the designated verifier for a specific pid are linked to this pid's signatures. Game 1.4 captures the definition of linkability. In this game, the adversary $\mathcal{A}$ tries to generate two unlinked signatures with *a single* secret key. $\mathcal{A}$ queries all the oracles according to any adaptive strategy to generate two signatures σ_1 and σ_2. The adversary can pick the public keys from two separate rings and select two different designated verifiers. $\mathcal{A}$ wins if both signatures verify and are not linked.

Definition 6. *Linkability. A UDVLRS scheme Π is linkable if for any* PPT *adversary $\mathcal{A}$ the following holds:*

$$\mathsf{Adv}_{\mathcal{A}}^{\mathsf{link}}(\lambda) \triangleq \Pr\left[\mathsf{Exp}_{\mathcal{A},\Pi}^{\mathsf{link}}(\lambda) = 1\right] \leq \mathsf{negl}(\lambda)$$

Note that our model in Game 1.4 differs from the corresponding one in [2] which stated that $\mathcal{A}$ could not produce $k + 1$ pairwise unlinked signatures by having access to k signing keys. In our case, $\mathcal{A}$ can only have access to a single private key instead of more. This restriction of linkability allows our scheme to have unconditional anonymity, but has the downside that the scheme is prone to linkability attacks, if there is collusion of signers or if the adversary has access to more than one private keys. As a result, this weaker linkability together with the unforgeability property do not imply non-slanderability as in [2]. In the next section we define the notion of non-slanderability.

Game 1.4: Linkability experiment $\mathsf{Exp}_{\mathcal{A},\Pi}^{\mathsf{link}}$

---

params $\leftarrow \Pi.\mathsf{Setup}(1^{\lambda})$

$\mathcal{U} \leftarrow \left\{(\mathsf{pk}_i, \mathsf{sk}_i) \leftarrow \Pi.\mathsf{KGen}()\right\}_{i=1}^{\mu(\lambda)}$

$(\sigma_1, \sigma_2, \mathsf{ev}, L_1 = \{\mathsf{pk}_i\}_{i=1}^{n_1}, L_2 = \{\mathsf{pk}_i\}_{i=1}^{n_2}, \mathsf{m}_1, \mathsf{m}_2, \mathsf{pk}_{D_1}, \mathsf{pk}_{D_2}) \leftarrow$
$\mathcal{A}^{\mathcal{RO},\mathcal{JO},\mathcal{CO},\mathcal{SO},\mathcal{MO}}(\mathcal{U})$

return
$|\mathcal{CO}| = 1$ AND $\sigma_1, \sigma_2 \notin \mathcal{SO}$ AND $\mathsf{Vrfy}(\mathsf{ev}, \sigma_1, L_1, \mathsf{m}_1, \mathsf{pk}_{D_1}) = 1$ AND
$\mathsf{Vrfy}(\mathsf{ev}, \sigma_2, L_2, \mathsf{m}_2, \mathsf{pk}_{D_2}) = 1$ AND $\mathsf{Link}(\sigma_1, \mathsf{ev}, \sigma_2, \mathsf{ev}) = 0$

3.8 Non-slanderability

Non-slanderability prevents framing, by not allowing adversarial attempts to link signatures to a specific ring member. Therefore, if a signature is linked to an existing one, it is either generated by the same signer or by the corresponding designated verifier. This is captured by Game 1.5. The adversary queries all the oracles, according to any adaptive strategy, and chooses the signature parameters and the public key of a selected signer pk_{π}, and gives them to challenger $\mathcal{C}$. Then, $\mathcal{C}$ using the Sign algorithm for the private key sk_{π}, produces a signature σ_1. Note that pk_{π} - chosen by $\mathcal{A}$ - should not have been asked from $\mathcal{CO}$ or included as public key of any query to $\mathcal{SO}$. Then, $\mathcal{A}$ queries the oracles with the same restrictions for pk_{π} and produces new signature parameters (except for ev) and σ_2, different from σ_1. $\mathcal{A}$ wins if σ_2 verifies and σ_1 and σ_2 are linked.

Definition 7. *Non-Slanderability. A UDVLRS scheme Π is non-slanderable if for any* PPT *adversary $\mathcal{A}$ the following holds:*

$$\mathsf{Adv}^{\mathrm{sland}}_{\mathcal{A}}(\lambda) \triangleq \Pr\left[\mathsf{Exp}^{\mathrm{sland}}_{\mathcal{A},\Pi}(\lambda) = 1\right] \leq \mathsf{negl}(\lambda)$$

Game 1.5: Non-slanderability experiment $\mathsf{Exp}^{\mathrm{sland}}_{\mathcal{A},\Pi}$

$\mathsf{params} \leftarrow \Pi.\mathsf{Setup}(1^\lambda)$

$\mathcal{U} \leftarrow \left\{(\mathsf{pk}_i, \mathsf{sk}_i) \leftarrow \Pi.\mathsf{KGen}()\right\}_{i=1}^{\mu(\lambda)}$

$(\mathsf{ev}, L_1 = \{\mathsf{pk}_i\}_{i=1}^{n_1}, \mathsf{m}_1, \mathsf{pk}_{D_1}, \mathsf{pk}_\pi) \leftarrow \mathcal{A}^{\mathcal{RO},\mathcal{JO},\mathcal{CO},\mathcal{SO},\mathcal{MO}}(\mathcal{U})$

$\sigma_1 \leftarrow \Pi.\mathsf{Sign}(\mathsf{ev}, L_1, \mathsf{m}_1, \mathsf{pk}_{D_1}, \mathsf{sk}_\pi)$

$(\sigma_2, L_2 = \{\mathsf{pk}_i\}_{i=1}^{n_2}, \mathsf{m}_2, \mathsf{pk}_{D_2}) \leftarrow \mathcal{A}^{\mathcal{RO},\mathcal{JO},\mathcal{CO},\mathcal{SO},\mathcal{MO}}(\mathcal{U})$

return

$\sigma_2 \neq \sigma_1$ AND $\mathsf{Vrfy}(\mathsf{ev}, \sigma_2, L_2, \mathsf{m}_2, \mathsf{pk}_{D_2}) = 1$ AND $\sigma_2 \notin \mathcal{SO}$ AND $\sigma_2 \notin \mathcal{MO}$ AND $\pi \notin \mathcal{CO}$ AND $D_2 \notin \mathcal{CO}$ AND $\mathsf{Link}(\sigma_1, \mathsf{ev}, \sigma_2, \mathsf{ev}) = 1$

---

4   Our Construction

Our proposed construction that implements the functionalities of Sect. 3 is depicted in Fig. 1. Our signatures (excluding the linking tag) consist of $2n + 4$ elements in contrast with [2] that consist of $3n + 1$. Thus, while they remain linear to the size of the ring they are in practice shorter.

5   Security Analysis

The completeness of our construction as well as detailed proofs of its security properties can be found in the full version of the paper[1]. Here we provide only proof sketches of the corresponding theorems.

Theorem 1 (Unforgeability). *Our UDVLRS scheme is unforgeable in the random oracle model if* DLOG *is hard in* $\mathbb{G}$.

Proof Sketch. Assume an adversary $\mathcal{A}$ that produces a forged signature in polynomial time with non-negligible probability. We construct another adversary $\mathcal{B}$ that simulates the oracles used by $\mathcal{A}$ in Game 1.1 and uses the forgery to solve DLOG for the 'x' part of pk_D or MDLR for the 'x' part of the keys of L in polynomial time and with non-negligible probability. As we assumed in Fig. 1, KGen returns a group where DLOG is hard, which is a contradiction. The proof uses the forking lemma [13] and techniques of [2,9].

---

[1] The full version of the paper appears in the IACR Cryptology ePrint Archive, Paper 2022/1138, https://eprint.iacr.org/2022/1138. Here we provide only proof sketches of the corresponding theorems.

- params ← Setup(λ):
  Return $\mathbb{G}$ of order q, where the DLOG problem is hard, two random generators $g, h \in \mathbb{G}$ and two hash functions $H_{\mathbb{G}} : \{0,1\}^* \to \mathbb{G}$ and $H_q : \{0,1\}^* \to \mathbb{Z}_q$. Also return $\mathcal{MSG} = \{0,1\}^*$, $\mathcal{SG} = \mathbb{G} \times \mathbb{Z}_q^{2n+4}$, $\mathcal{PID} = \mathbb{G}$, $\mathcal{EID} = \{0,1\}^*$. Note that the signature space is related to the size of the ring and that the relative discrete logarithms of g, h should not be known.
- (sk, pk) ← KGen():
  Each user i samples $x_i, y_i \xleftarrow{\$} \mathbb{Z}_q$, computes $Z_i \leftarrow g^{x_i} h^{y_i}$ and sets $\mathsf{sk}_i \leftarrow (x_i, y_i)$, $\mathsf{pk}_i = Z_i$. The designated verifier sets $\mathsf{sk}_D \leftarrow (x_D, y_D), \mathsf{pk}_D \leftarrow Z_D$.
- $\sigma \leftarrow$ Sign(ev, L, m, pk_D, sk_π):
  The signer π picks $r_x, r_y, r, s, \{c_i\}_{\substack{i \in [n] \\ i \neq \pi}}, \{w_i\}_{i \in [n]} \xleftarrow{\$} \mathbb{Z}_q$ and computes:

$$e \leftarrow H_{\mathbb{G}}(\mathsf{ev}), \qquad t \leftarrow e^{x_\pi}, \qquad K \leftarrow g^{r_x} h^{r_y} \cdot \prod_{\substack{i \in [n] \\ i \neq \pi}} Z_i^{c_i + w_i},$$

$$K' \leftarrow e^{r_x} \cdot t^{\sum_{\substack{i \in [n] \\ i \neq \pi}} c_i + w_i}, \qquad K'' \leftarrow h^s \mathsf{pk}_D^r \cdot \prod_{i=1}^{n} g^{w_i}$$

Then it computes c_π such that $\sum_{i=1}^{n} c_i \mod q = H_q(\mathsf{m}, L, \mathsf{ev}, t, K, K', K'')$ and computes:

$$\tilde{x} \leftarrow (r_x - (c_\pi + w_\pi)x_\pi) \mod q, \qquad \tilde{y} \leftarrow (r_y - (c_\pi + w_\pi)y_\pi) \mod q$$

The signature is the tuple $\sigma = (t, \tilde{x}, \tilde{y}, r, s, \{c_i\}_{i=1}^n, \{w_i\}_{i=1}^n)$.
- pid ← Extract(σ)
  Parse σ as the tuple $(t, \tilde{x}, \tilde{y}, r, s, \{c_i\}_{i=1}^n, \{w_i\}_{i=1}^n)$ and return t.
- $\sigma \leftarrow$ Sim(ev, L, m, pk_D, sk_D, pid): The designated verifier picks $\chi, \psi, \alpha, \beta, \gamma, \{c_i\}_{i=2}^n$, $\{w_i\}_{i=2}^n \xleftarrow{\$} \mathbb{Z}_q$, and $t = \mathsf{pid}$ as pseudoidentity and then computes:

$$K_D \leftarrow g^\chi \cdot h^\psi \cdot Z_1^\alpha \cdot \prod_{i=2}^{n} Z_i^{c_i + w_i},$$

$$K_D' \leftarrow e^\chi t^{\alpha + \sum_{i=2}^n c_i + w_i}, \qquad K_D'' \leftarrow g^\beta h^\gamma \cdot \prod_{i=2}^{n} g^{w_i}$$

Computes c_1 such that $\sum_{i=1}^{n} c_i \mod q = H_q(\mathsf{m}, L, \mathsf{ev}, t, K_D, K_D', K_D'')$ and sets:

$$\tilde{x} \leftarrow \chi, \qquad \tilde{y} \leftarrow \psi, \qquad w_1 \leftarrow \alpha - c_1 \mod q,$$

$$r \leftarrow (\beta - w_1)x_D^{-1} \mod q, \qquad s \leftarrow \gamma - ry_D \mod q$$

The simulated signature is the tuple $\sigma = (t, \tilde{x}, \tilde{y}, r, s, \{c_i\}_{i=1}^n, \{w_i\}_{i=1}^n)$.
Note that Sim can use any c_k, w_k for any $k \in [n]$ instead of c_1, w_1.
- $\{0,1\} \leftarrow$ Vrfy(ev, L, m, pk_D, σ):
  Parse σ as the tuple $(t, \tilde{x}, \tilde{y}, r, s, \{c_i\}_{i=1}^n, \{w_i\}_{i=1}^n)$ and compute the value:

$$c_0 \leftarrow H_q(\mathsf{m}, L, \mathsf{ev}, t, g^{\tilde{x}} h^{\tilde{y}} \cdot \prod_{i=1}^{n} Z_i^{c_i + w_i}, e^{\tilde{x}} \cdot t^{\sum_{i=1}^n c_i + w_i}, h^s \mathsf{pk}_D^r \cdot \prod_{i=1}^{n} g^{w_i}) \qquad (1)$$

Return 1 (valid) if $c_0 = \sum_{i=1}^{n} c_i \pmod{q}$ else return 0 (invalid).
- $\{0,1\} \leftarrow$ Link(σ_1, ev_1, σ_2, ev_2):
  Return 1 (linked) if $\mathsf{ev}_1 = \mathsf{ev}_2$ AND Extract(σ_1) = Extract(σ_2) and both signatures verify otherwise return 0 (unlinked).

Fig. 1. The UDVLRS construction

Theorem 2 (Anonymity). *Our UDVLRS scheme is unconditionally anonymous.*

Proof Sketch. The proof is based on the idea that the linking tag cannot reveal the signer even for an unbounded adversary. As a result, even if $\mathcal{A}$ is capable of solving the DLOG problem and get the first component 'x' of the secret key $\mathsf{sk} = (x, y)$ of a signer, it still cannot find, without extra information, to which pk the signatures it is attacking belongs to, since there are q different public keys that are equally likely to have the same linking tag. The full proof builds upon the techniques of [9].

Theorem 3 (Non-Transferability). *Our UDVLRS scheme is perfectly non-transferable in the random oracle model.*

Proof Sketch. We argue that the distributions of the outputs of Sign and Sim are identical, when the input is the same signature parameters.

Next, we state Lemma 1 which will be used in the proof of Theorem 4.

Lemma 1. *If an adversary $\mathcal{A}$ knows only one private key $\mathsf{sk}_\pi = (x_\pi, y_\pi)$ where $\pi \in [n]$ and produces a valid signature $\sigma = (t, \tilde{x}, \tilde{y}, r, s, \{c_i\}_{i=1}^n, \{w_i\}_{i=1}^n)$ for an event ev, then $t = e^{x_\pi}$, where $e = \mathsf{H}_{\mathbb{G}}(\mathsf{ev})$, provided that DLOG is hard, in the random oracle model.*

Theorem 4 (Linkability). *Our UDVLRS scheme is linkable in the random oracle model if DLOG is hard in $\mathbb{G}$.*

Proof. We assume that $\mathcal{A}$, knowing only one private key, can produce two valid signatures that are unlinked, that is, for the linking tags of the two signatures t_1 and t_2 it holds that $t_1 \neq t_2$. Then, by Lemma 1, it should hold that $t_i = e^{x_i}, i = 1, 2$ where $e = \mathsf{H}_{\mathbb{G}}(\mathsf{ev})$. This means that $\mathcal{A}$ knows two different private keys or that it can solve the DLOG problem, both of which are a contradiction. □

Theorem 5 (Non-Slanderability). *Our UDVLRS scheme is non-slanderable in the random oracle model if DLOG is hard in $\mathbb{G}$.*

Proof Sketch. We assume an adversary $\mathcal{A}$ that given a signature of a user with public key pk_π and without knowing sk_π, produces in polynomial time with non-negligible probability a signature with the same linking tag as the signature of the user. Similarly to the proof of Theorem 1, we construct another adversary $\mathcal{B}$ that simulates the oracles used by $\mathcal{A}$ in Game 1.5. The adversary $\mathcal{B}$ uses the forgery of $\mathcal{A}$ to solve in polynomial time and with non-negligible probability the DLOG problem for the 'x' part of either pk_D or pk_π. The KGen function of Fig. 1 returns a group where the DLOG is considered hard, which makes this a contradiction.

6   Conclusion and Future Work

In this work we defined UDVLRS, a signature construction that provides public verifiability, unconditional anonymity and non-transferability, unforgeability, non-slanderability, and linkability - the latter conditional to the hardness of DLOG. This is an improvement on its predecessor, DVLRS [2], the anonymity of which was computational. We formally defined the security model of our new scheme and proved its properties. While a UDVLR signature consists of fewer elements than a DVLR signature, the size of both schemes is linear to the number of ring members. As a result, the goal of shorter signatures set in [2] remains an open problem.

References

1. Kılınç Alper, H., Burdges, J.: Two-round trip schnorr multi-signatures via delinearized witnesses. In: Malkin, T., Peikert, C. (eds.) CRYPTO 2021. LNCS, vol. 12825, pp. 157–188. Springer, Cham (2021). https://doi.org/10.1007/978-3-030-84242-0_7

2. Behrouz, P., Grontas, P., Konstantakatos, V., Pagourtzis, A., Spyrakou, M.: Designated-verifier linkable ring signatures. In: Park, J.H., Seo, S.H. (eds.) ICISC 2021. LNCS, vol. 13218, pp. 51–70. Springer, Cham (2022). https://doi.org/10.1007/978-3-031-08896-4_3

3. Bellare, M., Rogaway, P.: Random oracles are practical: a paradigm for designing efficient protocols. In: CCS 1993, pp. 62–73. ACM (1993)

4. Bender, A., Katz, J., Morselli, R.: Ring signatures: stronger definitions, and constructions without random oracles. J. Cryptol. 22(1), 114–138 (2007). https://doi.org/10.1007/s00145-007-9011-9

5. Jakobsson, M., Sako, K., Impagliazzo, R.: Designated verifier proofs and their applications. In: Maurer, U. (ed.) EUROCRYPT 1996. LNCS, vol. 1070, pp. 143–154. Springer, Heidelberg (1996). https://doi.org/10.1007/3-540-68339-9_13

6. Lee, J.S., Chang, J.H.: Strong designated verifier ring signature scheme. In: Sobh, T. (ed.) Innovations and Advanced Techniques in Computer and Information Sciences and Engineering, pp. 543–547. Springer, Dordrech (2007). https://doi.org/10.1007/978-1-4020-6268-1_95

7. Li, J., Wang, Y.: Universal designated verifier ring signature (proof) without random oracles. In: Zhou, X., et al. (eds.) EUC 2006. LNCS, vol. 4097, pp. 332–341. Springer, Heidelberg (2006). https://doi.org/10.1007/11807964_34

8. Lipmaa, H., Wang, G., Bao, F.: Designated verifier signature schemes: attacks, new security notions and a new construction. In: Caires, L., Italiano, G.F., Monteiro, L., Palamidessi, C., Yung, M. (eds.) ICALP 2005. LNCS, vol. 3580, pp. 459–471. Springer, Heidelberg (2005). https://doi.org/10.1007/11523468_38

9. Liu, J.K., Au, M.H., Susilo, W., Zhou, J.: Linkable ring signature with unconditional anonymity. IEEE Trans. Knowl. Data Eng. 26(1), 157–165 (2014)

10. Liu, J.K., Wei, V.K., Wong, D.S.: Linkable spontaneous anonymous group signature for ad hoc groups. In: Wang, H., Pieprzyk, J., Varadharajan, V. (eds.) ACISP 2004. LNCS, vol. 3108, pp. 325–335. Springer, Heidelberg (2004). https://doi.org/10.1007/978-3-540-27800-9_28

11. Liu, J.K., Wong, D.S.: Enhanced security models and a generic construction approach for linkable ring signature. Int. J. Found. Comput. Sci. **17**(06), 1403–1422 (2006)
12. Liu, J.K., Wong, D.S.: Linkable ring signatures: security models and new schemes. In: Gervasi, O., et al. (eds.) ICCSA 2005. LNCS, vol. 3481, pp. 614–623. Springer, Heidelberg (2005). https://doi.org/10.1007/11424826_65
13. Pointcheval, D., Stern, J.: Security arguments for digital signatures and blind signatures. J. Cryptology **13**(3), 361–396 (2000)
14. Rivest, R.L., Shamir, A., Tauman, Y.: How to leak a secret. In: Boyd, C. (ed.) ASIACRYPT 2001. LNCS, vol. 2248, pp. 552–565. Springer, Heidelberg (2001). https://doi.org/10.1007/3-540-45682-1_32
15. Steinfeld, R., Bull, L., Wang, H., Pieprzyk, J.: Universal designated-verifier signatures. In: Laih, C.-S. (ed.) ASIACRYPT 2003. LNCS, vol. 2894, pp. 523–542. Springer, Heidelberg (2003). https://doi.org/10.1007/978-3-540-40061-5_33
16. Tsang, P.P., Wei, V.K.: Short linkable ring signatures for E-voting, E-cash and attestation. In: Deng, R.H., Bao, F., Pang, H.H., Zhou, J. (eds.) ISPEC 2005. LNCS, vol. 3439, pp. 48–60. Springer, Heidelberg (2005). https://doi.org/10.1007/978-3-540-31979-5_5

Finding Points on Elliptic Curves with Coppersmith's Method

Virgile Dossou-Yovo[1], Abderrahmane Nitaj[2(✉)], and Alain Togbé[3]

[1] Institut de Mathématiques et de Sciences Physiques, Dangbo, Bénin
[2] Normandie Univ, UNICAEN, CNRS, LMNO, 14000 Caen, France
abderrahmane.nitaj@unicaen.fr
[3] Department of Mathematics and Statistics, Purdue University Northwest, Westville, USA
atogbe@pnw.edu

Abstract. Several cryptosystems based on Elliptic Curve Cryptography such as ElGamal and KMOV process the message as a point $M = (x_0, y_0)$ of an elliptic curve with an equation of the form $y^2 = x^3 + ax + b$ over a finite field or a finite ring. In this paper, we present a method to find the small solutions of the former elliptic curve equation. Our method is based on Coppersmith's technique and enables one to find the solutions (x_0, y_0) when $|x_0|^3|y_0|^2$ is smaller than the modulus.

Keywords: Elliptic curve cryptography · Coppersmith's method · Lattice basis reduction · Cryptanalysis

1 Introduction

In 1985, Koblitz [10] and Miller [14] independently suggested the use of elliptic curves in public key cryptography. The benefit of using elliptic curves in cryptography is that the keys are much smaller than the keys in other systems based on factorization such as RSA [16], or based on finite fields such as the Diffie-Hellman key exchange protocol [5] and the ElGamal cryptosystem [7]. Therefore, the use of Elliptic Curve Cryptography (ECC) allows faster encryption, decryption, key exchange, signatures, and faster verification of transactions on blockchain. Several cryptographic systems are based on elliptic curves such as the Elliptic Curve Diffie-Hellman (ECDH) key agreement protocol, the Elliptic Curve Digital Signature Algorithm (ECDSA), the Edwards-curve Digital Signature Algorithm (EdDSA), the Elliptic Curve Integrated Encryption Scheme (ECIES), KMOV [11], Demytko [4], and the systems presented in [2,6]. Most of the arithmetic of ECC use short Weierstrass forms over finite fields. Typically, let p be a prime number and $\mathbb{F}_p$ be the finite field with p elements. Let $a, b \in \mathbb{F}_p$. In the short Weierstrass form, the elliptic curve $E_p(a, b)$ is the set of solutions of the modular equation $y^2 \equiv x^3 + ax + b \pmod{p}$ together with a special point $\mathcal{O}$, called the point at infinity. With the chord-and-tangent rule addition, the set $E_p(a, b)$ is a finite group. In some systems, the prime number is replaced by a

© The Author(s), under exclusive license to Springer Nature Switzerland AG 2022
D. Poulakis and G. Rahonis (Eds.): CAI 2022, LNCS 13706, pp. 69–80, 2022.
https://doi.org/10.1007/978-3-031-19685-0_6

composite integer of the form $n = pq$. In this case, the chord-and-tangent rule addition works as well, and $E_n(a, b)$ is a ring.

In some situations, it is required to use a solution $P_0 = (x_0, y_0)$ of the modular equation $y^2 \equiv x^3 + ax + b \pmod{p}$ where both $|x_0|$ and $|y_0|$ are small. Such solutions are often used as generators for cryptographic applications such as ECDH and ECDSA, or are used to represent plain messages as in EC-ElGamal, KMOV, and other systems. When the solution $P_0 = (x_0, y_0)$ is not public and $|x_0|$ and $|y_0|$ are sufficiently small, an attacker can find (x_0, y_0) by finding the small solutions by some algebraic method. In general, the known methods are devoted to solve the equation $y^2 = x^3 + ax + b$ over the integers [17]. The situation is different for the modular elliptic equation $y^2 \equiv x^3 + ax + b \pmod{n}$, especially when the prime factorization of n is not known.

In this paper, we use Coppersmith's method [3] to find the small solutions (x, y) of the modular elliptic curve equation

$$(Y_0 + y)^2 \equiv (X_0 + x)^3 + a(X_0 + x) + b \equiv 0 \pmod{n}, \tag{1}$$

where n is a positive integer with unknown factorization, $a, b \in \mathbb{Z}/n\mathbb{Z}$ are fixed, and X_0, Y_0 are known parameters. The method is valid for $X_0 = Y_0 = 0$ with the equation $y^2 \equiv x^3 + ax + b \pmod{n}$. Coppersmith's method was first presented in 1996 to compute small roots of bivariate polynomials $f(x, y)$ over the integers, and small solutions of univariate modular polynomial equations $f(x) \equiv 0 \pmod{n}$ using lattice reduction algorithms. We adapt Coppersmith's method to find the small solutions of the modular elliptic curve equation (1). We show that this equation can be solved in polynomial time if $|x|^3 y^2 < n$. When $X_0 = 0$ and $Y_0 = 0$, our method can be applied to find the plain message $M = (x_0, y_0)$ in EC-ElGamal, KMOV, Demytko and other related systems if $|x_0|^3 y_0^2 < n$. We note that our method works even when the equation $y^2 \equiv x^3 + ax + b \pmod{n}$ does not represent an elliptic curve, that is even if $4a^3 + 27b^2 \equiv 0 \pmod{n}$.

The rest of the paper is organized as follows. In Sect. 2, we present three ECC cryptosystems that use the message as a point of an elliptic curve, and review lattice basis reduction and Coppersmith's method. In Sect. 3, we present our method to find the small solutions of an elliptic curve equation over a finite field or a finite ring. In Sect. 4, we propose a numerical example, and conclude the paper in Sect. 5.

2   Preliminaries

In this section, we present some ECC cryptosystems where the message is represented as a point on an elliptic curve. We also present lattice basis reduction, and Coppersmith's method for solving modular polynomial equations.

2.1   Some ECC Cryptosystems

The original ElGamal cryptosystem [7] can easily be adapted for elliptic curves. First, a modulus p, an elliptic curve E over $\mathbb{Z}/p\mathbb{Z}$, and a point P on E are chosen. Then a private key a, and a public key $Q = aP$ are generated. To send a message, one matches it with a point $M = (m_1, m_2)$ on E, and encrypt it as $C = kQ + M$ and $K = kP$ where k is a random integer. To decrypt the message, one computes $M = C - aK$ on E.

In 1991, Koyama et al. [11] presented an elliptic curve RSA cryptosystem using a modulus of the form $n = pq$ where p and q are prime numbers satisfying $p \equiv q \equiv 2 \pmod{3}$. A message $M = (m_1, m_2) \in \mathbb{Z}/n\mathbb{Z} \times \mathbb{Z}/n\mathbb{Z}$ is used to form an elliptic curve $E_n(b)$ with the equation $y^2 = x^3 + b$ in $\mathbb{Z}/n\mathbb{Z}$ where $b \equiv m_2^2 - m_1^3 \pmod{n}$. The message M is encrypted by computing $C = eM$ on $E_n(b)$ where e is the public key. To decrypt, one just computes $M = dC$ on $E_n(b)$ where d is the private key which satisfies $ed \equiv 1 \pmod{(p+1)(q+1)}$.

In 1994, Demytko [4] proposed an RSA based cryptosystem using elliptic curves. The modulus is also in the form $n = pq$, and the elliptic curve $E_n(a, b)$ has the equation $y^2 = x^3 + ax + b$ in the ring $\mathbb{Z}/n\mathbb{Z}$. The public and the private keys are two integers e and d satisfying $ed \equiv 1 \pmod{(p + 1 \pm t_p)(q + 1 \pm t_q)}$ where $t_p = p + 1 - |E_p(a, b)|$, $t_q = q + 1 - |E_q(a, b)|$, and $|E_p(a, b)|$, $|E_q(a, b)|$ are the orders of the curve $y^2 = x^3 + ax + b$ in $\mathbb{Z}/p\mathbb{Z}$ and $\mathbb{Z}/q\mathbb{Z}$ respectively. To encrypt a message m, one transforms it to a point $M = (m, y)$ on $E_n(a, b)$, and computes $C = eM = (c, v)$. To decrypt M one computes $M = dC$ where $ed \equiv 1 \pmod{(p + 1 \pm t_p)(q + 1 \pm t_q)}$, depending on the values of the Legendre symbols $\left(\frac{v}{p}\right)$ and $\left(\frac{v}{q}\right)$ (see [4,15] for more details).

2.2   Lattice Basis Reduction and Coppersmith's Method

Let $u_1, \ldots, u_m \in \mathbb{R}^n$ be m linearly independent vectors where m and n are two positive integers satisfying $m \leq n$. The lattice $\mathcal{L}$ spanned by $\{u_1, \ldots, u_m\}$ is the set of linear combinations of the vectors $u_1, \cdots, u_m$ using integer coefficients

$$\mathcal{L} = \left\{ \sum_{i=1}^m a_i u_i \mid a_i \in \mathbb{Z} \right\}.$$

The set $\{u_1, \ldots, u_m\}$ form a basis for $\mathcal{L}$, and $\dim(\mathcal{L}) = m$ is its dimension. When $m = n$, the determinant is equal to the absolute value of the determinant of the matrix whose rows are the basis vectors $u_1, \ldots, u_m$, that is

$$\det(\mathcal{L}) = |\det(u_1, \ldots, u_m)|.$$

If $u = \sum_{i=1}^m a_i u_i$ is a vector of $\mathcal{L}$, then the Euclidean norm of u is

$$\|u\| = \left(\sum_{i=1}^m a_i^2 \right)^{\frac{1}{2}}.$$

A lattice has infinitely many bases with the same determinant and it is useful to find a basis with vectors of small Euclidean norms. However, finding the shortest nonzero vector in a lattice is very hard in general. It is called the

Shortest Vector Problem (SVP), and is known to be NP hard under randomized reductions [1]. In 1982, Lenstra, Lenstra and Lovász [12] invented the so-called LLL algorithm to reduce a basis and to approximate a shortest lattice vector in time polynomial in the bit-length of the entries of the basis matrix and in the dimension of the lattice. In the following theorem, we state a general result on the size of the individual reduced basis vectors. A proof can be found in [12,13].

Theorem 1 (LLL). *Let $\mathcal{L}$ be a lattice of dimension m. In polynomial time, the LLL-algorithm outputs a reduced basis $\{b_1, \ldots, b_m\}$ that satisfies, after rearranging the norms*

$$\|b_1\| \leq \|b_2\| \leq 2^{\frac{m}{4}} \det(\mathcal{L})^{\frac{1}{m-1}}.$$

Using lattice reduction techniques, Coppersmith [3] proposed in 1996 a rigorous technique to compute small roots of bivariate polynomials over the integers, and univariate modular polynomials modulo an integer with unknown factorization. Since then, Coppersmith's method has been extended to more variables. Howgrave-Graham [8] reformulated Coppersmith's technique and proved the following result.

Theorem 2 (Howgrave-Graham). *Let $f(x,y) \in \mathbb{Z}[x,y]$ be a polynomial which is a sum of at most ω monomials. Let N, x_0, y_0, X, and Y be integers such that*

$$f(x_0, y_0) \equiv 0 \pmod{N},$$
$$|x_0| < X, |y_0| < Y,$$
$$\|f(Xx, Yy)\| < \frac{N}{\sqrt{\omega}}.$$

Then $f(x_0, y_0) = 0$ holds over the integers.

If we know two algebraically independent polynomials $f_1(x,y)$ and $f_2(x,y)$ such that $f_1(x_0, y_0) = f_2(x_0, y_0) = 0$, then using resultant computation or Gröbner technique, it is possible to find the solution (x_0, y_0).

We note that multivariate Coppersmith's method requires the following heuristic assumption.

Assumption 1. *At least two polynomial equations determined by the short vectors in the LLL-reduced lattice are algebraically independent.*

3   Small Solutions of the Elliptic Curve Equation

In this section, we present a method for finding the small solutions of the elliptic curve equation $(Y_0 + y)^2 \equiv (X_0 + x)^3 + a(X_0 + x) + b \pmod{n}$ when $|x|$ and $|y|$ are suitably small, and X_0, Y_0 are known.

Theorem 3. *Let X_0, Y_0, a, b and n be integers. If the equation*

$$(Y_0 + y)^2 \equiv (X_0 + x)^3 + a(X_0 + x) + b \pmod{n},$$

has a solution (x_0, y_0) with $|x_0|^3 y_0^2 < n$, then, under Assumption 1, one can find x_0 and y_0 in polynomial time.

Proof. Suppose that

$$(Y_0 + y)^2 \equiv (X_0 + x)^3 + a(X_0 + x) + b \pmod{n},$$

has a solution (x_0, y_0). Consider the polynomial

$$
\begin{aligned}
f(x, y) &= (X_0 + x)^3 + a(X_0 + x) + b - (Y_0 + y)^2 \\
&= x^3 + 3X_0 x^2 + (3X_0^2 + a)x - y^2 - 2Y_0 y + X_0^3 - Y_0^2 + aX_0 + b.
\end{aligned}
$$

Then $f(x_0, y_0) \equiv 0 \pmod{n}$. Suppose that

$$|x_0| < X = n^\delta, \quad y_0 < Y = n^\gamma. \tag{2}$$

We mix Coppersmith's method [3] with the extended strategy of Jochemsz and May [9]. Let m and t be two positive integers to be optimized later. Define the set

$$M_k = \bigcup_{0 \le h \le t} \left\{ x^i y^{j+h} \mid x^i y^j \text{ is a monomial of } f^m(x, y) \right.$$

$$\left. \text{and} \quad \frac{x^i y^j}{x^{3k}} \text{ is a monomial of } f^{m-k} \right\}.$$

A straightforward calculation shows that

$$
f^m(x, y) = \sum_{i_1=0}^{m} \sum_{i_2=0}^{m-i_1} \sum_{i_3=0}^{m-i_1-i_2} \sum_{j_1=0}^{m-i_1-i_2-i_3} \sum_{j_2=0}^{m-i_1-i_2-i_3-j_1} \binom{m}{i_1} \binom{m-i_1}{i_2}
$$

$$
\binom{m-i_1-i_2}{i_3} \binom{m-i_1-i_2-i_3}{j_1} \binom{m-i_1-i_2-i_3-j_1}{j_2} (-1)^{j_1} (3X_0)^{i_2}
$$

$$
(3X_0^2 + a)^{i_3} (-2Y_0)^{j_2} (X_0^3 - Y_0^2 + aX_0 + b)^{m-i_1-i_2-i_3-j_1-j_2}
$$

$$
x^{3i_1+2i_2+i_3} y^{2j_1+j_2}.
$$

This gives the following properties

$$x^i y^j \in f^m \text{ if } \begin{cases} i = 3i_1 + 2i_2 + i_3, \\ j = 2j_1 + j_2. \end{cases}$$

where

$$
\begin{cases}
i_1 = 0, \ldots, m, \\
i_2 = 0, \ldots, m - i_1, \\
i_3 = 0, \ldots, m - i_1 - i_2, \\
j_1 = 0, \ldots, m - i_1 - i_2 - i_3, \\
j_2 = 0, \ldots, m - i_1 - i_2 - i_3 - j_1.
\end{cases}
$$

This implies that

$$x^i y^j \in f^{m-k} \text{ if } \begin{cases} i = 3i_1 + 2i_2 + i_3, \\ j = 2j_1 + j_2. \end{cases}$$

where

$$\begin{cases} i_1 = 0, \ldots, m - k, \\ i_2 = 0, \ldots, m - k - i_1, \\ i_3 = 0, \ldots, m - k - i_1 - i_2, \\ j_1 = 0, \ldots, m - k - i_1 - i_2 - i_3, \\ j_2 = 0, \ldots, m - k - i_1 - i_2 - i_3 - j_1. \end{cases}$$

If we assume that $x^i y^j$ is a monomial of $f^m(x, y)$, then $\frac{x^i y^j}{x^{3k}}$ is a monomial of $f^{m-k}(x, y)$ if

$$\begin{cases} i - 3k = 3i_1 + 2i_2 + i_3, \\ j = 2j_1 + j_2. \end{cases}$$

with

$$\begin{cases} i_1 = 0, \ldots, m - k, \\ i_2 = 0, \ldots, m - k - i_1, \\ i_3 = 0, \ldots, m - k - i_1 - i_2, \\ j_1 = 0, \ldots, m - k - i_1 - i_2 - i_3 \\ j_2 = 0, \ldots, m - k - i_1 - i_2 - i_3 - j_1. \end{cases}$$

For $0 \leq k \leq m$, we obtain

$$x^i y^j \in M_k \quad \text{if} \quad i = 3k + 3i_1 + 2i_2 + i_3, \quad j = 2j_1 + j_2 + h,$$

with

$$\begin{cases} i_1 = 0, \ldots, m - k, \\ i_2 = 0, \ldots, m - k - i_1, \\ i_3 = 0, \ldots, m - k - i_1 - i_2, \\ j_1 = 0, \ldots, m - k - i_1 - i_2 - i_3, \\ j_2 = 0, \ldots, m - k - i_1 - i_2 - i_3 - j_1. \\ h = 0, \ldots, t. \end{cases}$$

Also, we have

$$x^i y^j \in M_{k+1} \quad \text{if} \quad i = 3k + 3 + 3i_1 + 2i_2 + i_3, \quad j = 2j_1 + j_2 + h,$$

with

$$\begin{cases} i_1 = 0, \ldots, m - k - 1, \\ i_2 = 0, \ldots, m - k - 1 - i_1, \\ i_3 = 0, \ldots, m - k - 1 - i_1 - i_2, \\ j_1 = 0, \ldots, m - k - 1 - i_1 - i_2 - i_3, \\ j_2 = 0, \ldots, m - k - 1 - i_1 - i_2 - i_3 - j_1. \\ h = 0, \ldots, t. \end{cases}$$

Using M_k and M_{k+1}, we get

$$x^i y^j \in M_k \backslash M_{k+1} \quad \text{if} \quad i = \{3k, 3k+1, 3k+2\} + 3i_1 + 2i_2 + i_3, \quad j = 2j_1 + j_2 + h,$$

with

$$\begin{cases} i_1 = 0, \ldots, m - k, \\ i_2 = 0, \ldots, m - k - i_1, \\ i_3 = 0, \ldots, m - k - i_1 - i_2, \\ j_1 = 0, \ldots, m - k - i_1 - i_2 - i_3, \\ j_2 = 0, \ldots, m - k - i_1 - i_2 - i_3 - j_1. \\ h = 0, \ldots, t. \end{cases}$$

We can observe the following facts

- For $i = 3k + 3i_1 + 2i_2 + i_3$, the maximal value for j is $j = 2(m - k) + t$ and is attained for $(i_1, i_2, i_3) = (0, 0, 0)$ and $(j_1, j_2) = (m - k, 0)$.
- For $i = 3k + 1 + 3i_1 + i_2 + 2i_3$, the maximal value for j is $j = 2(m - k) + t - 2$ and is attained for $(i_1, i_2, i_3) = (0, 1, 0)$ and $(j_1, j_2) = (m - k - 1, 0)$.
- For $i = 3k + 2 + 3i_1 + i_2 + 2i_3$, the maximal value for j is $j = 2(m - k) + t - 2$ and is attained for $(i_1, i_2, i_3) = (0, 0, 1)$ and $(j_1, j_2) = (m - k - 1, 0)$.

This enables us to simplify the monomials of $M_k \backslash M_{k+1}$ as follows

$$x^i y^j \in M_k \backslash M_{k+1} \quad \text{if} \quad \begin{cases} i = 3k, \ j = 0, 1, \ldots, 2(m - k) + t, \\ i = 3k + 1, \ j = 0, 1, \ldots, 2(m - k) + t - 2, \\ i = 3k + 2, \ j = 0, 1, \ldots, 2(m - k) + t - 2. \end{cases}$$

For $0 \leq k \leq m$, define the polynomials

$$g_{k,i,j}(x, y) = \frac{x^i y^j}{x^{3k}} f(x, y)^k n^{m-k} \quad \text{with} \quad x^i y^j \in M_k \backslash M_{k+1}.$$

Then the polynomials $g_{k,i,j}(x, y)$ can be transformed into two forms $G_{k,i,j}(x, y)$ and $H_{k,i,j}(x, y)$ with

$$G_{k,i,j}(x, y) = y^j f(x, y)^k n^{m-k} \quad \text{with} \quad \begin{cases} i = 3k, \\ j = 0, 1, \ldots, 2(m - k) + t, \end{cases}$$

$$H_{k,i,j}(x, y) = x^{i-3k} y^j f(x, y)^k n^{m-k} \quad \text{with} \quad \begin{cases} i = 3k + 1, 3k + 2, \\ j = 0, 1, \ldots, 2(m - k) + t - 2, \end{cases}$$

Let $\mathcal{L}$ be the lattice spanned by the coefficients of the vectors $G_{k,i,j}(xX, yY)$ and $H_{k,i,j}(xX, yY)$ where X, Y are the bounds defined in (2). The rows are ordered following the natural way of (k, i, j), and the monomials $x^i y^j$ are ordered following the natural way of (i, j). For $m = 1$ and $t = 2$, the matrix of $\mathcal{L}$ is represented in Table 1. The non-zero elements are marked with an '$\circledast$'.

Table 1. The matrix of the lattice $\mathcal{L}$ for the case $m = 1$, $t = 2$.

$G_{k,i_1,i_2}(xX,yY)$ $H_{k,i_1,i_2}(xX,yY)$	1	y	y^2	y^3	y^4	x	xy	xy^2	x^2	x^2y	x^2y^2	x^3	x^3y	x^3y^2	x^4	x^5
$G_{0,0,0}(xX,yY)$	n	0	0	0	0	0	0	0	0	0	0	0	0	0	0	0
$G_{0,0,1}(xX,yY)$	0	Yn	0	0	0	0	0	0	0	0	0	0	0	0	0	0
$G_{0,0,2}(xX,yY)$	0	0	Y^2n	0	0	0	0	0	0	0	0	0	0	0	0	0
$G_{0,0,3}(xX,yY)$	0	0	0	Y^3n	0	0	0	0	0	0	0	0	0	0	0	0
$G_{0,0,4}(xX,yY)$	0	0	0	0	Y^4n	0	0	0	0	0	0	0	0	0	0	0
$H_{0,1,0}(xX,yY)$	0	0	0	0	0	Xn	0	0	0	0	0	0	0	0	0	0
$H_{0,1,1}(xX,yY)$	0	0	0	0	0	0	XYn	0	0	0	0	0	0	0	0	0
$H_{0,1,2}(xX,yY)$	0	0	0	0	0	0	0	XY^2n	0	0	0	0	0	0	0	0
$H_{0,2,0}(xX,yY)$	0	0	0	0	0	0	0	0	X^2n	0	0	0	0	0	0	0
$H_{0,2,1}(xX,yY)$	0	0	0	0	0	0	0	0	0	X^2Yn	0	0	0	0	0	0
$H_{0,2,2}(xX,yY)$	0	0	0	0	0	0	0	0	0	0	X^2Y^2n	0	0	0	0	0
$G_{1,3,0}(xX,yY)$	⊛	⊛	⊛	0	0	⊛	0	0	⊛	0	0	X^3	0	0	0	0
$G_{1,3,1}(xX,yY)$	0	⊛	⊛	⊛	0	0	⊛	0	0	⊛	0	0	X^3Y	0	0	0
$G_{1,3,2}(xX,yY)$	0	0	⊛	⊛	⊛	0	0	⊛	0	0	⊛	0	0	X^3Y^2	0	0
$H_{1,4,0}(xX,yY)$	0	0	0	0	0	⊛	⊛	⊛	⊛	0	0	⊛	0	0	X^4	0
$H_{1,5,0}(xX,yY)$	0	0	0	0	0	0	0	0	⊛	⊛	⊛	⊛	0	0	⊛	X^5

Since the matrix is triangular, then the values marked with the symbol ⊛ do not contribute in the computation of the determinant which is in the form

$$\det(\mathcal{L}) = n^{e_n} X^{e_X} Y^{e_Y}. \tag{3}$$

We have

$$e_n = \sum_{k=0}^{m} \sum_{i=3k}^{3k} \sum_{j=0}^{2m-2k+t} (m-k) + \sum_{k=0}^{m} \sum_{i=3k+1}^{3k+2} \sum_{j=0}^{2m-2k+t-2} (m-k)$$

$$= \frac{1}{2} m(m+1)(4m+3t+1).$$

Similarly, we get

$$e_X = \sum_{k=0}^{m} \sum_{i=3k}^{3k} \sum_{j=0}^{m-2k+t} i + \sum_{k=0}^{m} \sum_{i=3k+1}^{3k+2} \sum_{j=0}^{2m-2k+t-2} i$$

$$= \frac{3}{2}(m+1)\left(2m^2 + 3mt - m + 2t - 2\right)$$

and

$$e_Y = \sum_{k=0}^{m} \sum_{i=3k}^{3k} \sum_{j=0}^{2m-2k+t} j + \sum_{k=0}^{m} \sum_{i=3k+1}^{3k+2} \sum_{j=0}^{2m-2k+t-2} j$$

$$= \frac{1}{2}(m+1)\left(4m^2 + 6mt + 3t^2 - 3m - 5t + 4\right).$$

Let ω be the dimension of the lattice. Then

$$\omega = \sum_{k=0}^{m}\sum_{i=3k}^{3k}\sum_{j=0}^{2m-2k+t}1 + \sum_{k=0}^{m}\sum_{i=3k+1}^{3k+2}\sum_{j=0}^{2m-2k+t-2}1$$

$$= (m+1)(3m+3t-1).$$

When we apply the LLL algorithm to the lattice $\mathcal{L}$, we get a reduced basis $b_1(xX, yY), b_2(xX, yY), \ldots, b_\omega(xX, yY)$. Using Theorem 1, this basis satisfies

$$\|b_1(xX, yY)\| \le \|b_2(xX, yY)\| \le 2^{\frac{\omega}{4}}\det(\mathcal{L})^{\frac{1}{\omega-1}}.$$

Since $G_{k,i,j}(x_0, y_0) \equiv 0 \pmod{n^m}$ and $H_{k,i,j}(x_0, y_0) \equiv 0 \pmod{n^m}$ for any (k, i, j), to apply Theorem 2, we need $\|b_2(xX, yY)\| \le \frac{n^m}{\sqrt{\omega}}$. This can be satisfied if

$$2^{\frac{\omega}{4}}\det(\mathcal{L})^{\frac{1}{\omega-1}} < \frac{n^m}{\sqrt{\omega}},$$

which implies

$$\det(\mathcal{L}) < \frac{1}{\left(2^{\frac{\omega}{4}}\sqrt{\omega}\right)^{\omega-1}}n^{m(\omega-1)}.$$

Using (3), we get the inequality

$$n^{e_n}X^{e_X}Y^{e_Y} < \frac{1}{\left(2^{\frac{\omega}{4}}\sqrt{\omega}\right)^{\omega-1}}n^{m(\omega-1)}.$$

Using the values (2) and taking logarithms, we get

$$e_n + e_X\delta + e_Y\gamma < -\frac{(\omega-1)\log\left(2^{\frac{\omega}{4}}\sqrt{\omega}\right)}{\log(n)} + m(\omega-1). \qquad (4)$$

Set $t = \tau m$. Then, for sufficiently large m, $\omega - 1$ and the exponents e_n, e_X and e_Y reduce to

$$\omega - 1 = 3(\tau+1)m^2 + o(m^2),$$

$$e_n = \tfrac{1}{2}(3\tau+4)m^3 + o(m^3),$$

$$e_X = \tfrac{3}{2}(3\tau+2)m^3 + o(m^3),$$

$$e_Y = \tfrac{1}{2}\left(3\tau^2+6\tau+4\right)m^3 + o(m^3).$$

Plugging the former values in (4) and neglecting low order terms, we get

$$\frac{1}{2}(3\tau+4)m^3 + \frac{3}{2}(3\tau+2)\delta m^3 + \frac{1}{2}\left(3\tau^2+6\tau+4\right)\gamma m^3$$

$$< -\frac{(\omega-1)\log\left(2^{\frac{\omega}{4}}\sqrt{\omega}\right)}{\log(n)} + 3(\tau+1)m^3.$$

Simplifying by m^3, we get

$$\frac{1}{2}(3\tau + 4) + \frac{3}{2}(3\tau + 2)\delta + \frac{1}{2}\left(3\tau^2 + 6\tau + 4\right)\gamma < 3(\tau + 1) - \varepsilon_0,$$

where

$$\varepsilon_0 = \frac{(\omega - 1)\log\left(2^{\frac{\omega}{4}}\sqrt{\omega}\right)}{m^3\log(n)}.$$

Transforming this inequality, we get

$$3\gamma\tau^2 + (9\delta + 6\gamma - 3)\tau + 6\delta + 4\gamma - 2 < -2\varepsilon_0.$$

The left hand side is optimized with the value $\tau_0 = \frac{1 - 3\delta - 2\gamma}{2\gamma}$ under the condition $1 - 3\delta - 2\gamma \geq 0$. Plugging τ_0 in the former inequality, we get

$$-27\delta^2 + (18 - 12\gamma)\delta + 4\gamma^2 + 4\gamma - 3 < -8\gamma\varepsilon_0.$$

This is satisfied if

$$\delta < \frac{1}{3} - \frac{2}{3}\gamma - \varepsilon, \tag{5}$$

where ε is a small positive constant. From the two vectors $b_1(xX, yY)$ and $b_2(xX, yY)$, we obtain two polynomials $h_1(x, y)$, $h_2(x, y)$ with the common root (x_0, y_0). Assuming that $h_1(x, y)$, $h_2(x, y)$ are algebraically independent as in Assumption 1, we can find this common root by resultant computation or by Gröbner basis method. Neglecting ε, the condition (5) can be rewritten as $3\delta + 2\gamma < 1$, and finally $x_0^3 y_0^2 < n$. This terminates the proof. □

4   A Numerical Example

We have implemented our method using Maple on a computer with Windows 11 environment, and Intel(R) Core(TM) i5-8250U CPU 1.60 GHZ, 8.0 GO. Let us present the whole details of the method with a numerical example. Consider the following parameters:

$$n = 3650174173313416490006734778062821233,$$
$$a = 75668315480797829587618405552246706,$$
$$b = 1296686456476938429625387225487816755,$$
$$X_0 = 46222544894878157179178395056,$$
$$Y_0 = 32863537713312844398312781203967.$$

Our goal is to find the small solutions $(x, y) = (x_0, y_0)$ of the elliptic curve equation

$$(Y_0 + y)^2 \equiv (X_0 + x)^3 + a(X_0 + x) + b \pmod{n}.$$

We take $m = 4$, and $t = 1$, and form the lattice $\mathcal{L}$, with dimension $\omega = 70$. Also, we take

$$X = Y = \left\lfloor n^{\frac{1}{5}} \right\rfloor = 20533487.$$

By applying the LLL algorithm, we get a new basis with ω polynomials. We use the Gröbner basis method and find the solution

$$x = x_0 = 2114853, \quad y = y_0 = 329043.$$

Both lattice reduction phase and Gröbner basis technique took less than 283 seconds. We note that $x_0^3 y_0^2 < n$, as required by the method.

5 Conclusion

We have presented a method to find the small solutions (x, y) of the elliptic curve equation $(Y_0 + y)^2 = (X_0 + x)^3 + a(X_0 + x) + b$ in a finite field $\mathbb{Z}/p\mathbb{Z}$ where p is a prime number, or over a finite ring $\mathbb{Z}/n\mathbb{Z}$ with a composite integer n with unknown factorization. In the former equation, X_0, Y_0, a and b be fixed integers. Our method is based on Coppersmith's method and finds the solutions (x, y) if $|x|^3 y^2 < p$ or $|x|^3 y^2 < n$. By applying our method with $X_0 = 0$ and $Y_0 = 0$, this enables us to find the encrypted message in certain cryptosystems based on elliptic curves such as ElGamal, KMOV, Demytko and others, if the message is suitably small.

References

1. Ajtai, M.: The shortest vector problem in L_2 is NP-hard for randomized reductions (extended abstract). In: Vitter, J.S. (ed.) Proceedings of the Thirtieth Annual ACM Symposium on the Theory of Computing, Dallas, Texas, USA, 23–26 May 1998, pp. 10–19. ACM (1998)
2. Boudabra, M., Nitaj, A.: A new public key cryptosystem based on edwards curves. IACR Cryptology ePrint Archive, p. 1051 (2019)
3. Coppersmith, D.: Small solutions to polynomial equations, and low exponent RSA vulnerabilities. J. Cryptol. **10**(4), 233–260 (1997)
4. Demytko, N.: A new elliptic curve based analogue of RSA. In: Helleseth, T. (ed.) EUROCRYPT 1993. LNCS, vol. 765, pp. 40–49. Springer, Heidelberg (1994). https://doi.org/10.1007/3-540-48285-7_4
5. Diffie, W., Hellman, M.E.: New directions in cryptography. IEEE Trans. Inf. Theory **22**(6), 644–654 (1976)
6. Galindo, D., Molleví, S.M., Morillo, P., Villar, J.L.: An efficient semantically secure elliptic curve cryptosystem based on KMOV. IACR Cryptology ePrint Archive, p. 37 (2002)
7. El Gamal, T.: A public key cryptosystem and a signature scheme based on discrete logarithms. IEEE Trans. Inf. Theory **31**(4), 469–472 (1985)
8. Howgrave-Graham, N.: Finding small roots of univariate modular equations revisited. In: Darnell, M. (ed.) Cryptography and Coding 1997. LNCS, vol. 1355, pp. 131–142. Springer, Heidelberg (1997). https://doi.org/10.1007/BFb0024458

9. Jochemsz, E., May, A.: A strategy for finding roots of multivariate polynomials with new applications in attacking RSA variants. In: Lai, X., Chen, K. (eds.) ASIACRYPT 2006. LNCS, vol. 4284, pp. 267–282. Springer, Heidelberg (2006). https://doi.org/10.1007/11935230_18

10. Koblitz, N.: Elliptic curve cryptosystems. Math. Comput. **48**(177), 203–209 (1987)

11. Koyama, K., Maurer, U.M., Okamoto, T., Vanstone, S.A.: New public-key schemes based on elliptic curves over the ring Z_n. In: Feigenbaum, J. (ed.) CRYPTO 1991. LNCS, vol. 576, pp. 252–266. Springer, Heidelberg (1992). https://doi.org/10.1007/3-540-46766-1_20

12. Lenstra, A.K., Lenstra, H.W., Lovász, L.: Factoring polynomials with rational coefficients. Math. Ann. **261**(4), 515–534 (1982). Dec

13. May, A.: New RSA vulnerabilities using lattice reduction methods. Ph.D. thesis, University of Paderborn (2003)

14. Miller, V.S.: Use of elliptic curves in cryptography. In: Williams, H.C. (ed.) CRYPTO 1985. LNCS, vol. 218, pp. 417–426. Springer, Heidelberg (1986). https://doi.org/10.1007/3-540-39799-X_31

15. Nitaj, A., Fouotsa, E.: A new attack on RSA and demytko's elliptic curve cryptosystem. IACR Cryptology ePrint Archive, p. 1050 (2019)

16. Rivest, R.L., Shamir, A., Adleman, L.M.: A method for obtaining digital signatures and public-key cryptosystems. Commun. ACM **21**(2), 120–126 (1978)

17. Stroeker, R.J., de Weger, B.M.M.: Solving elliptic diophantine equations: the general cubic case. Acta Arithmetica **87**(4), 339–365 (1999)

Weighted Propositional Configuration Logic over De Morgan Algebras

Leonidas Efstathiadis[(✉)]

Department of Mathematics, Aristotle University of Thessaloniki,
54124 Thessaloniki, Greece
leonidase@math.auth.gr

Abstract. We introduce and investigate a weighted propositional configuration logic over De Morgan algebras. Our logic is intended to serve as a specification language for software architectures with quantitative features. We prove an efficient construction of full normal forms, the decidability of equivalence of formulas and the decidability of a partial order relation over polynomials derived from formulas in this logic. Moreover, we provide formulas of our logic which describe well-known architectures equipped with quantitative characteristics.

Keywords: Software architectures · Propositional configuration logic · Weighted propositional configuration logic · De Morgan algebras

1 Introduction

Architectures are an integral issue in design and development of complex software systems. They characterize coordination principles between components in order to build complex systems with practical management. A software system based on a well-designed architecture will satisfy most of the designer's requirements such as desired performance, functionality, quality, security etc. But, in order to tell apart the well-designed architectures from the rest, or attempt to construct new ones we will first need a formal treatment of architectures. A recent work towards this objective is [7], where the authors introduced the propositional configuration logic (PCL for short) where the meaning of every PCL formula is a configuration set, which represents the acceptable connections between components. PCL with its first-order and second-order extensions, presented in [7], was proven sufficient to describe the qualitative features of an architecture.

However, most practical applications of architectures require the description of some quantitative characteristics to ensure the functionality of software systems, such as the cost, the probability or the time needed for the interactions between system components. For instance, the Publish/Subscribe architecture, that has several IoT and cloud applications, requires the management of quantitative features especially when working with large scale systems [8,10,12]. In general, quantitative properties are crucial for the construction of efficient architectures.

© The Author(s), under exclusive license to Springer Nature Switzerland AG 2022
D. Poulakis and G. Rahonis (Eds.): CAI 2022, LNCS 13706, pp. 81–100, 2022.
https://doi.org/10.1007/978-3-031-19685-0_7

The authors of [9] introduced and investigated a weighted PCL (wPCL for short), which was an extension of PCL over a commutative semiring and serves as a specification language for software architectures with quantitative features such as the maximum cost of an architecture or the maximum priority of a component. Based on the direction of [9], the authors of [5] introduced and investigated a weighted PCL over product valuation monoids (w_{pvm}PCL for short) which could provide some quantitative characteristics that could not be represented by the framework of semirings, such as average cost of an architecture or the maximum most frequent priority of a component.

In this paper we extend the work of [7]. Specifically, we introduce and investigate an extension of PCL over De Morgan algebras (w_{DM}PCL for short), which is proved sufficient to serve as a specification language for software architectures. The motivation of our work is as follows. The structure of De Morgan algebras is suitable for the development of tools for model checking techniques [4,11], as well as allowing the description of quantitative features of an architecture similarly to the case of the commutative semirings of [9]. Additionally, the partial order and the complement function which are implied by the structure of De Morgan algebras can provide ways to compare weighted architectures and give an interpretation to negative weights.

The main contributions of our work are the following. We introduce the syntax and the semantics of w_{DM}PCL. The semantics of w_{DM}PCL formulas are polynomials with values in a De Morgan algebra. Then, in our first main result, we prove that for every w_{DM}PCL formula we can effectively construct a unique equivalent one in full normal form, in doubly exponential time. In our second main result, we prove the decidability of equivalence of w_{DM}PCL formulas, in doubly exponential time. In our third main result, we prove the decidability of a partial order relation over polynomials of w_{DM}PCL formulas, in doubly exponential time. Lastly we present some examples of well-known architectures with w_{DM}PCL formulas to display the expressive power and the potential of w_{DM}PCL.

2   Related Work

In [7], the authors explored many useful properties of PCL, such as the existence of an equivalent full normal form which implied the decidability of equivalence of PCL formulas. Both the equivalent full normal form calculation and the automated decidability of equivalence procedures of PCL, were proven in [9] to take doubly exponential time. Our w_{DM}PCL is an extension of the PCL of [7], which can include weights of a De Morgan algebra in the formulas, similarly allows the construction of an equivalent full normal form and an automated decidability of equivalence of w_{DM}PCL formulas, with the same doubly exponential time complexity.

In [9], the authors introduced wPCL, an extension of PCL, which can include weights of a commutative semiring in the formulas. The existence of a full normal form and the decidability of equivalence were also proven in [9] for wPCL formulas, with both procedures needing doubly exponential time. Our $\mathrm{w}_{DM}\mathrm{PCL}$ is defined through De Morgan algebras, algebraic structures which also constitute commutative semirings with a complement function and a partial order over their elements (cf. [4,11]). It is therefore interesting to note that even with the additional properties that $\mathrm{w}_{DM}\mathrm{PCL}$ entails, there is no increase of the overall complexity of the corresponding weighted procedures.

3   Preliminaries

Let K be a non-empty set. A *partial order* $\leq$ is a binary relation over K that is reflexive, antisymmetric, and transitive. A *partially ordered set* (*poset* for short) is a pair $(K, \leq)$ where $\leq$ is a partial order. A *lattice* is a partially ordered set $(K, \leq)$ where the supremum $k_1 \oplus k_2$ and the infimum $k_1 \otimes k_2$ exist in K for every $k_1, k_2 \in K$ (cf. [6]). A lattice is *distributive* if for all $k_1, k_2, k_3 \in K$ it holds that $k_1 \otimes (k_2 \oplus k_3) = (k_1 \otimes k_2) \oplus (k_1 \otimes k_3)$. A lattice is *bounded* if there exist $0, 1 \in K$ with $0 \neq 1$, such that $0 \leq k \leq 1$ for all $k \in K$.

A *De Morgan algebra* is a bounded distributive lattice $(K, \leq)$ equipped with a mapping $\neg : K \to K$ which satisfies involution and De Morgan laws: $\neg(\neg k) = k$, $\neg(k \oplus k') = \neg k \otimes \neg k'$, $\neg(k \otimes k') = \neg k \oplus \neg k'$, for all $k, k' \in K$. We call $\neg$ a *complement mapping*. We then have that $k \leq k'$ implies $\neg k' \leq \neg k$ for every $k, k' \in K$. The most commonly used De Morgan algebra is the Fuzzy algebra $([0,1], \leq, \neg)$ where $\leq$ is the usual order of real numbers, the supremum and infimum operators are the maximum and minimum operators, respectively, and $\neg k = 1 - k$. Any Boolean algebra is also a De Morgan algebra. Every De Morgan algebra $(K, \leq, \neg)$ constitutes a semiring $(K, \oplus, \otimes, 0, 1)$ with complement mapping $\neg$.

Let Q be a non-empty set. A *series* s over Q and K is a mapping $s : Q \to K$. The *support of a series* s is the set $supp(s) = \{q \in Q \mid s(q) \neq 0\}$. A *polynomial* is a series with finite support. We denote by $K\langle\langle Q \rangle\rangle$ the class of all series over Q and K, and by $K\langle Q \rangle$ the class of all polynomials over Q and K. For any $s, r \in K\langle\langle Q \rangle\rangle, q \in Q$ and $k \in K$, we define the complement $\neg s$, the point-wise supremum $s \oplus r$, the point-wise infimum with scalars ks and sk, and the point-wise infimum $s \otimes r$, respectively by $\neg s(q) = \neg(s(q))$, $(s \oplus r)(q) = s(q) \oplus r(q)$, $ks(q) = k \otimes s(q)$, $sk(q) = s(q) \otimes k$, $(s \otimes r)(q) = s(q) \otimes r(q)$.

Throughout the paper $(K, \oplus, \otimes, 0, 1, \neg)$ will denote a De Morgan algebra where $\oplus$ and $\otimes$ are the supremum and infimum operators in K, respectively.

4   Unweighted PCL

In this section we recall the propositional configuration logic of [7]. Let P be a non-empty set of ports. We let $I(P) = \mathcal{P}(P) \setminus \{\emptyset\}$, where $\mathcal{P}(P)$ denotes the power set of P. Every set $\alpha \in I(P)$ is called an interaction. The syntax of the propositional interaction logic (PIL for short) over P is given by the grammar $\phi ::= true \mid p \mid \overline{\phi} \mid \phi \vee \phi$ where $p \in P$. The operators $\overline{}$ and $\vee$ are called *negation* and *disjunction*, respectively. We set $\overline{\overline{\phi}} = \phi$ and $false = \overline{true}$. The *conjunction* of two PIL formulas ϕ, ϕ' is defined as $\phi \wedge \phi' = \overline{\overline{\phi} \vee \overline{\phi'}}$. A monomial is a PIL formula of the form $p_1 \wedge \ldots \wedge p_n$, where $n > 0$, and $p_i = p_i'$ or $p_i = \overline{p_i'}$ for some $p_i' \in P$, for all $i : 1 \leq i \leq n$.

Let ϕ be a PIL formula and α an interaction. The satisfaction $\alpha \models_i \phi$ is defined in [7] by induction on the structure of ϕ as follows:

$$\alpha \models_i true, \qquad \alpha \models_i p \text{ iff } p \in \alpha,$$
$$\alpha \models_i \overline{\phi} \text{ iff } \alpha \not\models_i \phi, \alpha \models_i \phi_1 \vee \phi_2 \text{ iff } \alpha \models_i \phi_1 \text{ or } \alpha \models_i \phi_2.$$

We define for every interaction α its characteristic monomial $m_\alpha = \bigwedge_{p \in \alpha} p \wedge \bigwedge_{p \notin \alpha} \overline{p}$ such that for any $\alpha' \in I(P), \alpha' \models_i m_\alpha$ iff $\alpha' = \alpha$.

The propositional configuration logic is an extension of PIL given by the grammar $f ::= true \mid \phi \mid \neg_c f \mid f + f \mid f \sqcup f$, where ϕ is a PIL formula. The operators $\neg_c, +$ and $\sqcup$ are called *complementation*, *coalescing* and *union*, respectively. We also define the *intersection* and *implication* operators $f_1 \sqcap f_2 :=$ $\neg_c(\neg_c f_1 \sqcup \neg_c f_2), f_1 \Rightarrow f_2 := \neg_c f_1 \sqcup f_2$, respectively. We let $C(P) = \mathcal{P}(I(P)) \setminus \{\emptyset\}$. Every set $\gamma \in C(P)$ is called a *configuration*. Any PCL formula that is also a PIL formula will be called an *interaction formula*. Let f be a PCL formula and γ a configuration. The satisfaction $\gamma \models f$ is defined in [7] by induction on the structure of f as follows:

$$\gamma \models true,$$
$$\gamma \models \phi, \qquad \text{iff } \forall \alpha \in \gamma, \alpha \models_i \phi, \text{ where } \phi \text{ is an interaction formula,}$$
$$\gamma \models f_1 + f_2, \text{ iff } \exists \gamma_1, \gamma_2 \in C(P) : \gamma = \gamma_1 \cup \gamma_2, \gamma_1 \models f_1 \text{ and } \gamma_2 \models f_2,$$
$$\gamma \models f_1 \sqcup f_2, \text{ iff } \gamma \models f_1 \text{ or } \gamma \models f_2,$$
$$\gamma \models \neg_c f, \qquad \text{iff } \gamma \not\models f.$$

Trivially we can also obtain that $\gamma \models f_1 \sqcap f_2$ iff $\gamma \models f_1$ and $\gamma \models f_2$, and $\gamma \models f_1 \Rightarrow f_2$ iff $\gamma \not\models f_1$ or $\gamma \models f_2$. The closure operator $\sim_c$ for any PCL formula f, is defined in [7] as $\sim_c f := f + true$, and the disjunction of any two PCL formulas f_1, f_2 as $f_1 \vee f_2 := f_1 \sqcup f_2 \sqcup (f_1 + f_2)$.[1] Two PCL formulas f, f' are called equivalent, and we denote it by $f \equiv f'$, if $\gamma \models f$ iff $\gamma \models f'$ for every $\gamma \in C(P)$. For any interaction formulas ϕ, ϕ' it holds that $\phi \sqcap \phi' \equiv \phi \wedge \phi'$, $\phi + \phi \equiv \phi$ and $\overline{\phi} \equiv \neg_c \sim_c \phi$. Therefore, from now on we shall denote both the conjunction and intersection operators with the symbol $\wedge$. The coalescing operator is associative, commutative, has $false$ as its absorbing element and distributes over union and disjunction. We denote by $\sum_{j \in J} f_j$ the coalescing of all the formulas f_j for any index set J. A formula f is said to be in *normal form* if it is expressed as $f = \bigsqcup_{i \in I} \sum_{j \in J_i} \bigvee_{k \in K_{i,j}} m_{i,j,k}$ where $I, J_i, K_{i,j}$ are (finite) index sets and $m_{i,j,k}$ are monomials for all i, j and k. A *full monomial* is a monomial that can be expressed as $m = \bigwedge_{p \in P_+} p \wedge \bigwedge_{p \in P_-} \overline{p}$, where P_+ and P_- are subsets of P such that $P_+ \cup P_- = P$ and $P_+ \cap P_- = \emptyset$.

---

[1] The binding power of all PCL operators (in decreasing order) is: negation, complementation/closure, conjunction, disjunction, coalescing and union.

A formula $f = \sum_{i \in I} m_i$, where m_i are full monomials for all $i \in I$, is satisfied by exactly one configuration $\gamma = \{\alpha_i \mid i \in I\}$ where for all $i \in I$, a_i is the unique interaction such that $\alpha_i \models_i m_i$. The reverse also holds, i.e., for all configurations $\gamma \in C(P)$ there exists an index set I and full monomials m_i for all $i \in I$, such that only γ satisfies the formula $\sum_{i \in I} m_i$. A PCL formula f is said to be in *full normal form* if it is expressed as $f = \bigsqcup_{i \in I} \sum_{j \in J_i} m_{i,j}$, where I, J_i are (finite) index sets and $m_{i,j}$ are full monomials for all i, j.

We now present one of the main results of [9] regarding the PCL of [7].

Theorem 1 ([9]). *Let P be a set of ports. Then, for every PCL formula f over P we can effectively construct, in doubly exponential time, an equivalent PCL formula f' in full normal form. The best run time for the construction of f' is exponential. Furthermore, f' is unique up to equivalence relation.*

5   Weighted PCL over De Morgan Algebras

In this section, we introduce and investigate the weighted propositional configuration logic over De Morgan algebras (w_{DM}PCL for short).

Definition 1. *Let P be a set of ports and $\mathcal{K} = (K, \oplus, \otimes, 0, 1, \neg)$ a De Morgan algebra. The syntax of w_{DM}PCL over P and $\mathcal{K}$ is given by the grammar:*

$$\zeta ::= k \mid f \mid \zeta \oplus \zeta \mid \zeta \otimes \zeta \mid \neg\zeta \mid \zeta \uplus \zeta$$

where $k \in K$, f is a PCL formula over P and $\uplus$ is the coalescing operator among w_{DM}PCL formulas.

We denote by $PCL(\mathcal{K}, P)$ the set of w_{DM}PCL formulas over the set of ports P and the De Morgan algebra $\mathcal{K}$. The semantics of w_{DM}PCL formulas $\zeta \in PCL(\mathcal{K}, P)$ will be represented by polynomials $\|\zeta\| \in K\langle C(P) \rangle$.

Definition 2. *Let P be a non-empty set of ports and $\mathcal{K}$ a De Morgan algebra. For every $\zeta \in PCL(\mathcal{K}, P)$, the semantics of ζ is a polynomial $\|\zeta\| \in K\langle C(P) \rangle$ where for every configuration $\gamma \in C(P)$ the value of $\|\zeta\|(\gamma)$ is defined inductively as follows:*

- $\|k\|(\gamma) = k,$
- $\|f\|(\gamma) = \begin{cases} 1 & \text{if } \gamma \models f \\ 0 & \text{otherwise} \end{cases},$
- $\|\zeta_1 \oplus \zeta_2\|(\gamma) = \|\zeta_1\|(\gamma) \oplus \|\zeta_2\|(\gamma),$
- $\|\zeta_1 \otimes \zeta_2\|(\gamma) = \|\zeta_1\|(\gamma) \otimes \|\zeta_2\|(\gamma),$
- $\|\neg\zeta\|(\gamma) = \neg\|\zeta\|(\gamma),$
- $\|\zeta_1 \uplus \zeta_2\|(\gamma) = \bigoplus_{\gamma = \gamma_1 \cup \gamma_2} (\|\zeta_1\|(\gamma_1) \otimes \|\zeta_2\|(\gamma_2)).$

Two w_{DM}PCL formulas $\zeta_1, \zeta_2 \in PCL(\mathcal{K}, P)$ will be called *equivalent*, and denoted by $\zeta_1 \equiv \zeta_2$, if $\|\zeta_1\| = \|\zeta_2\|$, i.e., $\|\zeta_1\|(\gamma) = \|\zeta_2\|(\gamma)$, $\forall \gamma \in C(P)$. We now extend the partial order $\leq$ of any De Morgan algebra $(K, \oplus, \otimes, 0, 1, \neg)$ to polynomials of $K\langle C(P)\rangle$ as follows: Let $\zeta, \xi \in K\langle C(P)\rangle$, then $\|\zeta\| \leq \|\xi\|$ iff $\forall \gamma \in C(P) : \|\zeta\|(\gamma) \leq \|\xi\|(\gamma)$. It is trivial to show that the extended relation $\leq$ is also a partial order, and therefore that $(K\langle C(P)\rangle, \oplus, \otimes, \|0\|, \|1\|, \neg)$ is also a De Morgan algebra. For any w_{DM}PCL formula ζ we define the *weighted closure operator* as $\sim \zeta := \zeta \uplus 1$.

Lemma 1. *Let* $\zeta \in PCL(\mathcal{K}, P)$. *Then* $\| \sim \zeta\|(\gamma) = \bigoplus_{\emptyset \neq \gamma' \subseteq \gamma} \|\zeta\|(\gamma')$.

Proof. We compute:

$$
\begin{aligned}
\| \sim \zeta\|(\gamma) &= \|\zeta \uplus 1\|(\gamma) \\
&= \bigoplus_{\gamma = \gamma_1 \cup \gamma_2} (\|\zeta\|(\gamma_1) \otimes \|1\|(\gamma_2)) \\
&= \bigoplus_{\emptyset \neq \gamma_1 \subseteq \gamma} \left(\bigoplus_{\gamma_1 \cup \gamma_2 = \gamma} (\|\zeta\|(\gamma_1) \otimes 1) \right) \\
&= \bigoplus_{\emptyset \neq \gamma_1 \subseteq \gamma} \left(\bigoplus_{\gamma_1 \cup \gamma_2 = \gamma} \|\zeta\|(\gamma_1) \right) \\
&= \bigoplus_{\emptyset \neq \gamma' \subseteq \gamma} \|\zeta\|(\gamma'),
\end{aligned}
$$

where the last equality holds due to the idempotence of the supremum operator $\oplus$.

Due to the definition of the weighted closure operator, for any formula ζ, $\| \sim \zeta\|(\gamma)$ equals the supremum of all the possible weights we can obtain from $\|\zeta\|(\gamma')$ for any configuration $\gamma' \subseteq \gamma$.

Proposition 1. *Let* $k \in K$ *and* $\zeta \in PCL(\mathcal{K}, P)$. *Then* $k \uplus \zeta \equiv \sim (k \otimes \zeta) \equiv k \otimes \sim \zeta$.

Proof. Let $\gamma \in C(P)$. For the first equivalence we have:

$$
\begin{aligned}
\|k \uplus \zeta\|(\gamma) &= \bigoplus_{\gamma_1 \cup \gamma_2 = \gamma} (\|k\|(\gamma_1) \otimes \|\zeta\|(\gamma_2)) \\
&= \bigoplus_{\gamma_1 \cup \gamma_2 = \gamma} (\|k \otimes \zeta\|(\gamma_2)) \\
&= \bigoplus_{\emptyset \neq \gamma_2 \subseteq \gamma} (\|k \otimes \zeta\|(\gamma_2)) \\
&= \| \sim (k \otimes \zeta)\|(\gamma),
\end{aligned}
$$

where the third equality holds due to the idempotence of the supremum operator $\oplus$. For the second equivalence we have:

$$
\begin{aligned}
\| \sim (k \otimes \zeta)\|(\gamma) &= \bigoplus_{\emptyset \neq \gamma' \subseteq \gamma} (\|k \otimes \zeta\|(\gamma')) \\
&= \bigoplus_{\emptyset \neq \gamma' \subseteq \gamma} (k \otimes \|\zeta\|(\gamma')) \\
&= k \otimes \bigoplus_{\emptyset \neq \gamma' \subseteq \gamma} \|\zeta\|(\gamma') \\
&= \|k \otimes \sim \zeta\|(\gamma).
\end{aligned}
$$

Lemma 2. *For any PCL formulas* f, g *the following* w_{DM}PCL *equivalences hold:* $\neg f \equiv \neg_c f$, $\sim f \equiv_c f$, $f \oplus g \equiv f \sqcup g$, $f \otimes g \equiv f \wedge g$, $f \uplus g \equiv f + g$.

Proof. 1. Equivalence of complementation operators:

$$
\begin{aligned}
\|\neg f\|(\gamma) &= \neg \|f\|(\gamma) \\
&= \begin{cases} \neg 1, & \text{if } \gamma \models f \\ \neg 0, & \text{otherwise} \end{cases} \\
&= \begin{cases} 0, & \text{if } \gamma \models f \\ 1, & \text{if } \gamma \not\models f \end{cases} \\
&= \begin{cases} 1, & \text{if } \gamma \models \neg_c f \\ 0, & \text{otherwise} \end{cases} \\
&= \|\neg_c f\|(\gamma).
\end{aligned}
$$

2. Equivalence of closure operators:

$$\| \sim f \|(\gamma) = \bigoplus_{\emptyset \neq \gamma' \subseteq \gamma} \| f \|(\gamma'),$$

$$= \begin{cases} 1, & \text{if } \exists \gamma' \subseteq \gamma : \| f \|(\gamma') = 1 \\ 0, & \text{otherwise} \end{cases}$$

$$= \begin{cases} 1, & \text{if } \exists \gamma' \subseteq \gamma : \gamma' \models f \\ 0, & \text{otherwise} \end{cases}$$

$$= \begin{cases} 1, & \text{if } \gamma \models \sim_c f \\ 0, & \text{otherwise} \end{cases}$$

$$= \| \sim_c f \|(\gamma).$$

3. Supremum extends union:

$$\| f \oplus g \|(\gamma) = \| f \|(\gamma) \oplus \| g \|(\gamma)$$

$$= \begin{cases} 1, & \text{if } \| f \|(\gamma) = 1 \text{ or } \| g \|(\gamma) = 1 \\ 0, & \text{otherwise} \end{cases}$$

$$= \begin{cases} 1, & \text{if } \gamma \models f \text{ or } \gamma \models g \\ 0, & \text{otherwise} \end{cases}$$

$$= \begin{cases} 1, & \text{if } \gamma \models f \sqcup g \\ 0, & \text{otherwise} \end{cases}$$

$$= \| f \sqcup g \|(\gamma),$$

4. Infimum extends intersection/conjunction:

$$\| f \otimes g \|(\gamma) = \| f \|(\gamma) \otimes \| g \|(\gamma)$$

$$= \begin{cases} 1, & \text{if } \| f \|(\gamma) = 1 \text{ and } \| g \|(\gamma) = 1 \\ 0, & \text{otherwise} \end{cases}$$

$$= \begin{cases} 1, & \text{if } \gamma \models f \text{ and } \gamma \models g \\ 0, & \text{otherwise} \end{cases}$$

$$= \begin{cases} 1, & \text{if } \gamma \models f \wedge g \\ 0, & \text{otherwise} \end{cases}$$

$$= \| f \wedge g \|(\gamma).$$

5. Equivalence of coalescing operators:

$$\| f \uplus g \|(\gamma) = \bigoplus_{\gamma = \gamma_1 \cup \gamma_2} (\| f \|(\gamma_1) \otimes \| g \|(\gamma_2))$$

$$= \begin{cases} 1, & \text{if } \exists \gamma_1, \gamma_2 : \gamma_1 \cup \gamma_2 = \gamma \text{ and } \| f \|(\gamma_1) = \| g \|(\gamma_2) = 1 \\ 0, & \text{otherwise} \end{cases}$$

$$= \begin{cases} 1, & \text{if } \exists \gamma_1, \gamma_2 : \gamma_1 \cup \gamma_2 = \gamma \text{ and } \gamma_1 \models f, \gamma_2 \models g \\ 0, & \text{otherwise} \end{cases}$$

$$= \begin{cases} 1, & \text{if } \gamma \models f + g \\ 0, & \text{otherwise} \end{cases}$$

$$= \| f + g \|(\gamma),$$

where the second equality holds since the De Morgan infimum $\| f \|(\gamma_1) \otimes \| g \|(\gamma_2)$ equals 1 instead of 0 iff $\| f \|(\gamma_1) = \| g \|(\gamma_2) = 1$.

Therefore, from now on we shall denote both the PCL and w_{DM}PCL complementation and closure operators with the symbols $\neg$ and $\sim$, respectively.[2]

Proposition 2. *The w_{DM}PCL closure operator distributes over supremum, coalescing and is idempotent.*

---

[2] The binding power of all w_{DM}PCL operators (in decreasing order) is: PCL operators, weighted complementation/closure, weighted coalescing, infimum and supremum.

Proof. For every $\gamma \in C(P)$ we have:

$$\| \sim (\zeta_1 \oplus \zeta_2)\|(\gamma) = \bigoplus_{\emptyset \neq \gamma' \subseteq \gamma} \|\zeta_1 \oplus \zeta_2\|(\gamma')$$
$$= \bigoplus_{\emptyset \neq \gamma' \subseteq \gamma} \|\zeta_1\|(\gamma') \oplus \bigoplus_{\emptyset \neq \gamma' \subseteq \gamma} \|\zeta_2\|(\gamma')$$
$$= \| \sim \zeta_1\|(\gamma) \oplus \| \sim \zeta_2\|(\gamma)$$
$$= \| \sim \zeta_1 \oplus \sim \zeta_2\|(\gamma),$$

$$\| \sim (\zeta_1 \uplus \zeta_2)\|(\gamma) = \bigoplus_{\emptyset \neq \gamma' \subseteq \gamma} \|\zeta_1 \uplus \zeta_2\|(\gamma')$$
$$= \bigoplus_{\emptyset \neq \gamma' \subseteq \gamma} \left(\bigoplus_{\gamma' = \gamma_1' \cup \gamma_2'} (\|\zeta_1\|(\gamma_1') \otimes \|\zeta_2\|(\gamma_2')) \right)$$
$$= \bigoplus_{\gamma = \gamma_1 \cup \gamma_2} \left(\bigoplus_{\gamma_1' \subseteq \gamma_1, \gamma_2' \subseteq \gamma_2} (\|\zeta_1\|(\gamma_1') \otimes \|\zeta_2\|(\gamma_2')) \right)$$
$$= \bigoplus_{\gamma = \gamma_1 \cup \gamma_2} \left(\left(\bigoplus_{\gamma_1' \subseteq \gamma_1} \|\zeta_1\|(\gamma_1') \right) \otimes \left(\bigoplus_{\gamma_2' \subseteq \gamma_2} \|\zeta_2\|(\gamma_2') \right) \right)$$
$$= \bigoplus_{\gamma = \gamma_1 \cup \gamma_2} (\| \sim \zeta_1\|(\gamma_1) \otimes \| \sim \zeta_2\|(\gamma_2))$$
$$= \| \sim \zeta_1 \uplus \sim \zeta_2\|(\gamma)$$

$$\| \sim \sim \zeta\|(\gamma) = \bigoplus_{\emptyset \neq \gamma_1 \subseteq \gamma} \| \sim \zeta\|(\gamma_1)$$
$$= \bigoplus_{\emptyset \neq \gamma_1 \subseteq \gamma} \left(\bigoplus_{\emptyset \neq \gamma_2 \subseteq \gamma_1} \|\zeta\|(\gamma_2) \right)$$
$$= \bigoplus_{\emptyset \neq \gamma' \subseteq \gamma} \|\zeta\|(\gamma')$$
$$= \| \sim \zeta\|(\gamma).$$

The third equality of the second proof and the third equality of the third proof hold due to the idempotence of $\oplus$.

Proposition 3. *The $w_{DM}PCL$ coalescing operator $\uplus$ is associative, commutative, has 0 as its absorbing element and distributes over the supremum $\oplus$.*

Proof. We will only prove the first and fourth properties since the rest are trivial. Let $\zeta_1, \zeta_2, \zeta_3 \in PCL(\mathcal{K}, P)$, then for any $\gamma \in C(P)$ we have:

$$\|\zeta_1 \uplus (\zeta_2 \uplus \zeta_3)\|(\gamma) = \bigoplus_{\gamma' \cup \gamma'' = \gamma} (\|\zeta_1\|(\gamma') \otimes \|\zeta_2 \uplus \zeta_3\|(\gamma''))$$
$$= \bigoplus_{\gamma' \cup \gamma'' = \gamma} (\|\zeta_1\|(\gamma') \otimes (\bigoplus_{\gamma_1 \cup \gamma_2 = \gamma''} \|\zeta_2\|(\gamma_1) \otimes \|\zeta_3\|(\gamma_2)))$$
$$= \bigoplus_{\gamma' \cup \gamma'' = \gamma} (\bigoplus_{\gamma_1 \cup \gamma_2 = \gamma''} (\|\zeta_1\|(\gamma') \otimes \|\zeta_2\|(\gamma_1) \otimes \|\zeta_3\|(\gamma_2)))$$
$$= \bigoplus_{\gamma' \cup \gamma_1 \cup \gamma_2 = \gamma} (\|\zeta_1\|(\gamma') \otimes \|\zeta_2\|(\gamma_1) \otimes \|\zeta_3\|(\gamma_2))$$
$$= \bigoplus_{\gamma'' \cup \gamma_2 = \gamma} ((\bigoplus_{\gamma' \cup \gamma_1 = \gamma''} \|\zeta_1\|(\gamma') \otimes \|\zeta_2\|(\gamma_1)) \otimes \|\zeta_3\|(\gamma_2))$$
$$= \bigoplus_{\gamma'' \cup \gamma_2 = \gamma} (\|\zeta_1 \uplus \zeta_2\|(\gamma'') \otimes \|\zeta_3\|(\gamma_2))$$
$$= \|(\zeta_1 \uplus \zeta_2) \uplus \zeta_3\|(\gamma),$$

$$\|\zeta_1 \uplus (\zeta_2 \oplus \zeta_3)\|(\gamma) = \bigoplus_{\gamma' \cup \gamma'' = \gamma} (\|\zeta_1\|(\gamma') \otimes \|\zeta_2 \oplus \zeta_3\|(\gamma''))$$
$$= \bigoplus_{\gamma' \cup \gamma'' = \gamma} (\|\zeta_1\|(\gamma') \otimes (\|\zeta_2\|(\gamma'') \oplus \|\zeta_3\|(\gamma'')))$$
$$= \bigoplus_{\gamma' \cup \gamma'' = \gamma} ((\|\zeta_1\|(\gamma') \otimes \|\zeta_2\|(\gamma'')) \oplus (\|\zeta_1\|(\gamma') \otimes \|\zeta_3\|(\gamma'')))$$
$$= \bigoplus_{\gamma' \cup \gamma'' = \gamma} (\|\zeta_1\|(\gamma') \otimes \|\zeta_2\|(\gamma''))$$
$$\oplus \bigoplus_{\gamma' \cup \gamma'' = \gamma} (\|\zeta_1\|(\gamma') \otimes \|\zeta_3\|(\gamma''))$$
$$= \|\zeta_1 \uplus \zeta_2\|(\gamma) \oplus \|\zeta_1 \uplus \zeta_3\|(\gamma)$$
$$= \|(\zeta_1 \uplus \zeta_2) \oplus (\zeta_1 \uplus \zeta_3)\|(\gamma).$$

Definition 3. *A $w_{DM}PCL$ formula $\zeta \in PCL(\mathcal{K}, P)$ is said to be in full normal form if either:*

- *$\zeta = k$ where $k \in K$, or*
- *there exist finite index sets I and J_i for every $i \in I$ such that*

$$\zeta = \bigoplus_{i \in I} (k_i \otimes \sum_{j \in J_i} m_{i,j}).$$

where $k_i \in K$ and $m_{i,j}$ are full monomials.

Proposition 4. *For any formula $\zeta \in PCL(\mathcal{K}, P)$ that is in full normal form, there exists an equivalent formula $\zeta' \in PCL(\mathcal{K}, P)$ in full normal form $\zeta' = \bigoplus_{i \in I}(k_i \otimes \sum_{j \in J_i} m_{i,j})$ such that it satisfies the following statements:*

- *For every $i \in I$ and $j, j' \in J_i$, $j \neq j'$ implies $m_{i,j} \not\equiv m_{i,j'}$.*
- *For every $i, i' \in I$, $i \neq i'$ implies $\sum_{j \in J_i} m_{i,j} \not\equiv \sum_{j \in J_{i'}} m_{i',j}$.*

Proof. Let $\zeta' = \bigoplus_{i \in I}(k_i \otimes \sum_{j \in J_i} m_{i,j})$.

1. For every pair of full monomials $m_{i,j} \equiv m_{i,j'}$ such that $j \neq j'$ we replace $m_{i,j} + m_{i,j'}$ by its equivalent $m_{i,j}$.
2. For every pair of indices $i \neq i'$ such that $\sum_{j \in J_i} m_{i,j} \equiv \sum_{j \in J_{i'}} m_{i',j}$, we replace the supremum $(k_i \otimes \sum_{j \in J_i} m_{i,j}) \oplus (k_{i'} \otimes \sum_{j \in J_{i'}} m_{i',j})$ with its equivalent $w_i \otimes \sum_{j \in J_i} m_{i,j}$, where $w_i \in K$ such that $w_i = k_i \oplus k_{i'}$.

At the end of the above replacements we are left with unique coalescings of pairwise non-equivalent full monomials.

Remark 1. For every w_{DM}PCL formula, the above proposition will help us define a unique full normal form. Notice that for any full monomial m and $k \in K$ the formulas $\zeta = k \otimes m$ and $\xi = (k \otimes m) \oplus (k \otimes (m + m))$ are not equal, but they are equivalent and they are both in full normal form. Only ζ satisfies the statements of Proposition 4. Any variation between full normal forms of the same formula, which satisfy the statements Proposition 4, would be due to the commutativity of the coalescing $+$ and the supremum $\oplus$ operators.

Lemma 3. *Let $k_1, k_2 \in K$ and $\zeta_1, \zeta_2 \in PCL(\mathcal{K}, P)$. Then:*

$$(k_1 \otimes \zeta_1) \uplus (k_2 \otimes \zeta_2) \equiv (k_1 \otimes k_2) \otimes (\zeta_1 \uplus \zeta_2).$$

Proof. For every $\gamma \in C(P)$ we have:
$$
\begin{aligned}
\|(k_1 \otimes \zeta_1) \uplus (k_2 \otimes \zeta_2)\|(\gamma) &= \bigoplus_{\gamma = \gamma_1 \cup \gamma_2}(\|k_1 \otimes \zeta_1\|(\gamma_1) \otimes \|k_2 \otimes \zeta_2\|(\gamma_2)) \\
&= \bigoplus_{\gamma = \gamma_1 \cup \gamma_2}(k_1 \otimes k_2 \otimes \|\zeta_1\|(\gamma_1) \otimes \|\zeta_2\|(\gamma_2)) \\
&= (k_1 \otimes k_2) \otimes \bigoplus_{\gamma = \gamma_1 \cup \gamma_2}(\|\zeta_1\|(\gamma_1) \otimes \|\zeta_2\|(\gamma_2)) \\
&= (k_1 \otimes k_2) \otimes \|\zeta_1 \uplus \zeta_2\|(\gamma) \\
&= \|(k_1 \otimes k_2) \otimes (\zeta_1 \uplus \zeta_2)\|(\gamma).
\end{aligned}
$$

Lemma 4. *Let J be an index set and m_j a full monomial for every index $j \in J$. Then, there exists a unique $\gamma' \in C(P)$ such that, for every $\gamma \in C(P)$:*

$$
\left\| \sum_{j \in J} m_j \right\|(\gamma) = \begin{cases} 1, & \text{if } \gamma = \gamma' \\ 0, & \text{otherwise} \end{cases}
$$

Proof. We define the configuration $\gamma' = \{a_j \mid m_j \text{ is the characteristic monomial of } a_j\}$. Notice that $\gamma' \models \sum_{j \in J} m_j$. For any $\gamma \in C(P)$, due to the semantics of the coalescing operator, $\|\sum_{j \in J} m_j\|(\gamma) = 1$ implies that for every m_j there exists a $\gamma_j \subseteq \gamma$ such that $\gamma_j \models m_j$. However m_j are full monomials, meaning that only the configuration $\{a_j\}$ satisfies each m_j, for all $j \in J$. Therefore $\gamma = \bigcup_{j \in J} \{a_j\} = \gamma'$. If $\|\sum_{j \in J} m_j\|(\gamma) = 0$ then $\gamma \not\models \sum_{j \in J} m_j$ and therefore cannot be γ'.

Lemma 5. *Let m_i, m'_j be full monomials for every $i \in I, j \in J$, then:*

$$\left(\sum_{i \in I} m_i \right) \otimes \left(\sum_{j \in J} m'_j \right) \equiv \begin{cases} \sum_{i \in I} m_i, & \text{if } \sum_{i \in I} m_i \equiv \sum_{j \in J} m'_j \\ 0, & \text{otherwise} \end{cases}.$$

Proof. By Lemma 4 there exist $\gamma', \gamma'' \in C(P)$ such that for any $\gamma \in C(P)$, $\|\sum_{i \in I} m_i\|(\gamma) = 1$ if $\gamma = \gamma'$ while $\|\sum_{i \in I} m_i\|(\gamma) = 0$ otherwise and similarly, $\|\sum_{j \in J} m'_j\|(\gamma) = 1$ if $\gamma = \gamma''$ while $\|\sum_{j \in J} m'_j\|(\gamma) = 0$ otherwise. We then compute:

$$\left\| \left(\sum_{i \in I} m_i \right) \otimes \left(\sum_{j \in J} m'_j \right) \right\| (\gamma) = \left\| \sum_{i \in I} m_i \right\| (\gamma) \otimes \left\| \sum_{j \in J} m'_j \right\| (\gamma)$$

$$= \begin{cases} 1, & \text{if } \|\sum_{i \in I} m_i\|(\gamma) = \|\sum_{j \in J} m'_j\|(\gamma) = 1 \\ 0, & \text{otherwise} \end{cases}$$

$$= \begin{cases} 1, & \text{if } \sum_{i \in I} m_i \equiv \sum_{j \in J} m'_j, \gamma \models \sum_{i \in I} m_i \\ 0, & \text{otherwise} \end{cases}$$

$$= \begin{cases} \|\sum_{i \in I} m_i\|(\gamma), & \text{if } \sum_{i \in I} m_i \equiv \sum_{j \in J} m'_j \\ 0, & \text{otherwise} \end{cases}.$$

We now present and prove our main results.

Theorem 2. *Let $\mathcal{K} = (K, \oplus, \otimes, 0, 1, \neg)$ be a De Morgan algebra and P a set of ports. Then, for every $w_{DM}PCL$ formula $\zeta \in PCL(\mathcal{K}, P)$ we can effectively construct an equivalent formula $\zeta' \in PCL(\mathcal{K}, P)$ in full normal form which is unique up to the equivalence relation, in doubly exponential time.*

Proof. We will prove our theorem by induction on the structure of $w_{DM}PCL$ formulas ζ over P and $\mathcal{K}$. We can assume that every full normal form in this proof satisfies both statements of Proposition 4. For the calculation of the full normal form ζ' of the formula ζ we distinguish five cases.

ζ *is an element of K*:
If $\zeta = k \in K$ then it is already in full normal form and therefore $\zeta' := k$.
In case an equivalent full normal form formula is needed for calculations then we use $k \equiv \bigoplus_{i \in I} \left(k \otimes \sum_{j \in J_i} m_{i,j} \right)$, where I is an index set such that $\{\gamma_i \mid i \in I\}$ is an enumeration of $C(P)$, and $m_{i,j}$ are full monomials such that $\forall i \in I, \gamma_i \models \sum_{j \in J_i} m_{i,j}$.

ζ *is a PCL formula*:
If $\zeta = f$ is a PCL formula then due to Theorem 1, it has an equivalent PCL full normal form $f' = \bigsqcup_{i \in I} \sum_{j \in J_i} m_{i,j}$ which due to Lemma 2, is equivalent to the $w_{DM}PCL$ full normal form $\zeta' := \bigoplus_{i \in I} (1 \otimes \sum_{j \in J_i} m_{i,j})$.

ζ *is a De Morgan supremum* $\oplus$ *of two* $w_{DM}PCL$ *formulas*:
Let $\zeta = \zeta_1 \oplus \zeta_2$ for some w_{DM}PCL formulas ζ_1, ζ_2 and let $\zeta_1' = \bigoplus_{i_1 \in I_1}(k_{i_1} \otimes \sum_{j_1 \in J_{i_1}} m_{i_1,j_1})$ and $\zeta_2' = \bigoplus_{i_2 \in I_2}(k_{i_2} \otimes \sum_{j_2 \in J_{i_2}} m_{i_2,j_2})$ be their equivalent full normal forms respectively. It obviously follows that $\zeta \equiv \zeta_1' \oplus \zeta_2'$. If both ζ_1, ζ_2 are elements of K then we can immediately set $\zeta' := \zeta_1 \oplus \zeta_2$. We now assume that at least one of ζ_1, ζ_2 is not an element of K, and if one of them is, we use the formula described in the first case of ζ to end up with its full normal form. If $\sum_{j_1 \in J_{i_1}} m_{i_1,j_1} \not\equiv \sum_{j_2 \in J_{i_2}} m_{i_2,j_2}$ for all indices $i_1 \in I_1, i_2 \in I_2$ then we can set $\zeta' := \zeta_1' \oplus \zeta_2'$ which will be in full normal form. If this is not the case, then we assume that $\sum_{j_1 \in J_{i_1'}} m_{i_1',j_1} \equiv \sum_{j_2 \in J_{i_2'}} m_{i_2',j_2}$ for some $i_1' \in I_1, i_2' \in I_2$, and we have: $\zeta_1' \oplus \zeta_2' \equiv (\bigoplus_{i_1 \in I_1 \setminus \{i_1'\}}(k_{i_1} \otimes \sum_{j_1 \in J_{i_1}} m_{i_1,j_1})) \oplus (\bigoplus_{i_2 \in I_2 \setminus \{i_2'\}}(k_{i_2} \otimes \sum_{j_2 \in J_{i_2}} m_{i_2,j_2})) \oplus ((k_{i_1'} \oplus k_{i_2'}) \otimes \sum_{j_1 \in J_{i_1'}} m_{i_1',j_1})$. We continue the same way for the first supremum of the right hand side of the equivalence, replacing every pair of equivalent coalescings with a new term outside the De Morgan supremum operator. In the end we will have the following form: $\zeta_1' \oplus \zeta_2' \equiv (\bigoplus_{i_1 \in I_1 \setminus I_1'}(k_{i_1} \otimes \sum_{j_1 \in J_{i_1}} m_{i_1,j_1})) \oplus (\bigoplus_{i_2 \in I_2 \setminus I_2'}(k_{i_2} \otimes \sum_{j_2 \in J_{i_2}} m_{i_2,j_2})) \oplus (\bigoplus_{i_1' \in I_1'}((k_{i_1'} \oplus k_{i_2'(i_1')}) \otimes \sum_{j_1 \in J_{i_1'}} m_{i_1',j_1}))$, where I_1' (resp. I_2') is the subset of I_1 (resp. I_2) whose elements are all the i_1' (resp. i_2') of the replacement process described above. The index $i_2'(i_1') \in I_2'$ of the last term is the corresponding index of i_1' that satisfies $\sum_{j_1 \in J_{i_1'}} m_{i_1',j_1} \equiv \sum_{j_2 \in J_{i_2'}} m_{i_2',j_2}$ for that specific i_1'. Lastly, we define ζ' as the above formula which is in full normal form and equivalent to ζ.

ζ *is a De Morgan infimum* $\otimes$ *of two* $w_{DM}PCL$ *formulas*:
Let $\zeta = \zeta_1 \otimes \zeta_2$ for some w_{DM}PCL formulas ζ_1, ζ_2 and let $\zeta_1' = \bigoplus_{i_1 \in I_1}(k_{i_1} \otimes \sum_{j_1 \in J_{i_1}} m_{i_1,j_1})$ and $\zeta_2' = \bigoplus_{i_2 \in I_2}(k_{i_2} \otimes \sum_{j_2 \in J_{i_2}} m_{i_2,j_2})$ be their equivalent full normal forms, respectively. If at least one of ζ_1, ζ_2 is constant then we can immediately calculate the full normal form ζ' using the distributivity of $\otimes$ over $\oplus$. In every case it follows that $\zeta \equiv \zeta_1' \otimes \zeta_2'$, and we compute:
$$\zeta_1' \otimes \zeta_2' \equiv (\bigoplus_{i_1 \in I_1}(k_{i_1} \otimes \sum_{j_1 \in J_{i_1}} m_{i_1,j_1})) \otimes (\bigoplus_{i_2 \in I_2}(k_{i_2} \otimes \sum_{j_2 \in J_{i_2}} m_{i_2,j_2}))$$
$$\equiv \bigoplus_{i_1 \in I_1} \bigoplus_{i_2 \in I_2}((k_{i_1} \otimes \sum_{j_1 \in J_{i_1}} m_{i_1,j_1}) \otimes (k_{i_2} \otimes \sum_{j_2 \in J_{i_2}} m_{i_2,j_2}))$$
$$\equiv \bigoplus_{i_1 \in I_1} \bigoplus_{i_2 \in I_2}((k_{i_1} \otimes k_{i_2}) \otimes (\sum_{j_1 \in J_{i_1}} m_{i_1,j_1} \otimes \sum_{j_2 \in J_{i_2}} m_{i_2,j_2})).$$
By Lemma 5, every $\sum_{j_1 \in J_{i_1}} m_{i_1,j_1} \otimes \sum_{j_2 \in J_{i_2}} m_{i_2,j_2}$ can be replaced either by its equivalent $\sum_{j_1 \in J_{i_1}} m_{i_1,j_1}$ if $\sum_{j_1 \in J_{i_1}} m_{i_1,j_1} \equiv \sum_{j_2 \in J_{i_2}} m_{i_2,j_2}$ or by 0 otherwise. After all these replacements we define ζ' as the formula we are left with which will be in the full normal form.

ζ *is a coalescing* $\uplus$ *of two* $w_{DM}PCL$ *formulas*:
Let $\zeta = \zeta_1 \uplus \zeta_2$ where $\zeta_1, \zeta_2 \in PCL(\mathcal{K}, P)$. If exactly one of them is constant we use the formula described in the first case of ζ to end up with its full normal form. If both of them are constant then $\zeta' := \zeta_1 \otimes \zeta_2 \equiv \zeta$. We can now let $\zeta_1' = \bigoplus_{i_1 \in I_1}(k_{i_1} \otimes \sum_{j_1 \in J_{i_1}} m_{i_1,j_1})$ and $\zeta_2' = \bigoplus_{i_2 \in I_2}(k_{i_2} \otimes \sum_{j_2 \in J_{i_2}} m_{i_2,j_2})$ be the equivalent full normal forms of ζ_1 and ζ_2 respectively. It obviously follows that

$\zeta \equiv \zeta_1' \uplus \zeta_2'$, and we compute:

$$\zeta_1' \uplus \zeta_2' \equiv \left(\bigoplus_{i_1 \in I_1}(k_{i_1} \otimes \sum_{j_1 \in J_{i_1}} m_{i_1,j_1})\right) \uplus \left(\bigoplus_{i_2 \in I_2}(k_{i_2} \otimes \sum_{j_2 \in J_{i_2}} m_{i_2,j_2})\right)$$
$$\equiv \bigoplus_{i_1 \in I_1}\left((k_{i_1} \otimes \sum_{j_1 \in J_{i_1}} m_{i_1,j_1}) \uplus \left(\bigoplus_{i_2 \in I_2}(k_{i_2} \otimes \sum_{j_2 \in J_{i_2}} m_{i_2,j_2})\right)\right)$$
$$\equiv \bigoplus_{i_1 \in I_1} \bigoplus_{i_2 \in I_2}\left((k_{i_1} \otimes \sum_{j_1 \in J_{i_1}} m_{i_1,j_1}) \uplus (k_{i_2} \otimes \sum_{j_2 \in J_{i_2}} m_{i_2,j_2})\right)$$
$$\equiv \bigoplus_{i_1 \in I_1} \bigoplus_{i_2 \in I_2}\left(k_{i_1} \otimes k_{i_2} \otimes \left((\sum_{j_1 \in J_{i_1}} m_{i_1,j_1}) \uplus (\sum_{j_2 \in J_{i_2}} m_{i_2,j_2})\right)\right)$$
$$\equiv \bigoplus_{i_1 \in I_1} \bigoplus_{i_2 \in I_2}\left((k_{i_1} \otimes k_{i_2}) \otimes \left((\sum_{j_1 \in J_{i_1}} m_{i_1,j_1}) + (\sum_{j_2 \in J_{i_2}} m_{i_2,j_2})\right)\right).$$

The second and third equivalences hold due to Proposition 3, the fourth one holds due to Lemma 3 and the last one holds due to Lemma 2. The last formula is in full normal form, therefore we can apply the replacing process of Proposition 4 to make sure it satisfies its statements and define ζ' as the formula we are left with.

ζ *is the De Morgan complement* $\neg$ *of a* $w_{DM}PCL$ *formula:*
Let $\zeta = \neg\xi$ where $\xi \in PCL(\mathcal{K}, P)$ and let $\xi' = \bigoplus_{i \in I}(k_i \otimes \sum_{j \in J_i} m_{i,j})$ be the equivalent full normal form of ξ. It obviously follows that $\zeta \equiv \neg\xi'$. For all $i \in I$ we define $k_i' := \neg k_i$, $f_i := \neg \sum_{j \in J_i} m_{i,j}$ and since f_i is a PCL formula for all $i \in I$ we denote f_i' their equivalent $w_{DM}PCL$ full normal forms. Lastly we define $\xi_i := k_i' \oplus f_i'$ and we compute:

$$\neg\xi' \equiv \neg \bigoplus_{i \in I}(k_i \otimes \sum_{j \in J_i} m_{i,j})$$
$$\equiv \bigotimes_{i \in I}(\neg(k_i \otimes \sum_{j \in J_i} m_{i,j}))$$
$$\equiv \bigotimes_{i \in I}((\neg k_i) \oplus (\neg \sum_{j \in J_i} m_{i,j}))$$
$$\equiv \bigotimes_{i \in I}(k_i' \oplus f_i')$$
$$\equiv \bigotimes_{i \in I} \xi_i,$$

where the second and third equivalences hold due to the De Morgan laws, and the fourth one due to Lemma 2. We now notice that ξ_i is a De Morgan supremum $\oplus$ of two $w_{DM}PCL$ formulas in full normal form, and therefore has its own equivalent full normal form ξ_i'. Lastly, the formula $\bigotimes_{i \in I} \xi_i'$ is a De Morgan infimum $\otimes$ of the full normal form $w_{DM}PCL$ formulas ξ_i' and therefore has an equivalent full normal form. We define ζ' as that formula and we have $\zeta' \equiv \zeta$.

We have shown that for every $\zeta \in PCL(\mathcal{K}, P)$ we can inductively construct an equivalent one $\zeta' \in PCL(\mathcal{K}, P)$ which will be in full normal form. The uniqueness, up to the equivalence relation, of ζ' is derived from Remark 1 using the statements of Proposition 4. We now present the rewriting algorithm:

Input: A $w_{DM}PCL$ formula ζ over a set of ports P and a De Morgan algebra $\mathcal{K} = (K, \oplus, \otimes, 0, 1, \neg)$.

1. *For* every PCL formula f present in ζ, we calculate its equivalent PCL full normal form and then its corresponding $w_{DM}PCL$ full normal form and replace f with it.
   Go To Step 3.

2. We *find* an operation between formulas in full normal form, calculate the equivalent full normal form as shown above and replace it with the result.
   Go To Step 3.

3. *If* we are left with a formula in full normal form *Return* it as ζ'.
   Else Go To Step 2.

Output: An equivalent w_{DM}PCL formula ζ'.
Notice that due to the syntax of w_{DM}PCL, after *Step 1* we are left with formulas in full normal form and w_{DM}PCL operators. Therefore, every time *Step 2* is executed, we are left with a formula equivalent to ζ that consists of formulas in full normal form and exactly one less w_{DM}PCL operator (between full normal form formulas) than before, which implies that the process will at some point finish, where we end up with the formula ζ', which will have to be in full normal form.

   Now for the time complexity, the calculation of a PCL formula's full normal form needed for *Step 1* takes at most doubly exponential time due to Theorem 1, with the best case being exponential time. The calculation of the corresponding w_{DM}PCL full normal forms of the PCL full normal forms of *Step 1* and, operations between full normal forms of *Step 2* and the application of Proposition 4, take polynomial time unless we have a constant k involved in a $\oplus$ or $\uplus$ operation with a non-constant formula, in which case we have to replace k with its equivalent full normal form formula, as described in the first case, whose computation takes doubly exponential time, since we need to compute every configuration over P and every unique, up to the equivalence relation, coalescing of full monomials over P, or an $\neg$ operator, in which case we have to calculate the full normal forms of the PCL complement of each coalescing of full monomials in the formula that it is applied on, and then end up with an $\oplus$ operator between a constant and a w_{DM}PCL formula, which both take doubly exponential time for the same reason.

   Therefore the best case for the time complexity of the above algorithm is exponential, while the worst case is doubly exponential.

Theorem 3. *Let $\mathcal{K}$ be a De Morgan algebra and P be a set of ports. Then, for every $\zeta, \xi \in PCL(\mathcal{K}, P)$, the equality $\|\zeta\| = \|\xi\|$ is decidable in doubly exponential time.*

Proof. By Theorem 2 we can effectively construct w_{DM}PCL formulas ζ', ξ' in full normal form, such that $\|\zeta\| = \|\zeta'\|$ and $\|\xi\| = \|\xi'\|$, in doubly exponential time. We now assume that, $\zeta' = \bigoplus_{i \in I}(k_i \otimes \sum_{j \in J_i} m_{i,j})$ and $\xi' = \bigoplus_{l \in L}(k'_l \otimes \sum_{r \in M_l} m'_{l,r})$ which satisfy both Statements of Proposition 4. We will prove that $\|\zeta'\| = \|\xi'\|$ iff the following 3 requirements hold:

1. $\text{card}(I) = \text{card}(L)$,
2. $\{k_i \mid i \in I\} = \{k'_l \mid l \in L\}$,
3. Depending on the cardinality of I (and L) we require one of the following:
   (a) if $\text{card}(I) = \text{card}(\{k_i \mid i \in I\})$, then
       for every pair of indices $i \in I$ and $l \in L$ such that $k_i = k'_l$,

$$\sum_{j \in J_i} m_{i,j} \equiv \sum_{r \in M_l} m'_{l,r},$$

(b) if $\text{card}(I) > \text{card}(\{k_i \mid i \in I\})$, then

we get $\zeta' \equiv \bigoplus_{i' \in I'} \left(k_{i'} \otimes \bigsqcup_{i \in R_{i'}} \sum_{j \in J_i} m_{i,j} \right)$ where $I' \subset I$ s.t. $\forall i_1' \neq i_2'$ in I', $k_{i_1'} \neq k_{i_2'}$, and $\forall i' \in I'$, $R_{i'} = \{i \in I \mid k_i = k_{i'}\}$. Similarly, $\xi' \equiv \bigoplus_{l' \in L'} \left(k_{l'}' \otimes \bigsqcup_{l \in R_{l'}} \sum_{r \in M_i} m_{l,r}' \right)$, where $L' \subset L$ s.t. $\forall l_1' \neq l_2'$ in L', $k_{l_1'} \neq k_{l_2'}$, and $\forall l' \in L'$, $R_{l'} = \{l \in I \mid k_i' = k_{l'}'\}$. Then, for every $i' \in I', l' \in L'$ such that $k_{i'} = k_{l'}'$,

$$\bigsqcup_{i \in R_{i'}} \sum_{j \in J_i} m_{i,j} \equiv \bigsqcup_{l \in R_{l'}} \sum_{r \in M_i} m_{l,r}'.$$

If (1) does not hold then there must exist a configuration γ such that only one of $\|\zeta'\|(\gamma), \|\xi'\|(\gamma)$ is non-zero. If (2) or (3) do not hold, then there must exist a configuration γ such that $\|\zeta'\|(\gamma) \neq \|\xi'\|(\gamma)$. Therefore all the above requirements are necessary. If all three hold, then obviously $\|\zeta'\|(\gamma) = \|\xi'\|(\gamma)$ for all $\gamma \in C(P)$ and therefore $\|\zeta'\| = \|\xi'\|$.

Using Lemma 4, we reduce the decidability of the equivalences in (3.a) to the decidability of equality of sets of interactions (configurations) corresponding to full monomials, and the decidability of the equivalences in (3.b) to the decidability of equality of sets whose elements are sets of interactions (configuration sets) corresponding to full monomials. All these full monomials and their corresponding interaction sets are computed during the construction of the equivalent full normal forms ζ' and ξ', which due to Theorem 2 takes doubly exponential time. Therefore, the decidability for all the required equalities and equivalences ((1), (2) and (3.a) or (3.b)) above takes at most polynomial time, and therefore doubly exponential time in total.

Theorem 4. *Let $\mathcal{K}$ be a De Morgan algebra and P be a set of ports. Then, for every $\zeta, \xi \in PCL(\mathcal{K}, P)$, the relation $\|\zeta\| \leq \|\xi\|$ is decidable in doubly exponential time.*

Proof. By Theorem 2 we can effectively construct w_{DM}PCL formulas ζ', ξ' in full normal form, such that $\|\zeta\| = \|\zeta'\|$ and $\|\xi\| = \|\xi'\|$, in doubly exponential time. We now assume that, $\zeta' = \bigoplus_{i \in I}(k_i \otimes \sum_{j \in J_i} m_{i,j})$ and $\xi' = \bigoplus_{l \in L}(k_l' \otimes \sum_{r \in M_l} m_{l,r}')$ which satisfy both Statements of Proposition 4, while $k_i, k_l' \neq 0$ for all $i \in I, l \in L$. We will prove that $\|\zeta\| \leq \|\xi\|$ iff the following requirement holds:

– For all $i \in I$ there exists an $l \in L$ such that:

$$\sum_{j \in J_i} m_{i,j} \equiv \sum_{r \in M_l} m_{l,r}' \text{ and } k_i \leq k_l' \text{ where } \leq \text{ is the partial order of } \mathcal{K}.$$

We first prove that $\|\zeta\| \leq \|\xi\|$ implies the above requirement. Let $\|\zeta\| \leq \|\xi\|$, then we have:

$$\|\zeta\| \leq \|\xi\| \text{ iff } \|\zeta'\| \leq \|\xi'\|$$
$$\text{iff } \forall \gamma \in C(P) : \|\zeta'\|(\gamma) \leq \|\xi'\|(\gamma)$$

We now define the configuration sets $C_\zeta = \{\gamma \in C(P) \mid \exists i \in I : \gamma \models \sum_{j \in J_i} m_{i,j}\}$ and $C_\xi = \{\gamma \in C(P) \mid \exists l \in L : \gamma \models \sum_{r \in M_l} m'_{l,r}\}$. It obviously holds that if $\gamma \notin C_\zeta$ (resp. $\gamma \notin C_\xi$) then $\|\zeta\|(\gamma) = 0$ (resp. $\|\xi\|(\gamma) = 0$), while if $\gamma \in C_\zeta$ (resp. $\gamma \in C_\xi$) then $\|\zeta\|(\gamma) \in \{k_i \mid i \in I\}$ (resp. $\|\xi\|(\gamma) \in \{k'_l \mid l \in L\}$). We also notice that since ζ' and ξ' satisfy both statements of Proposition 4 then $\mathrm{card}(C_\zeta) = \mathrm{card}(I)$ and $\mathrm{card}(C_\xi) = \mathrm{card}(L)$. Therefore we distinguish the 4 following cases for any configuration γ.

1. $\gamma \in C_\zeta$ and $\gamma \in C_\xi$, therefore $\|\zeta\|(\gamma) \in \{k_i \mid i \in I\}$ and $\|\xi\|(\gamma) \in \{k'_l \mid l \in L\}$
2. $\gamma \in C_\zeta$ and $\gamma \notin C_\xi$, therefore $\|\zeta\|(\gamma) \in \{k_i \mid i \in I\}$ and $\|\xi\|(\gamma) = 0$
3. $\gamma \notin C_\zeta$ and $\gamma \in C_\xi$, therefore $\|\zeta\|(\gamma) = 0$ and $\|\xi\|(\gamma) \in \{k'_l \mid l \in L\}$
4. $\gamma \notin C_\zeta$ and $\gamma \notin C_\xi$, therefore $\|\zeta\|(\gamma) = 0$ and $\|\xi\|(\gamma) = 0$

The second case implies that $\|\zeta\|(\gamma) \not\leq \|\xi\|(\gamma)$ meaning that $\|\zeta\| \not\leq \|\xi\|$ which contradicts our hypothesis. Therefore, for all $\gamma \in C(P)$, only cases 1,3 and 4 can occur if $\|\zeta\| \leq \|\xi\|$, and all three imply that $C_\zeta \subseteq C_\xi$. In cases 3 and 4, where $\|\zeta\|(\gamma) = 0$, it obviously holds that $\|\zeta\|(\gamma) \leq \|\xi\|(\gamma)$.

For all the configurations such that case 1 holds, we have:

$$\forall \gamma \in C_\zeta \cap C_\xi = C_\zeta : \|\zeta'\|(\gamma) \leq \|\xi'\|(\gamma) \text{ iff } k_{i'} \leq k'_{l'}$$

where $i' \in I$ and $l' \in L$, are the (unique due to Proposition 4) indices such that $\|\sum_{j \in J_{i'}} m_{i',j}\|(\gamma) = \|\sum_{r \in M_{l'}} m'_{l',r}\|(\gamma) = 1$.

Since $C_\zeta \cap C_\xi = C_\zeta$ then for each $i \in I$ there must exist an $l \in L$ such that $k_i \leq k'_l$ and $\|\sum_{j \in J_i} m_{i,j}\|(\gamma) = \|\sum_{r \in M_l} m'_{l,r}\|(\gamma)$ for all $\gamma \in C(P)$, which is the requirement we described we wanted to prove.

Now, for the reverse implication, we assume that for all $i \in I$ there exists an $l \in L$ such that $\sum_{j \in J_i} m_{i,j} \equiv \sum_{r \in M_l} m'_{l,r}$ and $k_i \leq k'_l$.

For any configuration γ we have:

$$\|\zeta\|(\gamma) = \|\zeta'\|(\gamma) = \begin{cases} k'_i & \text{if } \gamma \in C_\zeta \\ 0 & \text{if } \gamma \notin C_\zeta \end{cases}$$

where $i' \in I$ is the unique index such that $\gamma \models \sum_{j \in J_{i'}} m_{i',j}$, given that $\gamma \in C_\zeta$. If $\gamma \in C_\zeta$ then due to our hypothesis there also has to exist an index $l' \in L$ such that $\gamma \models \sum_{r \in M_{l'}} m'_{l',r}$ and $k_{i'} \leq k'_{l'}$, therefore $\|\zeta'\|(\gamma) \leq \|\xi'\|(\gamma)$. Otherwise, we have that $\gamma \notin C_\zeta$, therefore $\|\zeta'\|(\gamma) = 0$ which again implies that $\|\zeta'\|(\gamma) \leq \|\xi'\|(\gamma)$. Therefore, since for any configuration γ we have $\|\zeta'\|(\gamma) \leq \|\xi'\|(\gamma)$, then $\|\zeta\| \leq \|\xi\|$, which is what we wanted to prove.

Therefore $\|\zeta\| \leq \|\xi\|$ iff our requirement holds.

Now for the time complexity, using Lemma 4, we reduce the decidability of the equivalences in the above requirement to the decidability of equality of sets of interactions (configurations) corresponding to full monomials. All these full monomials and their corresponding interaction sets are computed during the construction of the equivalent full normal forms ζ' and ξ', a process which, due to Theorem 2, takes doubly exponential time. Therefore, the decidability for all the required order relations and equivalences take at most polynomial time, and therefore doubly exponential time in total.

6   Examples

In this section we provide $w_{DM}PCL$ formulas that describe some examples of well-known architectures with quantitative features.

It is important to note however that, even though our work is similar to [9] and [5], the corresponding weighted formula of an architecture will be calculated with a different approach. In wPCL of [9] for example, architectures were described mostly with full monomials, each of which had a corresponding weight attached to it through the multiplication operator of the commutative semiring. In the case of $w_{DM}PCL$, the complement function $\neg$ of the De Morgan algebra allows us to include negative weights in our calculations, meaning that such cases should be taken into consideration when defining our formulas.

In the sequel we will calculate the weighted formulas of architectures as follows. Let m be a monomial and $k_m \in K$ be its corresponding weight. The weighted formula of m will be $k_m \otimes m$. The weighted formula of its PIL negation $\overline{m}$ will be $\neg k_m \oplus \overline{m}$. For the qualitative aspects of an architecture we will use PCL formulas in which we describe the requirements with monomials and the corresponding restrictions with PIL negations of monomials. For the quantitative aspects, we replace all monomials and their negations with their respective weighted formulas, and each PCL operator with its equivalent $w_{DM}PCL$ one.

Remark 2. The reader should notice that the way we described that the formulas of architectures are built, implies that the complement $\neg k$ of any weight $k \in K$ will only be present in weighted restriction formulas. Therefore, if for any configuration $\gamma \in C(P)$ the weight of the architecture requires the calculation of $\neg k$, this means that $\|\neg k \oplus \overline{m}\|(\gamma) = \neg k$ for some restriction formula $\overline{m}$, which in turn implies that $\|\overline{m}\|(\gamma) = 0$, i.e., $\gamma \not\models \overline{m}$. This means that the configuration γ does not satisfy one of the restriction formulas and as a result the qualitative requirements of the architecture are not met with γ.

Example 1. The star architecture is an architecture defined for a set of components of the same type. One of the components, called the central component, is connected with every other through a binary interaction and there

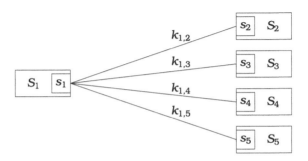

Fig. 1. The weighted star architecture.

are no other interactions. In Fig. 1 we see a star architecture with 5 components S_1, S_2, S_3, S_4, S_5, where S_1 is the central one. Let $I = \{1, 2, 3, 4, 5\}$ and $P = \{s_1, s_2, s_3, s_4, s_5\}$ be the set of ports of a star architecture with 5 components. We define the formulas

$$f_1 := s_1 s_2 + s_1 s_3 + s_1 s_4 + s_1 s_5 \text{ and } r_1 := \overline{s_2 s_3} \wedge \overline{s_2 s_4} \wedge \overline{s_2 s_5} \wedge \overline{s_3 s_4} \wedge \overline{s_3 s_5} \wedge \overline{s_4 s_5}.$$

The PCL formula that describes the star architecture with S_1 as the central component, is $g_1 = f_1 \wedge r_1$. In the weighted case, we denote by $k_{i,j}$ the weight of the monomial $s_i s_j$, for $i, j \in \{1, 2, 3, 4, 5\}, i \neq j$, i.e., the weight of an interaction between S_i and S_j. Here we also assume that any one of the components could become a central one. We generalise the formulas f_1, r_1 for any potential central component S_i with $f_i := \sum_{j \in I \setminus \{i\}} s_i s_j$ and $r_i := \bigwedge_{j_1, j_2 \in I \setminus \{i\}, j_1 \neq j_2} \overline{s_{j_1} s_{j_2}}$, respectively, and therefore $g = \bigsqcup_{i \in I} g_i$ describes this architecture. The corresponding weighted formulas of f_i and r_i are

$$\zeta_i := \biguplus_{j \in I \setminus \{i\}} (k_{i,j} \otimes s_i s_j) \text{ and } r'_i := \bigotimes_{j_1, j_2 \in I \setminus \{i\}, j_1 \neq j_2} (\neg k_{j_1, j_2} \oplus \overline{s_{j_1} s_{j_2}}),$$

respectively, for all $i \in I$. The formula that describes the weight of the architecture will be $\zeta = \sim \bigoplus_{i \in I} (\zeta_i \otimes r'_i)$. Let for example $\gamma = \{\{s_1, s_2\}, \{s_1, s_3\}, \{s_1, s_4\}, \{s_1, s_5\}, \{s_2, s_3\}, \{s_2, s_4\}, \{s_2, s_5\}, \{s_4, s_5\}\}$, and we compute $\|\zeta\|(\gamma) = \|\zeta_1 \otimes r'_1\|(\gamma_1) \oplus \|\zeta_2 \otimes r'_2\|(\gamma_2) = (k_{1,2} \otimes k_{1,3} \otimes k_{1,4} \otimes k_{1,5}) \oplus (k_{1,2} \otimes k_{2,3} \otimes k_{2,4} \otimes k_{2,5})$ where $\gamma_1 = \{\{s_1, s_2\}, \{s_1, s_3\}, \{s_1, s_4\}, \{s_1, s_5\}\}$, $\gamma_2 = \{\{s_1, s_2\}, \{s_2, s_3\}, \{s_2, s_4\}, \{s_2, s_5\}\}$. If however we let $\gamma = \{\{s_1, s_2, s_3\}, \{s_1, s_4\}, \{s_1, s_5\}\}$, then we compute $\|\zeta\|(\gamma) = \|\zeta_1 \otimes r'_1\|(\gamma) = k_{1,2} \otimes k_{1,3} \otimes k_{1,4} \otimes k_{1,5} \otimes \neg k_{2,3}$. As explained in Remark 2, the presence of the De Morgan complement function in our result implies that the configuration cannot satisfy the qualitative requirements of the star architecture, i.e., $\gamma \not\models_\sim g_1$.

Example 2. The blackboard architecture is most commonly used for the implementation of problem solving techniques through AI (cf. [3]). In a Blackboard

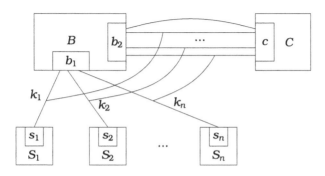

Fig. 2. A blackboard architecture for n knowledge source components.

architecture we have 3 types of components, knowledge sources, one blackboard and one control component. The blackboard component has 2 ports, while all other components have one. A knowledge source has to connect to the blackboard component through one port, but not with the other sources, while the control component is connected with the blackboard at all times through its other port. In Fig. 2 we see the general case of a blackboard architecture for n knowledge source components. We define the interaction formulas $\phi_i := b_1 s_i \wedge \bigwedge_{i' \in I \setminus \{i\}} \overline{b_1 s_{i'}}$, for all $i \in I = \{1, 2, ..., n\}$. Then the PCL formula that describes the architecture is $f = b_2 c \wedge \sum_{i \in I} \phi_i$. Now for the weighted case, we assign the weight $k_i \in K$ to the monomial $s_i b_1$ for all $i \in I$, while we also assign a weight of 1 to the monomial $b_2 c$, since the blackboard and the control components have to be connected. The corresponding weighted formula of ϕ_i is $\zeta_i := k_i \otimes b_1 s_i \otimes \bigotimes_{i' \in I \setminus \{i\}} \left(\neg k_{i'} \oplus \overline{b_1 s_{i'}} \right)$ for all $i \in I$, and the formula describing the weight of the architecture is $\zeta = b_2 c \otimes \biguplus_{i \in I} \zeta_i$. If for example we let $\gamma = \{\{b_2, c, b_1, s_i\} \mid i \in I\}$ then we compute $\|\zeta\|(\gamma) = \left\| \biguplus_{i \in I} \zeta_i \right\|(\gamma) = \bigotimes_{i \in I} \|\zeta_i\|(\{\{b_1, c, b_2, s_i\}\}) = \bigotimes_{i \in I} k_i$.

If instead we let $\gamma = \{\{b_2, c, b_1, s_1, s_2, s_3\}\} \cup \{\{b_2, c, b_1, s_i\} \mid i \in I \setminus \{1\}\}$ then we compute

$$\|\zeta\|(\gamma) = \|\zeta_1\|(\{\{b_1, c, b_2, s_1, s_2, s_3\}\}) \otimes \bigotimes_{i \in I \setminus \{1\}} \|\zeta_i\|(\{\{b_1, c, b_2, s_i\}\})$$
$$= (k_1 \otimes \neg k_2 \otimes \neg k_3) \otimes \bigotimes_{i \in I \setminus \{1\}} k_i,$$

which is an expected result due to Remark 2, since γ does not satisfy the qualitative requirements described by f.

7   Discussion

In our logic we considered the algebraic structure of De Morgan algebras. The formulas of w_{DM}PCL can be built with PCL formulas over a set of ports P, elements and operations of a De Morgan algebra $(K, \oplus, \otimes, 0, 1, \neg)$ and a weighted coalescing operator $\uplus$. Our goal was to describe the qualitative features of an architecture with PCL formulas, and when a quantitative approach was required, transition to the corresponding w_{DM}PCL formulas. In Lemma 2 however we prove that every PCL operator is equivalent to a w_{DM}PCL one, which basically implies that we could skip PCL and define w_{DM}PCL as an extension of PIL instead, with syntax $\zeta ::= k \mid \phi \mid \zeta \oplus \zeta \mid \zeta \otimes \zeta \mid \neg \zeta \mid \zeta \uplus \zeta$, where ϕ is a PIL formula and with semantics $\|\phi\|(\gamma) = 1$ if $\forall \alpha \in \gamma, \alpha \models_i \phi$, and $\|\phi\|(\gamma) = 0$ otherwise. This w_{DM}PCL is equivalent to the one of Definition 1 since any PCL formula f can be replaced by an equivalent one using PIL formulas and w_{DM}PCL operators, as shown in Lemma 2. In that case, if one wishes to use w_{DM}PCL the way PCL functions then they can let $\mathcal{K} = (\{0, 1\}, \vee, \wedge, 0, 1, \neg)$ be the usual 2-element Boolean algebra, and define satisfaction as follows, $\gamma \models \zeta$ iff $\|\zeta\|(\gamma) = 1$. Here, 0 and 1 represent the *false* and *true* of PCL, respectively, which means that PCL is a special case of w_{DM}PCL, i.e., w_{DM}PCL is a generalization of PCL.

In the definitions of interactions and configurations in Sect. 4, we excluded, following [7], the empty interaction and the empty configuration. Several properties of PCL and our weighted extension w_{DM}PCL do not hold if we accept that $\emptyset \in I(P)$. For example we would have that $\{\emptyset\} \models false$ which implies that $false \not\equiv 0$ and that $false$ is no longer an absorbing element of the coalescing operator, a property used in [7] for the computation of the equivalent full normal form of a PCL formula. The inclusion of the empty interaction to $C(P)$ would not help with the description of single architectures, even though it is used in architectural composition (cf. [1,2]) to represent the case where two architectures cannot be composed. Hence, one could rebuild the theory of PCL and of our w_{DM}PCL by considering the empty interaction and the empty configuration, but this is beyond the scope of this paper.

8 Conclusion

We introduced a weighted extension of PCL over a set of ports P and a De Morgan algebra $\mathcal{K}$, where the semantics of every w_{DM}PCL formula is a polynomial with values in $\mathcal{K}$. We proved that every w_{DM}PCL formula has an equivalent one in full normal form, which we can effectively construct in doubly exponential time. We also proved the decidability of equivalence of w_{DM}PCL formulas and the decidability of a partial order relation over polynomials of w_{DM}PCL formulas, in doubly exponential time. Lastly, we provided some examples of well-known architectures with w_{DM}PCL formulas to display the potential applications of our logic.

Future research could include a more specific rewriting system (similarly to [7], Fig. 6) for the full normal form calculation with the purpose of reducing the time complexity for special cases of formulas where it is possible, an application of w_{DM}PCL for the construction of model checking tools for large architectures, or the investigation of the first and second- order level of w_{DM}PCL for the description of architecture styles with quantitative features.

References

1. Attie, P., Baranov, E., Bliudze, S., Jaber, M., Sifakis, J.: A general framework for architecture composability. Form. Asp. Comput. **28**(2), 207–231 (2016). https://doi.org/10.1007/s00165-015-0349-8
2. Bozga, M., Iosif, R., Sifakis, J.: Local reasoning about parametric and reconfigurable component-based systems. CoRR abs/1908.11345 (2019). https://arxiv.org/abs/1908.11345
3. Corkill, D.D.: Blackboard Systems. Blackboard Technology Group Inc., New York (1991)
4. Droste, M., Kuich, W., Rahonis, G.: Multi-valued MSO logics over words and trees. Fundamenta Informaticae **84** (2008)
5. Karyoti, V., Paraponiari, P.: Weighted PCL over product valuation monoids. In: Bliudze, S., Bocchi, L. (eds.) COORDINATION 2020. LNCS, vol. 12134, pp. 301–319. Springer, Cham (2020). https://doi.org/10.1007/978-3-030-50029-0_19
6. Droste, M., Kuich, W.: Semirings and formal power series. In: Droste, M., Kuich, W., Vogler, H. (eds.) Handbook of Weighted Automata. EATCS, pp. 3–28. Springer, Heidelberg (2009). https://doi.org/10.1007/978-3-642-01492-5_1

7. Mavridou, A., Baranov, E., Bliudze, S., Sifakis, J.: Configuration logics: modeling architecture styles. J. Logical Algebraic Methods Program. **86**(1), 2–29 (2017). https://doi.org/10.1016/j.jlamp.2016.05.002

8. Olivieri, A.C., Rizzo, G., Morard, F.: A publish-subscribe approach to IoT integration: the smart office use case. In: 2015 IEEE 29th International Conference on Advanced Information Networking and Applications Workshops, pp. 644–651 (2015). https://doi.org/10.1109/WAINA.2015.28

9. Paraponiari, P., Rahonis, G.: Weighted propositional configuration logics: a specification language for architectures with quantitative features. Inf. Comput. 104647 (2020). https://doi.org/10.1016/j.ic.2020.104647

10. Patel, S., Jardosh, S., Makwana, A., Thakkar, A.: Publish/subscribe mechanism for IoT: a survey of event matching algorithms and open research challenges. In: Modi, N., Verma, P., Trivedi, B. (eds.) Proceedings of International Conference on Communication and Networks. AISC, vol. 508, pp. 287–294. Springer, Singapore (2017). https://doi.org/10.1007/978-981-10-2750-5_30

11. Rahonis, G.: Fuzzy languages. In: Droste, M., Kuich, W., Vogler, H. (eds.) Handbook of Weighted Automata. EATCS, pp. 481–517. Springer, Heidelberg (2009). https://doi.org/10.1007/978-3-642-01492-5_12

12. Yang, K., Zhang, K., Jia, X., Hasan, M.A., Shen, X.: Privacy-preserving attribute-keyword based data publish-subscribe service on cloud platforms. Inf. Sci. **387**, 116–131 (2017). https://doi.org/10.1016/j.ins.2016.09.020

Weighted Two-Way Transducers

Fan Feng[ID] and Andreas Maletti[✉][ID]

Faculty of Mathematics and Computer Science, Universität Leipzig,
P.O. box 100 920, 04009 Leipzig, Germany
{fanfeng,maletti}@informatik.uni-leipzig.de

Abstract. Weighted two-way transducers over complete commutative semirings are introduced and investigated. Their computed mappings have two-way definable support. Conversely, for every two-way definable relation R there exists a mapping computable by a weighted two-way transducer whose support is R. It is shown that the class of all such computed mappings is closed under sum and the subclass computed by deterministic weighted two-way transducers is closed under composition.

1 Introduction

Two-way transducers received a lot of interest recently [15]. They are a natural generalization of finite-state transducers [1] extended with the facility to move left and right along the input string like a Turing machine but without the Turing machine's facility to write onto that tape (or leave the area delimited by the input). They can similarly be considered as an extension of two-way automata [17,18] with the facility to generate output. Recently, two-way automata were generalized to weighted two-way automata [2,13], in which transitions carry weights in order to model success probabilities, multiplicities, or other quantitative aspects. In this contribution we introduce weighted two-way transducers that similarly extend two-way transducers with the facility to charge weights on each transition.

Our weighted extension follows the usual principles [6]. Transition weights along a run are multiplied, and multiple runs for the same input-output pair are added up. Since a single input-output pair can have an infinite number of runs (in the nondeterministic case), we require that our weight structures are complete commutative semirings [9,11], which permit infinite summations subject to the usual laws of commutativity, associativity, and distributivity. Infinite runs are not permitted, so no similar requirement is necessary for the product. For the deterministic case, in which the state and currently read input symbol uniquely determine the follow state, the movement direction, the generated output, and the charged weight, the completeness of the semiring is not required since there is at most one run for each input string. Nevertheless we limit ourselves to complete semirings in this contribution.

The seminal paper of Engelfriet and Hoogeboom [7] established celebrated logical characterizations of the two-way definable relations and mappings.

© The Author(s), under exclusive license to Springer Nature Switzerland AG 2022
D. Poulakis and G. Rahonis (Eds.): CAI 2022, LNCS 13706, pp. 101–114, 2022.
https://doi.org/10.1007/978-3-031-19685-0_8

Our eventual goal is to establish a similar characterization of the expressive power of weighted two-way transducers. To this end, we start an investigation of the expressive power and the basic properties of weighted two-way transducers. More specifically, we show that the support of every mapping computed by a weighted two-way transducer is two-way definable. This follows rather directly from the zero-sum freeness of complete semirings and a construction of [12], which unfortunately is not necessarily effective. Conversely, we can represent any two-way definable relation as the support of a mapping computed by a weighted two-way transducer. Several constructions for such weighted two-way transducers are available and the simplest variant relies on the infinite element of our complete semiring. These results establish a nice relationship to the unweighted case.

Next we investigate the standard closure properties of the class of mappings computed by weighted two-way transducers. We particularly focus on those closure properties, which traditionally are used to establish logical characterizations [5]. The class of computed mappings is closed under sums, which is entirely trivial to establish. It is not closed under (HADAMARD) products if support emptiness is decidable due to the straightforward embedding of PCP instances. The final closure that we consider is composition, which we initially consider only for reversible weighted two-way transducers [3]. These transducers are deterministic as well as co-deterministic and are the main reason for our specific syntactic variant of weighted two-way transducers. As in the unweighted case [3], we also prove that every deterministic weighted two-way transducer can be transformed into an equivalent reversible weighted two-way transducer, so our composition construction actually applies to all deterministic weighted two-way transducers. We thus prove that the class of all mappings computed by deterministic weighted two-way transducers is closed under composition.

Our composition construction is heavily inspired by the corresponding construction of [3], but adjustments were necessary to make sure that the weights of the first transducer in the composition are correctly accounted for. Roughly speaking, we split each run of the first transducer into two phases. In the initial phase the original run is simulated with the correct weights, but no outputs are generated in this phase. Once the first phase terminates in the final state at the end of the input, we rewind the reading head back to the beginning of the input and restart the transducer in the second phase, in which only unit weights are used, but now the correct outputs are produced. On those outputs we can now simulate the second transducer and even if an output is recalled multiple times the weight is no longer distorted since only unit weights are charged in the second phase.

2   Preliminaries

We use $\mathbb{N}$ for the set of nonnegative integers. The restriction $f|_B \colon B \to C$ of a mapping $f \colon A \to C$ to a subset $B \subseteq A$ is given by $f|_B(b) = f(b)$ for every $b \in B$. A *(commutative) semiring* [9,11] is an algebraic structure $(S, +, \cdot, 0, 1)$ such that $(S, +, 0)$ and $(S, \cdot, 1)$ are commutative monoids, $s \cdot 0 = 0$ for every $s \in S$, and the distributive law $s \cdot (s' + s'') = (s \cdot s') + (s \cdot s'')$ holds for all $s, s', s'' \in S$. An element $s \in S \setminus \{0\}$ is a *zero-divisor* if there exists $s' \in S \setminus \{0\}$ such that $s \cdot s' = 0$. The semiring is *zero-sum free* if for all $s, s' \in S$ with $s + s' = 0$ we have $s = 0 = s'$. Given a mapping $f \colon X \to S$, we let $\operatorname{supp}(f) = \{x \in X \mid f(x) \neq 0\}$ be its *support* and $\operatorname{ran}(f) = \{f(x) \mid x \in X\}$ be its *range*. A semiring $(S, +, \cdot, 0, 1)$ is *ordered* [9,11] if there exists a partial order $\leq$ on S such that for all $s, s', s'' \in S$ with $s' \leq s''$ we have (i) $0 \leq s$, (ii) $s + s' \leq s + s''$, and (iii) $s \cdot s' \leq s \cdot s''$. Thus $s \leq s + s'$ for all $s, s' \in S$. Hence an ordered semiring is necessarily zero-sum free because for all $s, s' \in S$ with $s + s' = 0$ we obtain $0 \leq s \leq s + s' = 0$, which proves $s = 0 = s'$.

The relation $\leqslant$ is defined for every $s, s' \in S$ by $s \leqslant s'$ if there exists $s'' \in S$ such that $s + s'' = s'$. This relation is reflexive and transitive, and if it is also anti-symmetric, then S is called *naturally ordered* [9,11]. Naturally ordered semirings are ordered because for every $s, s', s'' \in S$ with $s' \leqslant s''$ (i.e., there exists $\overline{s} \in S$ such that $s' + \overline{s} = s''$) we have (i) $0 \leqslant s$ since $0 + s = s$, (ii) $s + s' \leqslant s + s''$ since $(s + s') + \overline{s} = s + s''$, and (iii) $s \cdot s' \leqslant s \cdot s''$ since $(s \cdot s') + s \cdot \overline{s} = s \cdot (s' + \overline{s}) = s \cdot s''$. A *partition* Π of a set I is a mapping $\Pi \colon J \to \mathcal{P}(I)$ for some set J such that $I = \bigcup_{j \in J} \Pi(j)$ and $\Pi(j) \cap \Pi(j') = \emptyset$ for all $j, j' \in J$ with $j \neq j'$. A *complete semiring* $(S, +, \cdot, 0, 1, \sum)$ is a semiring $(S, +, \cdot, 0, 1)$ in which for any set I and mapping $\alpha \colon I \to S$ the infinite sum $\sum \alpha$ is defined and satisfies the following three axioms [9–11].

1. If $I = \{i, j\}$ with $i \neq j$, then $\sum \alpha = \alpha(i) + \alpha(j)$.
2. $\sum \alpha = \sum \beta$ for every partition $\Pi \colon J \to \mathcal{P}(I)$ of I, where $\beta \colon J \to S$ is such that $\beta(j) = \sum \alpha|_{\Pi(j)}$ for all $j \in J$.
3. $s \cdot \sum \alpha = \sum \alpha_s$ for every $s \in S$, where $\alpha_s \colon I \to S$ is given by $\alpha_s(i) = s \cdot \alpha(i)$ for every $i \in I$.

To avoid the explicit definition of α we often write $\sum_{i \in I} \alpha(i)$ instead of $\sum \alpha$. It follows from the last axiom that $\sum_{i \in I} 0 = 0$. For example, $(\mathbb{N} \cup \{\infty\}, +, \cdot, 0, 1, \sum)$ with the usual addition and multiplication extended to ∞ (i.e., $\infty \cdot 0 = 0$) is a complete semiring, where for every $\alpha \colon I \to S$ we have $\sum \alpha = \sum_{i \in \operatorname{supp}(\alpha)} \alpha(i)$ if $\operatorname{supp}(\alpha)$ is finite and $\sum \alpha = \infty$ otherwise. A semiring can be completed (i.e., embedded into a complete semiring) if and only if it is naturally ordered [10]. We note that each complete semiring is zero-sum free [9]. For example, the semiring $(\operatorname{Reg}(\Sigma), \cup, \cap, \emptyset, \Sigma^*)$ of regular languages for an alphabet Σ with union as addition and intersection as multiplication is naturally ordered. It is not complete because the union of countably many regular languages need not be regular. However, it can be embedded into the complete semiring $(\mathcal{P}(\Sigma^*), \cup, \cap, \emptyset, \Sigma^*)$.

3   Weighted Two-Way Transducers

For the rest of the contribution, let $(S, +, \cdot, 0, 1, \sum)$ be a complete semiring, and we let $A_{\vdash \dashv} = A \cup \{\vdash, \dashv\}$ for an alphabet A such that $A \cap \{\vdash, \dashv\} = \emptyset$. The symbols $\vdash$ and $\dashv$ are used as left- and right-end marker, respectively. For every alphabet B we let $B^{\leq 1} = B \cup \{\varepsilon\}$. Next we introduce the weighted two-way finite-state transducers as a straightforward generalization of two-way finite-state transducers [7] in a variant presented in [3]. We also immediately recall the notions 'deterministic' and 'reversible' from [3].

Definition 1. *A* weighted two-way finite-state transducer *(for short: w2fst) is a tuple* $\mathfrak{A} = (Q^{\rightarrow}, Q^{\leftarrow}, A, B, T, I, F)$, *in which* $Q^{\rightarrow}$ *and* $Q^{\leftarrow}$ *are disjoint finite sets of forward and backward states, resp., A and B are alphabets of input and output symbols, resp.,* $I, F \colon Q^{\rightarrow} \rightarrow S$ *are initial and final weights, resp., and* $T \colon Q \times A_{\vdash\dashv} \times B^{\leq 1} \times Q \rightarrow S$ *assigns weights to transitions, where* $Q = Q^{\rightarrow} \cup Q^{\leftarrow}$. *It is* deterministic *(for short: wd2fst) if* $|\text{supp}(I)| \leq 1$ *and for every* $q \in Q$ *and input symbol* $a \in A_{\vdash\dashv}$ *there exists at most one pair* $(b, q') \in B^{\leq 1} \times Q$ *with* $(q, a, b, q') \in \text{supp}(T)$. *Similarly, it is* co-deterministic *if* $|\text{supp}(F)| \leq 1$ *and for every* $q' \in Q$ *and input symbol* $a \in A_{\vdash\dashv}$ *there exists at most one pair* $(b, q) \in B^{\leq 1} \times Q$ *such that* $(q, a, b, q') \in \text{supp}(T)$. *The w2fst is* reversible *(for short: wr2fst) if it is deterministic as well as co-deterministic.*

Let us consider a small example, which is inspired by [7, Example 2]. It processes the input left-to-right and first outputs all seen letters a until it meets the first letter b, for which it charges weight 2. The direction then reverses and the just processed chain of letters a is reprocessed right-to-left and each letter a in the input yields an output letter b. Once the block of letters a ends, the direction is reversed again and the w2fst proceeds left-to-right over the letters a again and even passing the next letter b. It then resets and is ready to process the next block of letters a.

Example 2. We utilize the complete semiring $\left(\mathbb{N} \cup \{\infty\}, +, \cdot, 0, 1, \sum\right)$ in this example and consider the w2fst $\mathfrak{A} = (Q^{\rightarrow}, Q^{\leftarrow}, A, A, T, I, F)$ with input and output alphabet $A = \{a, b\}$, forward states $Q^{\rightarrow} = \{q_0, q_1, q_2, q_5, q_f\}$ and backward states $Q^{\leftarrow} = \{q_3, q_4\}$, and the following nonzero-weighted transitions, nonzero initial and final weights $I(q_0) = 1 = F(q_f)$. Thus $\text{supp}(I) = \{q_0\}$ and $\text{supp}(F) = \{q_f\}$.

$$T(q_1, b, \varepsilon, q_2) = 2$$
$$T(q_0, \vdash, \varepsilon, q_1) = T(q_1, a, a, q_1) = T(q_1, \dashv, \varepsilon, q_f) = T(q_2, a, \varepsilon, q_3) = 1$$
$$T(q_3, b, \varepsilon, q_4) = T(q_4, a, b, q_4) = T(q_4, b, \varepsilon, q_5) = T(q_4, \vdash, \varepsilon, q_5) = 1$$
$$T(q_5, a, \varepsilon, q_5) = T(q_5, b, \varepsilon, q_1) = 1$$

The w2fst $\mathfrak{A}$ is deterministic and co-deterministic and thus a wr2fst.   □

The input $w \in A^*$ to a w2fst is encoded as $\vdash w \dashv$. Intuitively, the w2fst can move its reading head on this input forward and backward and the head always stands between two letters of $A_{\vdash\dashv}$, completely in front of the input (i.e., left of the initial $\vdash$) or completely behind the input (i.e., right of the final $\dashv$). The move direction of the head is completely determined by the states in the executed transition. Given a transition $\tau = (q, a, b, q') \in \text{supp}(T)$, the head moves one step to the right if $q, q' \in Q^{\rightarrow}$. Analogously, it moves one step to the left if $q, q' \in Q^{\leftarrow}$.

Otherwise it does not move. We require a complete semiring because there may be infinitely many runs for a given input and output, which does not occur for wd2fst.

Definition 3. *Let* $\mathfrak{A} = (Q^{\rightarrow}, Q^{\leftarrow}, A, B, T, I, F)$ *be a w2fst,* $w \in A^*$ *be an input, and* $Q = Q^{\rightarrow} \cup Q^{\leftarrow}$. *A* configuration *on* w *is an element* $(\ell, q, r) \in A^*_{\vdash\dashv} \times Q \times A^*_{\vdash\dashv}$ *such that* $\ell r = \vdash w \dashv$. *The set of all configurations on* w *is denoted by* $\mathcal{C}_{\mathfrak{A}}(w)$. *Let* $\tau = (q, a, b, q') \in \mathrm{supp}(T)$ *be a transition. If* $q \in Q^{\rightarrow}$, *then for every configuration* $(\ell, q, ar) \in \mathcal{C}_{\mathfrak{A}}(w)$ *we let* $(\ell, q, ar) \xrightarrow{\tau}_{\mathfrak{A}} (\ell a, q', r)$ *if* $q' \in Q^{\rightarrow}$ *and otherwise* $(\ell, q, ar) \xrightarrow{\tau}_{\mathfrak{A}} (\ell, q', ar)$. *Similarly, if* $q \in Q^{\leftarrow}$, *then for every* $(\ell a, q, r) \in \mathcal{C}_{\mathfrak{A}}(w)$ *we let* $(\ell a, q, r) \xrightarrow{\tau}_{\mathfrak{A}} (\ell, q', ar)$ *if* $q' \in Q^{\leftarrow}$ *and* $(\ell a, q, r) \xrightarrow{\tau}_{\mathfrak{A}} (\ell a, q', r)$ *otherwise. Given configurations* $C_0, \ldots, C_n \in \mathcal{C}_{\mathfrak{A}}(w)$ *and transitions* $\tau_1, \ldots, \tau_n \in \mathrm{supp}(T)$ *such that* $C_0 \xrightarrow{\tau_1}_{\mathfrak{A}} C_1 \xrightarrow{\tau_2}_{\mathfrak{A}} \cdots \xrightarrow{\tau_n}_{\mathfrak{A}} C_n$, *we write* $C_0 \xrightarrow{\rho}_{\mathfrak{A}} C_n$ *with* $\rho = (\tau_1, \ldots, \tau_n)$, *call* ρ *a* run *from* C_0 *to* C_n, *and denote the set of all such runs by* $\mathrm{Run}_{\mathfrak{A}}(C_0, C_n)$. *The* weight $\mathrm{wt}_{\mathfrak{A}}(\rho)$ *and the* output $\mathrm{out}(\rho)$ *of the run* ρ *are* $\mathrm{wt}_{\mathfrak{A}}(\rho) = \prod_{i=1}^{n} T(\tau_i)$ *and* $\mathrm{out}(\rho) = \pi_3(\tau_1) \cdots \pi_3(\tau_n)$, *where* π_3 *is the projection to the third component as usual. The w2fst* $\mathfrak{A}$ *computes the mapping* $\|\mathfrak{A}\|: A^* \times B^* \to S$ *given for every* $w \in A^*$ *and* $v \in B^*$ *by*

$$\|\mathfrak{A}\|(w, v) = \sum_{\substack{q_0, q_{\mathrm{f}} \in Q^{\rightarrow} \\ C = (\varepsilon, q_0, \vdash w \dashv),\, C' = (\vdash w \dashv, q_{\mathrm{f}}, \varepsilon) \\ \rho \in \mathrm{Run}_{\mathfrak{A}}(C, C'),\, \mathrm{out}(\rho) = v}} I(q_0) \cdot \mathrm{wt}_{\mathfrak{A}}(\rho) \cdot F(q_{\mathrm{f}}) \ .$$

A mapping $f: A^* \times B^* \to S$ is *w2fst-computable* (resp. *wd2fst-computable* and *wr2fst-computable*) if there exists a w2fst $\mathfrak{A}$ (resp. wd2fst $\mathfrak{A}$ and wr2fst $\mathfrak{A}$) that computes f (i.e., $f = \|\mathfrak{A}\|$). If $S = \mathbb{B}$ is the BOOLEAN semiring

$$\mathbb{B} = \left(\{0, 1\}, \vee, \wedge, 0, 1, \bigvee\right) \ ,$$

then we can identify mappings $f: A^* \times B^* \to \mathbb{B}$ with relations $\mathrm{supp}(f) \subseteq A^* \times B^*$ and obtain the *two-way definable* (resp. *deterministic two-way definable* and *reversible two-way definable*) relations [7].

Example 4. Recall the wr2fst $\mathfrak{A}$ of Example 2. Let us illustrate a run without the transitions on top of $\to_{\mathfrak{A}}$ and configurations (ℓ, q, r) written simply as $\ell\, q\, r$.

$$q_0 \vdash aaba \dashv \to_{\mathfrak{A}} \vdash q_1\, aaba \dashv \to_{\mathfrak{A}} \vdash a\, q_1\, aba \dashv \to_{\mathfrak{A}} \vdash aa\, q_1\, ba \dashv \to_{\mathfrak{A}} \vdash aab\, q_2\, a \dashv$$
$$\to_{\mathfrak{A}} \vdash aab\, q_3\, a \dashv \to_{\mathfrak{A}} \vdash aa\, q_4\, ba \dashv \to_{\mathfrak{A}} \vdash a\, q_4\, aba \dashv \to_{\mathfrak{A}} \vdash q_4\, aaba \dashv \to_{\mathfrak{A}} \vdash q_5\, aaba \dashv$$
$$\to_{\mathfrak{A}} \vdash a\, q_5\, aba \dashv \to_{\mathfrak{A}} \vdash aa\, q_5\, ba \dashv \to_{\mathfrak{A}} \vdash aab\, q_1\, a \dashv \to_{\mathfrak{A}} \vdash aaba\, q_1 \dashv \to_{\mathfrak{A}} \vdash aaba \dashv q_{\mathrm{f}}$$

The output of this run is $aabba$ and its weight is 2 because each utilized transition has weight 1 except for the fourth transition, which has weight 2. The nonzero entries in $\|\mathfrak{A}\|$ are $\|\mathfrak{A}\|\left(a^{i_1} b a^{i_2} b \cdots a^{i_n} b a^{i_{n+1}}, a^{i_1} b^{i_1} a^{i_2} b^{i_2} \cdots a^{i_n} b^{i_n} a^{i_{n+1}}\right) = 2^n$ for every $n \in \mathbb{N}$ and $i_1, \ldots, i_{n+1} \in \mathbb{N}$.

4   Relation to Unweighted Case

In this section, we relate the supports of w2fst-computable mappings to the traditional two-way definable relations [7]. We already remarked that every complete semiring is zero-sum free, but zero-divisors can still interfere and render a run moot despite the individual transition weights being universally nonzero. Fortunately, we can avoid this anomaly with the help of a construction of [12]. We adjust the version of [14] of this construction to prove that the support of each w2fst-computable mapping is two-way definable.

Theorem 5. *For every w2fst-computable mapping $f: A^* \times B^* \to S$ the support $\mathrm{supp}(f)$ is two-way definable. This extends to the deterministic and reversible case.*

Proof. Let $\mathfrak{A} = (Q^{\to}, Q^{\leftarrow}, A, B, T, I, F)$ be a w2fst computing f. Moreover, let $Q = Q^{\to} \cup Q^{\leftarrow}$ and W be the finite set $W = \{0\} \cup \mathrm{ran}(T) \cup \mathrm{ran}(I) \cup \mathrm{ran}(F) \subseteq S$. We consider the monoid $(\mathbb{N}^W, +, \mathbf{0})$ with point-wise addition and the homomorphism $h: \mathbb{N}^W \to S$ into the multiplicative monoid $(S, \cdot, 1)$ of S given by

$$h(\varphi) = \prod_{s \in W} s^{\varphi(s)}$$

for every $\varphi: W \to \mathbb{N}$, where we assume that $0^0 = 1$. By DICKSON's lemma [4] the set $\min h^{-1}(0)$ is finite, where the partial order, for which the minimal elements are determined, is the standard pointwise order on $\mathbb{N}^W$. Consequently, there exists $u \in \mathbb{N}$ such that $\min h^{-1}(0) \subseteq \{0, \dots, u\}^W = U$. We define the mapping $\oplus: U^2 \to U$ by $(\varphi \oplus \varphi')(s) = \min(\varphi(s) + \varphi'(s), u)$ for every $\varphi, \varphi' \in U$ and $s \in W$. Moreover, for every $s \in W$ we let $\bar{s} \in U$ be such that $\bar{s}(s) = 1$ and $\bar{s}(s') = 0$ for all $s' \in S \setminus \{s\}$. Let $V = U \setminus h^{-1}(0)$. We construct the equivalent w2fst $\mathfrak{A}' = (Q^{\to} \times V, Q^{\leftarrow} \times V, A, B, T', I', F')$ with $P = Q \times V$ and

$$T'(\langle q, \varphi \rangle, a, b, \langle q', \varphi' \rangle) = \begin{cases} T(q, a, b, q') & \text{if } \varphi' = \varphi \oplus \overline{T(q, a, b, q')} \\ 0 & \text{otherwise} \end{cases}$$

$$I'(\langle q'', \varphi'' \rangle) = \begin{cases} I(q'') & \text{if } \varphi'' = \overline{I(q'')} \\ 0 & \text{otherwise} \end{cases}$$

$$F'(\langle q'', \varphi'' \rangle) = F(q'')$$

for every $\langle q, \varphi \rangle, \langle q', \varphi' \rangle \in P$, $\langle q'', \varphi'' \rangle \in Q^{\to} \times V$, $a \in A_{\vdash\dashv}$, and $b \in B^{\leq 1}$. Note that $\mathfrak{A}'$ is deterministic if $\mathfrak{A}$ is deterministic, but co-determinism is not necessarily preserved. Obviously, we track the run weight as long as necessary (to avoid zero-divisors) in the second component of the state $\langle q, \varphi \rangle$. By definition $h(\varphi) \neq 0$, which yields that any run ρ with universally nonzero transition weights will have nonzero weight. Together with the zero-sum freeness of the semiring S we consequently obtain that the two-way transducer $\mathrm{supp}(\mathfrak{A}') = (Q^{\to} \times V, Q^{\leftarrow} \times V, A, B, \mathrm{supp}(T'), \mathrm{supp}(I'), \mathrm{supp}(F'))$ accepts the desired relation $\mathrm{supp}(\|\mathfrak{A}'\|) = \mathrm{supp}(\|\mathfrak{A}\|) = \mathrm{supp}(f)$. If $\mathfrak{A}$ is reversible, then the obtained two-way transducer $\mathrm{supp}(\mathfrak{A}')$ is deterministic and by [3, Corollary 5] there exists an equivalent reversible two-way transducer as the classes of deterministic and reversible two-way definable relations coincide.   □

For the converse, let $\mathfrak{A} = (Q^{\rightarrow}, Q^{\leftarrow}, A, B, T, I, F)$ be a w2fst and suppose that $\rho = (\tau_1, \ldots, \tau_n) \in \mathrm{Run}_{\mathfrak{A}}(C, C')$ is a run of $\mathfrak{A}$ on some input $w \in A^*$; i.e., $C, C' \in C_{\mathfrak{A}}(w)$. The run ρ is *accepting* if (i) $C = (\varepsilon, q_0, \vdash w \dashv)$ is an initial configuration with $q_0 \in \mathrm{supp}(I)$, (ii) $\tau_i \in \mathrm{supp}(T)$ for every $1 \leq i \leq n$, and (iii) $C' = (\vdash w \dashv, q_{\mathrm{f}}, \varepsilon)$ is a final configuration with $q_{\mathrm{f}} \in \mathrm{supp}(F)$. For every set $X \subseteq X'$ and $s \in S$ we define the mapping $s_X \colon X' \to S$ given for every $x \in X'$ by $s_X(x) = s$ if $x \in X$ and $s_X(x) = 0$ otherwise. It follows from [19, Theorem 2.1] that there exists a (semiring) homomorphism $h \colon S \to \mathbb{B}$ from S into the BOOLEAN semiring $\mathbb{B}$ because S is certainly not a ring (due to its zero-sum freeness). By [8, Theorem 6.2] this homomorphism even preserves countable sums, which are all sums that occur in the semantics of a weighted two-way transducer. Finally, there is an element $\infty = \sum_{s \in S} (\sum_{i \in \mathbb{N}} s) \in S \setminus \{0\}$ such that $s + \infty = \infty$ for every $s \in S$ and $\sum_{i \in \mathbb{N}} \infty = \infty$ [11, Theorem IV.2.6]. Now we are ready to demonstrate that every two-way definable relation is the support of a w2fst-computable mapping.

Theorem 6. *For every two-way definable relation $R \subseteq A^* \times B^*$ there exists a w2fst-computable mapping $f \colon A^* \times B^* \to S$ such that $\mathrm{supp}(f) = R$. More specifically, (i) ∞_R is w2fst-computable, (ii) 1_R is wd2fst-computable if R is deterministic two-way definable, and (iii) 1_R is wr2fst-computable if R is reversible two-way definable.*

Proof. Statement (i) proves the general statement since $\mathrm{supp}(\infty_R) = R$. We show that ∞_R is w2fst-computable. Let $\mathfrak{A} = (Q^{\rightarrow}, Q^{\leftarrow}, A, B, T, I, F)$ be a two-way transducer accepting R. Consider the w2fst

$$\mathfrak{A}' = (Q^{\rightarrow}, Q^{\leftarrow}, A, B, 1_T, 1_I, \infty_F) \quad \text{and} \quad \mathfrak{A}'' = (Q^{\rightarrow}, Q^{\leftarrow}, A, B, 1_T, 1_I, 1_F),$$

of which $\mathfrak{A}'$ trivially computes ∞_R because each run weight is 1, which together with the final weight yields ∞, so the total weight assigned by $\mathfrak{A}'$ to each input-output pair $(w, v) \in A^* \times B^*$ is a (countable) sum of summands ∞. Thus, the weight is ∞ if there exists at least one accepting run of $\mathfrak{A}'$ (equivalently $\mathfrak{A}$) and 0 otherwise. Similarly, each run of $\mathfrak{A}''$ has weight 1 and the total weight is thus a sum of (countably) many summands 1, which is nonzero if there is at least one summand of 1. Thus $\mathrm{supp}(\|\mathfrak{A}''\|) = R$ as well. Finally, if $\mathfrak{A}$ is deterministic (resp. reversible), then $\mathfrak{A}''$ is also deterministic (resp. reversible) and computes 1_R.   □

5   Closure Properties

In this section we look at the closure properties of various classes of w2fst-computable mappings. We start with sum, which for every set X and all mappings $f, g \colon X \to S$ is $(f + g) \colon X \to S$ and defined by $(f + g)(x) = f(x) + g(x)$

108 F. Feng and A. Maletti

for every $x \in X$. The class of all w2fst-computable mappings is obviously closed under sum, which is entirely trivial.

Proposition 7. *The sum* $(f + g) \colon A^* \times B^* \to S$ *of two w2fst-computable mappings* $f, g \colon A^* \times B^* \to S$ *is again w2fst-computable.*

Proof. Let $\mathfrak{A} = (Q^{\to}, Q^{\leftarrow}, A, B, T, I, F)$ and $\mathfrak{A}' = (P^{\to}, P^{\leftarrow}, A, B, T', I', F')$ be w2fst that compute f and g, respectively. Without loss of generality, suppose that $(Q^{\to} \cup Q^{\leftarrow}) \cap (P^{\to} \cup P^{\leftarrow}) = \emptyset$. The w2fst $\mathfrak{A} + \mathfrak{A}'$

$$\mathfrak{A} + \mathfrak{A}' = (Q^{\to} \cup P^{\to}, Q^{\leftarrow} \cup P^{\leftarrow}, A, B, T \cup T', I \cup I', F \cup F')$$

obviously computes $f + g$. $\qquad\square$

Clearly the construction in the proof of Proposition 7 does not preserve determinism. Only in special cases, we can achieve closure under sum for wd2fst-computable mappings (as already in the unweighted case the union of two two-way deterministically definable functions is generally only a two-way definable relation).

As a second closure property we consider the point-wise (or HADAMARD) product. As usual for finite-state transducers [15] (not even necessarily two-way) there is essentially no chance for closure under intersection (or products). Let us first introduce products formally. For every set X and all mappings $f, g \colon X \to S$, their point-wise product $(f \cdot g) \colon X \to S$ is defined by $(f \cdot g)(x) = f(x) \cdot g(x)$ for every $x \in X$. The emptiness problem for w2fst asks whether $\mathrm{supp}(\|\mathfrak{A}\|) = \emptyset$ for a given w2fst $\mathfrak{A}$. By Theorem 5 the support $\mathrm{supp}(\|\mathfrak{A}\|)$ is two-way definable and emptiness of two-way definable relations is decidable, but since the construction of the two-way transducer in the proof of Theorem 5 is not necessarily effective, we cannot use it to conclude that w2fst emptiness is universally decidable.

Proposition 8. *If w2fst emptiness is decidable (for the semiring S), then the class of w2fst-computable mappings is not closed under products.*

Proof. We consider an instance (h, h') of the undecidable Post Correspondence Problem (PCP) [16], so $h, h' \colon A \to B^*$ are mappings for some alphabets A and B that extend to homomorphisms $h, h' \colon A^* \to B^*$. This instance is solvable if there exists $w \in A^* \setminus \{\varepsilon\}$ such that $h(w) = h'(w)$. By Theorem 6 the mappings 1_h and $1_{h'}$ are wd2fst-computable (because homomorphisms are deterministic rational transductions). Their product $1_h \cdot 1_{h'}$ contains only solutions in the support. If there were a w2fst $\mathfrak{A}$ computing $1_h \cdot 1_{h'}$, then we would be able to decide whether the PCP instance has a solution by deciding emptiness of $\mathfrak{A}$. This contradicts the fact that PCP is undecidable. $\qquad\square$

The final closure property we consider is composition for classes of wd2fst-computable mappings. Let $f \colon A^* \times B^* \to S$ and $g \colon B^* \times C^* \to S$ be two mappings. Their composition $(f \, ; g) \colon A^* \times C^* \to S$ is defined for every $w \in A^*$ and $u \in C^*$ by $(f \, ; g)(w, u) = \sum_{v \in B^*} f(w, v) \cdot g(v, u)$. We note that for wd2fst-computable mappings f and g the sum in the definition of $f \, ; g$ is always finite.

We start with wr2fst-computable mappings. We plan to utilize an approach of [3], but a minor difference to the two-way transducers of [3] concerns the transition output. We allow only a single letter or the empty word ε as output in each transition, whereas the definition in [3] allows an arbitrary sequence of letters as output on each transition. However, this difference is immaterial as we can simulate a transition $(q, a, v_1 \cdots v_n, q')$ with $v_1, \ldots, v_n \in B$ by the following sequence of transitions. If n is even, $q \in Q^{\rightarrow}$, and $q' \in Q^{\leftarrow}$, then the sequence is

$$(q, a, v_1, q_1), (q_1, a', v_2, q_2), (q_2, a, v_3, q_3), \ldots, (q_{n-1}, a', v_n, q_n), (q_n, a, \varepsilon, q')$$

for every $a' \in A_{\vdash\dashv}$ and fresh states $q_1, \ldots, q_n$ such that the direction strictly alternates in the sequence $q, q_1, q_2, \ldots, q_n$ (i.e., $q_1 \in Q^{\leftarrow}$, $q_2 \in Q^{\rightarrow}$, etc.). The sequences are similar in the remaining cases. We note that determinism and co-determinism (and thus reversibility) are preserved in this construction. Obviously a wr2fst $\mathfrak{A} = (Q^{\rightarrow}, Q^{\leftarrow}, A, B, T, I, F)$ has a single initial state $q_0 \in \mathrm{supp}(I)$ and a single final state $q_{\mathrm{f}} \in \mathrm{supp}(F)$. It is *normalized* if (i) $(q_0, \vdash, b, q) \in \mathrm{supp}(T)$ implies $b = \varepsilon$ and (ii) $(q, \dashv, b, q_{\mathrm{f}}) \in \mathrm{supp}(T)$ implies $b = \varepsilon$. In other words, the initial and final transition produce no output. The wr2fst of Example 2 is normalized. Without loss of generality, we can assume that a wr2fst is normalized. To prepare the composition construction, we need to separate the output and weight generation in order to avoid charging weights multiple times. We achieve this by essentially repeating each run. In the first phase we only charge the weights, but do not actually produce any output. We then rewind back to the beginning of the input and restart the transducer in the second phase, in which we do not charge weights (i.e., use only weight 1), but instead produce the correct output. This separation is possible for wd2fst since the second phase will reproduce the first-phase run due to determinism.

Definition 9. *Let $\mathfrak{A} = (Q^{\rightarrow}, Q^{\leftarrow}, A, B, T, I, F)$ be a normalized wr2fst using the states $Q = Q^{\rightarrow} \cup Q^{\leftarrow}$, and $g \colon Q \to P$ be a bijection for some set P such that $P \cap Q = \emptyset$. Additionally, let $P^{\rightarrow} = \{g(q) \mid q \in Q^{\rightarrow}\}$, $P^{\leftarrow} = \{g(q) \mid q \in Q^{\leftarrow}\}$, and $\bot \notin Q \cup P$ be a fresh state. Finally, let $q_0 \in \mathrm{supp}(I)$ be the unique initial and $q_{\mathrm{f}} \in \mathrm{supp}(F)$ the unique final state. We construct the wr2fst*

$$S(\mathfrak{A}) = \left(Q^{\rightarrow} \cup P^{\rightarrow}, Q^{\leftarrow} \cup P^{\leftarrow} \cup \{\bot\}, A, B_{\vdash\dashv}, T', I', F'\right)$$

with initial state $\mathrm{supp}(I') = \{q_0\}$ with initial weight $I'(q_0) = I(q_0)$, final state $\mathrm{supp}(F') = \{g(q_{\mathrm{f}})\}$ with final weight $F'(g(q_{\mathrm{f}})) = 1$, and (potentially nonzero) weighted transitions such that for every $(q, a, b, q') \in \mathrm{supp}(T)$

- $T'(q, a, \varepsilon, q') = \sum_{b' \in B^{\leq 1}} T(q, a, b', q')$ *if $a \neq \dashv$,*
- $T'(q, \dashv, \varepsilon, q') = \sum_{b' \in B^{\leq 1}} T(q, \dashv, b', q')$ *if $q' \in Q^{\leftarrow}$,*
- $T'(q, \dashv, \varepsilon, \bot) = \sum_{b' \in B^{\leq 1}} T(q, \dashv, b', q_{\mathrm{f}}) \cdot F(q_{\mathrm{f}})$,
- $T'(\bot, a, \varepsilon, \bot) = 1$ *if $a \neq \vdash$,*
- $T'(\bot, \vdash, \vdash, g(q')) = 1$ *if $q = q_0$ and $a = \vdash$,*
- $T'(g(q), \vdash, b, g(q')) = 1$ *if $a = \vdash$ and $q \neq q_0$,*
- $T'(g(q), a, b, g(q')) = 1$ *if $a \in A$,*

- $T'\big(g(q), \dashv, b, g(q')\big) = 1$ *if* $a = \dashv$ *and* $q' \neq q_{\mathrm{f}}$, *and*
- $T'\big(g(q), \dashv, \dashv, g(q_{\mathrm{f}})\big) = 1$ *if* $a = \dashv$ *and* $q' = q_{\mathrm{f}}$.

Note that due to determinism all the sums in the definition of T' consist of at most one summand. We also note that $S(\mathfrak{A})$ is reversible since $q_0, q_{\mathrm{f}} \in Q^{\rightarrow}$, so we must have $q' = q_{\mathrm{f}}$ if $q' \notin Q^{\leftarrow}$ (3rd item).

The wr2fst $S(\mathfrak{A})$ computes essentially the same mapping as $\mathfrak{A}$, but with added delimiters, so for every $w \in A^*$ and $v \in B^*$ we have $\|S(\mathfrak{A})\|(w, \vdash v \dashv) = \|\mathfrak{A}\|(w, v)$ and $\|S(\mathfrak{A})\|$ is 0 otherwise. This property relies crucially on the determinism of $\mathfrak{A}$, but is otherwise straightforward to establish.

Example 10. Recall the normalized wr2fst $\mathfrak{A}$ of Example 2. The wr2fst $S(\mathfrak{A})$ is given by $S(\mathfrak{A}) = (P^{\rightarrow}, P^{\leftarrow}, A, A, T', I', F')$ with states $P^{\leftarrow} = \{q_3, q_4, p_3, p_4, \bot\}$ and $P^{\rightarrow} = \{q_0, q_1, q_2, q_5, q_{\mathrm{f}}, p_0, p_1, p_2, p_5, p_{\mathrm{f}}\}$, and the following nonzero weighted transitions, initial states $\mathrm{supp}(I') = \{q_0\}$ with $I'(q_0) = 1$, and final states given by $\mathrm{supp}(F') = \{p_{\mathrm{f}}\}$ with $F'(p_{\mathrm{f}}) = 1$.

$$T'(q_1, b, \varepsilon, q_2) = 2$$
$$T'(q_0, \vdash, \varepsilon, q_1) = T'(q_1, a, \varepsilon, q_1) = T'(q_1, \dashv, \varepsilon, \bot) = T'(q_2, a, \varepsilon, q_3) = 1$$
$$T'(q_3, b, \varepsilon, q_4) = T'(q_4, a, \varepsilon, q_4) = T'(q_4, b, \varepsilon, q_5) = T'(q_4, \vdash, \varepsilon, q_5) = 1$$
$$T'(q_5, a, \varepsilon, q_5) = T'(q_5, b, \varepsilon, q_1) = T'(\bot, a, \varepsilon, \bot) = T'(\bot, b, \varepsilon, \bot) = 1$$

$$T'(\bot, \vdash, \vdash, p_1) = T'(p_1, a, a, p_1) = T'(p_1, b, \varepsilon, p_2) = T'(p_1, \dashv, \dashv, p_{\mathrm{f}}) = 1$$
$$T'(p_2, a, \varepsilon, p_3) = T'(p_3, b, \varepsilon, p_4) = T'(p_4, a, b, p_4) = T'(p_4, b, \varepsilon, p_5) = 1$$
$$T'(p_4, \vdash, \varepsilon, p_5) = T'(p_5, a, \varepsilon, p_5) = T'(p_5, b, \varepsilon, p_1) = 1$$

All transitions in the first part are copies of the original transitions that produce no output but keep the weight. Instead of switching to the original final state q_{f} we switch to $\bot$ to rewind. The second part contains copies of the original transitions with unit weight but keep the output.   □

Now we can present our composition construction, which follows the construction of [3, Theorem 1].

Theorem 11. *The composition $(f \,;g)\colon A^* \times C^* \to S$ of two wr2fst-computable mappings $f\colon A^* \times B^* \to S$ and $g\colon B^* \times C^* \to S$ is again wr2fst-computable.*

Proof. Let $\mathfrak{A}$ and $\mathfrak{A}' = (P^{\rightarrow}, P^{\leftarrow}, B, C, T', I', F')$ be wr2fst that compute the mappings f and g, respectively. Moreover, let $S(\mathfrak{A}) = (Q^{\rightarrow}, Q^{\leftarrow}, A, B, T, I, F)$, $Q = Q^{\rightarrow} \cup Q^{\leftarrow}$ and $P = P^{\rightarrow} \cup P^{\leftarrow}$. Finally, let $q_0 \in \mathrm{supp}(I)$ and $q_{\mathrm{f}} \in \mathrm{supp}(F)$ be the unique initial and final state of $S(\mathfrak{A})$, respectively, and $p_0 \in \mathrm{supp}(I')$ as well as $p_{\mathrm{f}} \in \mathrm{supp}(F')$ be the corresponding states of $\mathfrak{A}'$. To simplify the construction, we let $T'(p, \varepsilon, c, p') = 1$ if $c = \varepsilon$ and $p' = p$ and $T'(p, \varepsilon, c, p') = 0$ otherwise for every $p, p' \in P$ and $c \in C^{\leq 1}$. In other words, we extend T' such that on empty input $\mathfrak{A}'$ can only make a transition such that the state remains the same, no output is generated, and no weight (i.e., the neutral element 1) is charged. This is no actual transition, but simplifies the following definitions. At this point, we can construct the wr2fst

$$\mathfrak{B} = \Big((Q^\to \times P^\to) \cup (Q^\gets \times P^\gets), (Q^\to \times P^\gets) \cup (Q^\gets \times P^\to), A, C, T'', I'', F''\Big)$$

with initial and final weights

$$I''(\langle q, p\rangle) = I(q) \cdot I'(p) \quad \text{and} \quad F''(\langle q, p\rangle) = F(q) \cdot F'(p)$$

for every $q \in Q$ and $p \in P$, and for every $q, q' \in Q$, $p, p' \in P$, and $a \in A_{\vdash\dashv}$

$$T''\Big(\langle q, p\rangle, a, c, \langle q', p'\rangle\Big)$$
$$= \begin{cases} \sum_{b \in B^{\leq 1}} T(q, a, b, q') \cdot T'(p, b, c, p') & \text{if } p, p' \in P^\to \\ \sum_{b \in B^{\leq 1}} T(q', a, b, q) \cdot T'(p, b, c, p') & \text{if } p, p' \in P^\gets \\ \sum_{b \in B, q'' \in Q} T(q, a, b, q'') \cdot T'(p, b, c, p') & \text{if } p \in P^\to, p' \in P^\gets, q = q' \\ \sum_{b \in B, q'' \in Q} T(q'', a, b, q) \cdot T'(p, b, c, p') & \text{if } p \in P^\gets, p' \in P^\to, q = q' \\ 0 & \text{otherwise} \end{cases}$$

Again we note that the sums in the definition of T'' are for notational convenience only as they all have at most one summand. The wr2fst $\mathfrak{B}$ computes the mapping $f \, ; g$ by simulating transitions of $S(\mathfrak{A})$ as well as the behavior of $\mathfrak{A}'$ on the output of those transitions. If $\mathfrak{A}'$ moves left, then the transitions of $S(\mathfrak{A})$ are reversed to reobtain the outputs produced earlier. Due to the separation of weights and outputs in $S(\mathfrak{A})$ revisiting an earlier output symbol does not cause duplication of the corresponding weight. □

Let us illustrate the construction on an example. We first introduce the wr2fst that we will compose to our running example wr2fst.

Example 12. The wr2fst $\mathfrak{R} = (\{r_0, r_f\}, \{r_1\}, A, A, T', I', F')$ with input and output alphabet $A = \{a, b\}$ and the following nonzero weighted transitions, initial state $\text{supp}(I') = \{r_0\}$ with $I'(r_0) = 1$, and final states $\text{supp}(F') = \{r_f\}$ with $F'(r_f) = 1$ computes the mapping 1_R with $R = \{(w, w^R w) \mid w \in A^*\}$.

$$T'(r_0, \vdash, \varepsilon, r_0) = T'(r_0, a, \varepsilon, r_0) = T'(r_0, b, \varepsilon, r_0) = T'(r_0, \dashv, \varepsilon, r_1) = 1$$
$$T'(r_1, a, a, r_1) = T'(r_1, b, b, r_1) = T'(r_1, \vdash, \varepsilon, r_f) = 1$$
$$T'(r_f, a, a, r_f) = T'(r_f, b, b, r_f) = T'(r_f, \dashv, \varepsilon, r_f) = 1 \qquad □$$

Example 13. Recall the wr2fst $\mathfrak{A}$ and $\mathfrak{R}$ of Examples 2 and 12. The construction in the proof of Theorem 11 yields the wr2fst $\mathfrak{B}$ with the following nonzero weighted (relevant) transitions. The initial state with weight 1 is $\langle q_0, r_0\rangle$ and the final state with weight 1 is $\langle p_f, r_f\rangle$. To save space we write a state $\langle q, r\rangle$ simply as qr.

$$T''(q_1r_0, b, \varepsilon, q_2r_0) = 2$$
$$T''(q_0r_0, \vdash, \varepsilon, q_1r_0) = T''(q_1r_0, a, \varepsilon, q_1r_0) = T''(q_1r_0, \dashv, \varepsilon, \perp r_0) = 1$$
$$T''(q_2r_0, a, \varepsilon, q_3r_0) = T''(q_3r_0, b, \varepsilon, q_4r_0) = T''(q_4r_0, a, \varepsilon, q_4r_0) = 1$$
$$T''(q_4r_0, b, \varepsilon, q_5r_0) = T''(q_4r_0, \vdash, \varepsilon, q_5r_0) = T''(q_5r_0, a, \varepsilon, q_5r_0) = 1$$

$$T''(q_5r_0, b, \varepsilon, q_1r_0) = T''(\perp r_0, a, \varepsilon, \perp r_0) = T''(\perp r_0, b, \varepsilon, \perp r_0) = 1$$

$$T''(\perp r_0, \vdash, \varepsilon, p_1r_0) = T''(p_1r_0, a, \varepsilon, p_1r_0) = T''(p_1r_0, b, \varepsilon, p_2r_0) = 1$$
$$T''(p_1r_0, \dashv, \varepsilon, p_1r_1) = T''(p_2r_0, a, \varepsilon, p_3r_0) = T''(p_3r_0, b, \varepsilon, p_4r_0) = 1$$
$$T''(p_4r_0, a, \varepsilon, p_4r_0) = T''(p_4r_0, b, \varepsilon, p_5r_0) = T''(p_4r_0, \vdash, \varepsilon, p_5r_0) = 1$$

$$T''(p_5r_0, a, \varepsilon, p_5r_0) = T''(p_5r_0, b, \varepsilon, p_1r_0) = T''(p_1r_1, a, a, p_1r_1) = 1$$
$$T''(p_1r_1, b, \varepsilon, p_5r_1) = T''(p_1r_1, \vdash, \varepsilon, p_1r_f) = T''(p_1r_f, a, a, p_1r_f) = 1$$
$$T''(p_1r_f, b, \varepsilon, p_2r_f) = T''(p_1r_f, \dashv, \varepsilon, p_fr_f) = T''(p_5r_1, a, \varepsilon, p_5r_1) = 1$$
$$T''(p_5r_1, b, \varepsilon, p_4r_1) = T''(p_5r_1, \vdash, \varepsilon, p_4r_1) = T''(p_4r_1, a, b, p_4r_1) = 1$$

$$T''(p_4r_1, b, \varepsilon, p_3r_1) = T''(p_3r_1, a, \varepsilon, p_2r_1) = T''(p_2r_1, b, \varepsilon, p_1r_1) = 1$$
$$T''(p_2r_f, a, \varepsilon, p_3r_f) = T''(p_3r_f, b, \varepsilon, p_4r_f) = T''(p_4r_f, a, b, p_4r_f) = 1$$
$$T''(p_4r_f, b, \varepsilon, p_5r_f) = T''(p_4r_f, \vdash, \varepsilon, p_5r_f) = T''(p_5r_f, a, \varepsilon, p_5r_f) = 1$$
$$T''(p_5r_f, b, \varepsilon, p_1r_f) = 1 \qquad \square$$

Finally, we show that, as in the unweighted case, the classes of wd2fst-computable mappings and wr2fst-computable mappings coincide by demonstrating that every wd2fst can be transformed into an equivalent wr2fst. This yields closure under composition also for wd2fst-computable mappings.

Theorem 14. *For every wd2fst there exists an equivalent wr2fst.*

Proof. Let $\mathfrak{A} = (Q^\rightarrow, Q^\leftarrow, A, B, T, I, F)$ be the given wd2fst. We construct the wd2fst $\mathfrak{A}' = (Q^\rightarrow, Q^\leftarrow, A, \operatorname{supp}(T), T', I, F)$ with the transitions $\operatorname{supp}(T)$ as output alphabet and

$$T'(q, a, \tau, q') = \begin{cases} 1 & \text{if } \exists v \in B^{\leq 1} : \tau = (q, a, v, q') \\ 0 & \text{otherwise} \end{cases}$$

for every $q, q' \in Q^\rightarrow \cup Q^\leftarrow$, $a \in A_{\vdash \dashv}$, and $\tau \in \operatorname{supp}(T)$. In other words, on input $w \in A^*$, the wd2fst $\mathfrak{A}'$ outputs with weight 1 the transition sequence used by $\mathfrak{A}$ to process the input w. Additionally, let

$$\mathfrak{M} = (\{p\}, \emptyset, \operatorname{supp}(T), B, T'', \{p\}, \{p\})$$

be the wr2fst such that for every $\tau \in \operatorname{supp}(T)_{\vdash \dashv}$ and $v' \in B^{\leq 1}$

$$T''(p, \tau, v', p) = \begin{cases} T(\tau) & \text{if } \tau = (q, a, v', q') \in \operatorname{supp}(T) \\ 1 & \text{if } \tau \in \{\vdash, \dashv\} \text{ and } v' = \varepsilon \\ 0 & \text{otherwise.} \end{cases}$$

Clearly, $\mathfrak{M}$ processes transition sequences and generates the corresponding output and weight, so obviously $\|\mathfrak{A}\| = \|\mathfrak{A}'\| ; \|\mathfrak{M}\|$. By Theorem 5 $\mathrm{supp}(\|\mathfrak{A}'\|)$ is reversible two-way definable, so there exists a (unweighted) reversible two-way finite state transducer $\mathfrak{A}''$ that computes $\|\mathfrak{A}''\| = \mathrm{supp}(\|\mathfrak{A}'\|)$. Moreover,

$$\|\mathfrak{A}'\| = 1_{\mathrm{supp}(\|\mathfrak{A}'\|)} = 1_{\|\mathfrak{A}''\|}$$

and $1_{\|\mathfrak{A}''\|}$ is wr2fst-computable by Theorem 6. Hence $\|\mathfrak{A}\| = 1_{\|\mathfrak{A}''\|} ; \|\mathfrak{M}\|$ and thus $\|\mathfrak{A}\|$ is wr2fst-computable by Theorem 11 since all mappings on the right-hand side are wr2fst-computable. □

Corollary 15 (of Theorems 11 and 14). *The classes of wd2fst-computable and wr2fst-computable mappings coincide and they are both closed under composition.*

References

1. Berstel, J.: Transductions and Context-free Languages. Teubner-Verlag (1979). https://doi.org/10.1007/978-3-663-09367-1
2. Carnino, V., Lombardy, S.: On determinism and unambiguity of weighted two-way automata. In: Ésik, Z., Fülöp, Z. (eds.) Proceedings 14th International Conference Automata and Formal Languages. Electronic Proceedings of Theoretical Computer Science, vol. 151, pp. 188–200. arXiv (2014). https://doi.org/10.4204/EPTCS.151.13
3. Dartois, L., Fournier, P., Jecker, I., Lhote, N.: On reversible transducers. In: Chatzigiannakis, I., Indyk, P., Kuhn, F., Muscholl, A. (eds.) Proceedings– Leibniz-Zentrum für Informatik (2017). https://doi.org/10.4230/LIPIcs.ICALP.2017.113
4. Dickson, L.E.: Finiteness of the odd perfect and primitive abundant numbers with n distinct prime factors. Am. J. Math. **35**(4), 413–422 (1913). https://doi.org/10.2307/2370405
5. Droste, M., Gastin, P.: Weighted automata and weighted logics. In: Caires, L., Italiano, G.F., Monteiro, L., Palamidessi, C., Yung, M. (eds.) ICALP 2005. LNCS, vol. 3580, pp. 513–525. Springer, Heidelberg (2005). https://doi.org/10.1007/11523468_42
6. Droste, M., Kuich, W., Vogler, H.: Handbook of Weighted Automata. EATCS Monographs in Theoretical Computer Science, Springer (2009). https://doi.org/10.1007/978-3-642-01492-5
7. Engelfriet, J., Hoogeboom, H.J.: MSO definable string transductions and two-way finite state transducers. ACM Trans. Comput. Log. **2**(2), 216–254 (2001). https://doi.org/10.1145/371316.371512
8. Fülöp, Z., Maletti, A., Vogler, H.: Weighted extended tree transducers. Fund. Inform. **111**(2), 163–202 (2011). https://doi.org/10.3233/FI-2011-559
9. Golan, J.S.: Semirings and their Applications. Kluwer Academic, Dordrecht (1999). https://doi.org/10.1007/978-94-015-9333-5
10. Goldstern, M.: Completion of semirings. Master's thesis, TU Wien (1985). https://doi.org/10.48550/arXiv.math/0208134
11. Hebisch, U., Weinert, H.J.: Semirings – algebraic theory and applications in computer science. World Sci. (1998). https://doi.org/10.1142/3903

12. Kirsten, D.: The support of a recognizable series over a zero-sum free, commutative semiring is recognizable. Acta Cybernetica **20**(2), 211–221 (2011). https://doi.org/10.14232/actacyb.20.2.2011.1

13. Lombardy, S.: Two-way representations and weighted automata. RAIRO Theor. Inf. Appl. **50**(4), 331–350 (2016). https://doi.org/10.1051/ita/2016026

14. Maletti, A., Nász, A.: Weighted tree automata with constraints. In: Diekert, V., Volkov, M.V. (eds.) Proceedings 26th International Conference Developments in Language Theory. Lecture Notes in Computer Science, vol. 13257, pp. 226–238. Springer (2022). https://doi.org/10.1007/978-3-031-05578-2_18

15. Muscholl, A., Puppis, G.: The many facets of string transducers (invited talk). In: Niedermeier, R., Paul, C. (eds.) Proceedings 36th International Symposium Theoretical Aspects of Computer Science. LIPIcs, vol. 126, pp. 2:1–2:21. Schloss Dagstuhl – Leibniz-Zentrum für Informatik (2019). https://doi.org/10.4230/LIPIcs.STACS.2019.2

16. Papadimitriou, C.H.: Computational Complexity. Addison Wesley (1994)

17. Rabin, M.O., Scott, D.: Finite automata and their decision problems. IBM J. Res. Dev. **3**(2), 114–125 (1959). https://doi.org/10.1147/rd.32.0114

18. Shepherdson, J.C.: The reduction of two-way automata to one-way automata. IBM J. Res. Dev. **3**(2), 198–200 (1959). https://doi.org/10.1147/rd.32.0198

19. Wang, H.: On characters of semirings. Houston J. Math. **23**(3), 391–405 (1997). https://www.math.uh.edu/~hjm/restricted/archive/v023n3/0391WANG.pdf

A Formal Algebraic Approach for the Quantitative Modeling of Connectors in Architectures

Christina Chrysovalanti Fountoukidou and Maria Pittou[(✉)]

Department of Mathematics, Aristotle University of Thessaloniki,
54124 Thessaloniki, Greece
mpittou@math.auth.gr

Abstract. In this paper, we propose an algebraic formalization of connectors in the quantitative setting. We firstly present a weighted Algebra of Interactions over a set of ports and a commutative and idempotent semiring, which is proved sufficient for modeling well-known coordination schemes in the weighted setup. In turn, we study a weighted Algebra of Connectors over a set of ports and a commutative and idempotent semiring, which extends the former algebra with types that describe Rendezvous and Broadcast synchronization. We show the expressiveness of the algebra by modeling several weighted connectors, and we provide conditions for proving a concept of congruence relation between weighted connectors.

Keywords: Weighted Algebra of Interactions · Weighted Algebra of Connectors · Coordination schemes · Architectures

1 Introduction

Coordination plays a prominent role in component-based design, where systems are constructed by multiple interacting components [3,6]. In such a setting, architectures have been proved fine-grained models for defining communication of components, that is implemented by the concept of the so-called connectors [2,14]. Connectors are architectural entities that regulate the synchronization mode among the permissible interactions, where the latter are specified by the imposed coordination scheme [7,13]. For instance, in an architecture with a sender and two receiver components, a coordination scheme may forbid any interaction between the receivers only. In turn, a connector may impose Rendezvous synchronization mode requiring that all the components should interact simultaneously or Broadcast mode where a component, namely the sender, should initiate the interactions with some of the receivers. Architectures place significant importance on the rigorous formalization of connectors, which in turn, is crucial for the efficient modeling of coordinated systems [7,15].

ⓒ The Author(s), under exclusive license to Springer Nature Switzerland AG 2022
D. Poulakis and G. Rahonis (Eds.): CAI 2022, LNCS 13706, pp. 115–135, 2022.
https://doi.org/10.1007/978-3-031-19685-0_9

The formal modeling of connectors is a well-known problem that has been studied mainly in the qualitative setting, with alternative approaches, including process algebras [2,7] and category theory [6,7]. Connectors have been also supported by architectural description languages in order to facilitate the specification of coordination [1,11,15]. However, well-founded design of architectures should incorporate not only the required qualitative properties but also the related non-functional aspects (cf. [11,14,15]). Such features include available resources, energy consumption, probabilities, etc., for implementing the interactions in architectures. The quantitative modeling of connectors has been addressed in the setting of architectural description languages and mainly deals with their probabilistic behavior (cf. [11,15]). On the other hand, there is a lack of a formal algebraic framework for connectors in the weighted setup, which is the main contribution of this work. In particular, we extend the results of [3] in the weighted setting, in order to encode the quantitative features of connectors. In [3], the authors introduced two algebras for modeling the interactions and connectors in component based-systems, and studied their properties. In this paper, we extend these algebras in the weighted framework, and we show that the key results from [3] still hold.

Our weighted algebras do not require knowledge on the behavior of components, and the only necessary information lies on the ports that perform the communication. In particular, we associate each port with a weight from a commutative and idempotent semiring K, expressing the "cost" of its participation in the interactions, which are described as sets of ports. Also, we define an equivalence relation for the elements of the algebras and by their equivalence classes we derive several properties. In turn, we introduce a concept of congruence relation for weighted connectors and we extend the respective results from [3] in the weighted setup. Congruences are important for connectors since they allow to use them interchangeably without affecting the architecture [2,3,7]. Specifically, the contributions of the current paper are the following:

(i) We introduce the *weighted Algebra of Interactions* over a set of ports P and a commutative and idempotent semiring K ($wAI(P)$ for short). The syntax of $wAI(P)$ is over P, contains the symbols "0" and "1" such that $0, 1 \notin P$, and allows a weighted union operator "$\oplus$" and a weighted synchronization operator "$\otimes$", encoding the weight of independent and simultaneous interactions, respectively. We refer to the elements of the algebra simply as $wAI(P)$ elements and given a set of interactions we interpret their semantics as polynomials over P and K. We define the quotient set $wAI(P)/ \equiv$ of "$\equiv$" on $wAI(P)$, where "$\equiv$" is the equivalence relation of two $wAI(P)$ elements, i.e., elements with the same weight on the same set of interactions. We show that $wAI(P)/ \equiv$ is a commutative and idempotent semiring with operations $\oplus$ and $\otimes$, and constant elements the equivalence classes of 0 and 1, respectively, a result that is used in our weighted algebra of connectors. In turn, we apply $wAI(P)$ for encoding the coordination schemes Rendezvous, Broadcast, Atomic Broadcast, and Causality Chain, in the weighted setup.

(ii) We introduce the *weighted Algebra of Connectors* over a set of ports P and a commutative and idempotent semiring K ($wAC(P)$ for short). The syntax of $wAC(P)$ includes two typing operators that characterize the type of synchronization applied to ports, namely triggers "$[\cdot]'$" that can initiate an interaction and synchrons "$[\cdot]$" that need synchronization with other ports in order to interact, and allows two binary operators "$\oplus$" and "$\otimes$" called weighted union operator and weighted fusion operator, respectively. Weighted fusion is a generalization of the weighted synchronization in $wAI(P)$, while weighted union has the same meaning in both algebras. We define the semantics of $wAC(P)$ connectors as $wAI(P)$ elements, and then, by the semantics of $wAI(P)$ we can derive the weight of the $wAC(P)$ connectors on a set of interactions over P. We prove some nice properties for $wAC(P)$ and then we apply the algebra for describing several connectors in the weighted setup.

(iii) We show that the weighted fusion operator does not preserve the equivalence of $wAC(P)$ connectors, i.e., of connectors with the same $wAI(P)$ elements. For this, we are interested in a congruence relation for $wAC(P)$ connectors. We explain why we cannot derive a congruence relation between $wAC(P)$ connectors in general, and in turn, we introduce a concept of congruence applied to specific $wAC(P)$ connectors occurring in our examples, the so-called *fusion-$wAC(P)$* connectors. Finally, we derive two theorems that provide conditions for proving such a congruence for fusion-$wAC(P)$ connectors, by extending the results of [3] in our weighted framework.

The detailed proofs and results of the current paper can be found in [9].

2   Related Work

In [2], the authors formalized architectural types by a process algebra based on an architectural description language that supported dynamic interactions. In turn, they presented a weak bisimulation equivalence technique for verifying architectural compatibility and conformity. In [13], the authors proposed an architectural metamodel for describing connectors in message passing and remote procedure call mechanisms, formalized by Alloy. Then the authors verified conformance of these communication styles at the model level. In contrast to [2,13], our framework does not deal with dynamic interactions. On the other hand, we model the quantitative aspects of connectors that were not addressed in [2,13].

In [6], the authors developed an algebra for connectors based on category theory, that supported symmetry, synchronization, mutual exclusion, hiding and inaction connectors. The authors provided the operational, observational, and denotational semantics of connectors and presented a complete normal-form axiomatization for the algebra. In contrast to our framework, the work of [6] studied the qualitative modeling of connectors. On the other hand, in [6], connectors were specified as entities with behavior and interactions, where concurrent actions were permitted. Modeling the behavior of connectors in the weighted setup is left as future work.

In [1], Reo was used for the coordination of concurrent processes and for the compositional construction of connectors among component-based systems. Some work has also investigated Reo connectors in the probabilistic setup (cf. [11, 15]). In [15], the authors studied Reo connectors with probabilistic behavior as timed data distribution streams implemented in Coq. Moreover, in [11], the authors formalized Reo connectors with random and probabilistic behavior in PVS. In contrast to [11,15], we propose a general algebraic modeling framework in the weighted setup. Another difference is that Reo treats input and output ports separately, while in our framework, we use bidirectional ports.

In [3], the authors introduced two algebras, namely the Algebra of Interactions and the Algebra of Connectors in order to formalize connectors in BIP framework. Also, the authors studied a congruence relation for connectors and presented applications for improving the language and the execution engine of BIP. Our paper extends most of the results from [3] in the weighted setup. A main difference results from the fact that in [3], there was a clear distinction between syntactic equality and semantic equivalence. In particular, the properties of the algebras were defined as axioms using syntactic equality. On the contrary, in the weighted framework, we need to define an equivalence relation for the elements of the algebras and then by their equivalence classes we can derive the respective results. In turn, this difference makes the congruence problem more difficult in the weighted setup. Specifically, the investigation of a congruence relation for any $wAC(P)$ connector is an open problem (see also Sect. 6).

3   Preliminaries

For every natural number $n \geq 1$ we denote by $[n]$ the set $\{1, \ldots, n\}$. We recall the basic notions for semirings and series from [8]. A semiring $(K, +, \cdot, \hat{0}, \hat{1})$ consists of a set K, two binary operations $+$ and $\cdot$, and two constant elements $\hat{0}$ and $\hat{1}$ in K, such that $(K, +, \hat{0})$ is a commutative monoid, $(K, \cdot, \hat{1})$ is a monoid, $\cdot$ distributes over $+$, i.e., $(k_1 + k_2) \cdot k_3 = (k_1 \cdot k_3) + (k_2 \cdot k_3)$ and $k_1 \cdot (k_2 + k_3) = (k_1 \cdot k_2) + (k_1 \cdot k_3)$ for every $k_1, k_2, k_3 \in K$, and $\hat{0} \cdot k = k \cdot \hat{0} = \hat{0}$ for every $k \in K$. If $(K, \cdot, \hat{1})$ is commutative, then the semiring is called commutative. The semiring is denoted simply by K if the operations and the constant elements are understood. Also, K is called idempotent if $k + k = k$ for every $k \in K$. Well-known commutative semirings are: the semiring of natural numbers $(\mathbb{N}, +, \cdot, 0, 1)$, the Boolean semiring $B = (\{0, 1\}, +, \cdot, 0, 1)$, the max-plus semiring $\mathbb{R}_{\max} = (\mathbb{R}_+ \cup \{-\infty\}, \max, +, -\infty, 0)$ where $\mathbb{R}_+ = \{r \in \mathbb{R} \mid r \geq 0\}$, the min-plus semiring $\mathbb{R}_{\min} = (\mathbb{R}_+ \cup \{\infty\}, \min, +, \infty, 0)$, the Viterbi semiring $([0, 1], \max, \cdot, 0, 1)$, and the Fuzzy semiring $F = ([0, 1], \max, \min, 0, 1)$. All the above but the first one are idempotent.

Let K be a semiring and P be a non-empty set. A formal series (or simply series) over P and K is a mapping $s : P \to K$. The support of s is the set $\mathrm{supp}(s) = \{p \in P \mid s(p) \neq \hat{0}\}$. A series with finite support is called a polynomial. We denote by $K \langle P \rangle$ the class of all polynomials over P and K. Let $s, r \in K \langle P \rangle$ and $k \in K$. The sum $s \oplus r$, the product with scalars ks and sk, and the Hadamard product $s \otimes r$ are polynomials in $K \langle P \rangle$, and are defined elementwise by $s \oplus r(p) =$

$s(p) + r(p), (ks)(p) = k \cdot s(p), (sk)(p) = s(p) \cdot k$ and $s \otimes r(p) = s(p) \cdot r(p)$ for every $p \in P$, respectively.

In our setting, communication is performed by a set of ports and is defined by interactions. In turn, the permissible set of interactions is specified by the coordination scheme implemented in the architecture.

Let P be a finite set of ports. Then an *interaction* a is a set of ports over P, i.e., $a \in 2^P$. Also, an *interactions set* γ is a set of interactions over P, i.e., $\gamma \in 2^{2^P}$. We let $I(P) = 2^P$ and $\Gamma(P) = 2^{I(P)}$. Furthermore, in the sequel, we assign to each port $p \in P$ a unique weight $k_p \in K$.

In the rest of the paper, K denotes a commutative and idempotent semiring.

4   The Weighted Algebra of Interactions

In this section, we study the weighted Algebra of Interactions over a set of ports P and a commutative and idempotent semiring K, in order to encode the weight of interactions. We prove several properties for the algebra, and then we apply the algebra for modeling concrete coordination schemes in the weighted setup.

Definition 1. *Let P be a set of ports such that $0, 1 \notin P$. The syntax of the weighted Algebra of Interactions (wAI(P) for short) over P and K is given by*

$$z ::= 0 \mid 1 \mid p \mid z \oplus z \mid z \otimes z \mid (z)$$

where $p \in P$, "$\oplus$" is the weighted union operator and "$\otimes$" is the weighted synchronization operator that binds stronger than "$\oplus$".

We call z a $wAI(P)$ element over P and K. When the latter are understood we simply refer to z as a $wAI(P)$ element. We interpret the semantics of z as polynomials $\|z\| \in K \langle \Gamma(P) \rangle$. Hence, given an interactions set $\gamma \in \Gamma(P)$ we can derive the weight of implementing γ in a given architecture.

Definition 2. *Let z be a $wAI(P)$ element over P and K. The semantics of z is a polynomial $\|z\| \in K \langle \Gamma(P) \rangle$. For every interactions set $\gamma \in \Gamma(P)$, the value $\|z\| (\gamma)$ is defined inductively on z as follows:*

- $\|0\| (\gamma) = \hat{0}$,
- $\|1\| (\gamma) = \begin{cases} \hat{1} \ if \ \ \emptyset \in \gamma \\ \hat{0} \ otherwise \end{cases}$,
- $\|p\| (\gamma) = \begin{cases} k_p \ if \ \ \exists \ a \in \gamma \ such \ that \ p \in a \\ \hat{0} \ \ \ otherwise \end{cases}$,
- $\|z_1 \oplus z_2\| (\gamma) = \sum_{a \in \gamma} \left(\|z_1\| (\{a\}) + \|z_2\| (\{a\}) \right)$,
- $\|z_1 \otimes z_2\| (\gamma) = \sum_{a \in \gamma} \left(\sum_{a = a_1 \cup a_2} \left(\|z_1\| (\{a_1\}) \cdot \|z_2\| (\{a_2\}) \right) \right)$,
- $\|(z)\| (\gamma) = \|z\| (\gamma)$.

Recall that an element of the Algebra of Interactions from [3] served to encode a specific interactions set $\gamma \in \Gamma(P)$. On the other hand, a $wAI(P)$ element can be interpreted for any $\gamma \in \Gamma(P)$, inducing a weight from K. This difference results from the semantics given to $p \in P$. In [3], a term p was associated only with $\gamma = \{\{p\}\}$, while in $wAI(P)$, a weighted term p returns its weight k_p whenever it occurs in some interaction $a \in \gamma$, which implies that the port is "activated" with its assigned weight.

Observe that if $\gamma = \emptyset$, then $\|z\|(\gamma) = \hat{0}$ for every $z \in wAI(P)$. We say that $z_1, z_2 \in wAI(P)$ are equivalent and we write $z_1 \equiv z_2$, when $\|z_1\|(\gamma) = \|z_2\|(\gamma)$ for every interactions set $\gamma \in \Gamma(P)$. Obviously, the relation "$\equiv$" is an equivalence relation. We define the quotient set $wAI(P)/\equiv$ of "$\equiv$" on $wAI(P)$ and for every $z \in wAI(P)$ we denote its equivalence class by $\bar{z}$. We define on $wAI(P)/\equiv$ the operations $\overline{z_1} \oplus \overline{z_2} = \overline{z_1 \oplus z_2}$ and $\overline{z_1} \otimes \overline{z_2} = \overline{z_1 \otimes z_2}$ for every $z_1, z_2 \in wAI(P)$. Trivially, these two operations are well-defined, since, if $\overline{z_1} = \overline{z_3}$ and $\overline{z_2} = \overline{z_4}$, then $\overline{z_1} \oplus \overline{z_2} = \overline{z_3} \oplus \overline{z_4}$ and $\overline{z_1} \otimes \overline{z_2} = \overline{z_3} \otimes \overline{z_4}$ for every $z_1, z_2, z_3, z_4 \in wAI(P)$.

In [3], the authors derived the properties of the Algebra of Interactions as axioms obtained directly by syntactic equality. On the contrary, in the weighted setup, we derive an equivalence relation induced by the semantics of the algebra. In turn, we use the equivalence classes of $wAI(P)$ for proving the respective properties.

Next proposition states several nice properties satisfied by $wAI(P)$. As a consequence we get that $(wAI(P)/\equiv, \oplus, \otimes, \bar{0}, \bar{1})$ is a commutative and idempotent semiring, a result which is used in our weighted algebra for connectors. Firstly, we prove the subsequent lemma which is needed for proving the equalities (iii), (iv) and (vii) of Proposition 1.

Lemma 1. *Let $z \in wAI(P)$. Then for every $\gamma \in \Gamma(P)$ it holds that*

$$\sum_{a \in \gamma} \Big(\|z\|(\{a\}) \Big) = \|z\|(\gamma).$$

Proof. For $z = 0$, $z = 1$, $z = p$, and $z = (z_1)$ the proof is direct. For $z = z_1 \oplus z_2$ we have that:

$$\sum_{a \in \gamma} \Big(\|z_1 \oplus z_2\|(\{a\}) \Big) = \sum_{a \in \gamma} \Big(\sum_{a' \in \{a\}} \big(\|z_1\|(\{a'\}) + \|z_2\|(\{a'\}) \big) \Big)$$

$$= \sum_{a \in \gamma} \big(\|z_1\|(\{a\}) + \|z_2\|(\{a\}) \big)$$

$$= \|z_1 \oplus z_2\|(\gamma).$$

Finally, for $z = z_1 \otimes z_2$ we have that:

$$\sum_{a \in \gamma} \Big(\|z_1 \otimes z_2\|(\{a\}) \Big) = \sum_{a \in \gamma} \Big(\sum_{a' \in \{a\}} \Big(\sum_{a' = a_1 \cup a_2} \big(\|z_1\|(\{a_1\}) \cdot \|z_2\|(\{a_2\}) \big) \Big) \Big)$$

$$= \sum_{a \in \gamma} \Big(\sum_{a = a_1 \cup a_2} \big(\|z_1\|(\{a_1\}) \cdot \|z_2\|(\{a_2\}) \big) \Big)$$

$$= \|z_1 \otimes z_2\|(\gamma),$$

and we are done.

Proposition 1. *Let* $\overline{z_1}, \overline{z_2}, \overline{z_3} \in wAI(P)/\equiv$. *Then*

(i) $(\overline{z_1} \oplus \overline{z_2}) \oplus \overline{z_3} = \overline{z_1} \oplus (\overline{z_2} \oplus \overline{z_3})$ *(vi)* $\overline{z_1} \otimes \overline{z_2} = \overline{z_2} \otimes \overline{z_1}$
(ii) $\overline{z_1} \oplus \overline{z_2} = \overline{z_2} \oplus \overline{z_1}$ *(vii)* $\overline{z_1} \otimes \overline{1} = \overline{z_1}$
(iii) $\overline{z_1} \oplus \overline{z_1} = \overline{z_1}$ *(viii)* $\overline{z_1} \otimes \overline{0} = \overline{0}$
(iv) $\overline{z_1} \oplus \overline{0} = \overline{z_1}$ *(ix)* $\overline{z_1} \otimes (\overline{z_2} \oplus \overline{z_3}) = (\overline{z_1} \otimes \overline{z_2}) \oplus (\overline{z_1} \otimes \overline{z_3})$
(v) $(\overline{z_1} \otimes \overline{z_2}) \otimes \overline{z_3} = \overline{z_1} \otimes (\overline{z_2} \otimes \overline{z_3})$ *(x)* $(\overline{z_1} \oplus \overline{z_2}) \otimes \overline{z_3} = (\overline{z_1} \otimes \overline{z_3}) \oplus (\overline{z_2} \otimes \overline{z_3})$.

Proof. For (iii) we have that:

$$\overline{z_1} \oplus \overline{z_1} = \overline{z_1 \oplus z_1}$$
$$= \{z \in wAI(P) \mid z \equiv z_1 \oplus z_1\}$$
$$= \{z \in wAI(P) \mid \|z\|(\gamma) = \|z_1 \oplus z_1\|(\gamma) \text{ for every } \gamma \in \Gamma(P)\}$$
$$= \{z \in wAI(P) \mid \|z\|(\gamma) = \sum_{a \in \gamma} \big(\|z_1\|(\{a\}) + \|z_1\|(\{a\}) \big) \text{ for every } \gamma \in \Gamma(P)\}$$
$$= \{z \in wAI(P) \mid \|z\|(\gamma) = \sum_{a \in \gamma} (\|z_1\|(\{a\})) \text{ for every } \gamma \in \Gamma(P)\}$$
$$= \{z \in wAI(P) \mid \|z\|(\gamma) = \|z_1\|(\gamma) \text{ for every } \gamma \in \Gamma(P)\}$$
$$= \{z \in wAI(P) \mid z \equiv z_1\}$$
$$= \overline{z_1},$$

where the fifth equality holds since K is idempotent and sixth equality holds by a direct application of Lemma 1. The rest of the cases are proved similarly.

Corollary 1. *The structure* $(wAI(P)/\equiv, \oplus, \otimes, \overline{0}, \overline{1})$ *is a commutative and idempotent semiring.*

Next we apply $wAI(P)$ for encoding several coordination schemes in the weighted setup. We choose to compute the weight of each $wAI(P)$ element on $\gamma \in \Gamma(P)$ that contains only the interactions of the scheme. Obviously, by Definition 2, for any other $\gamma \in \Gamma(P)$ our $wAI(P)$ algebra returns the expected weight.

In the following example, we consider the Broadcast scheme implemented in a simple architecture. This allows us to present the detailed computations and the respective tables required for deriving the weight of the resulting $wAI(P)$ element over K.

Example 1. Consider an architecture with a sender and a receiver. The sender and the receiver have a unique port s and r_1 with weight k_s and k_{r_1}, for sending and receiving messages, respectively. Hence, $P = \{s, r_1\}$. If we apply the Broadcast scheme, then we allow all interactions involving the sender and any subset of receivers, possibly the empty one. Thus, the allowed interactions are $\{s\}$ and $\{s, r_1\}$. The $wAI(P)$ element describing the weight of Broadcast is $z = s \otimes (1 \oplus r_1)$. We let $\gamma = \{\{s\}, \{s, r_1\}\} \in \Gamma(P)$. According to Definition 2, we derive the weight of z on $a = \{s\}$ and $a = \{s, r_1\}$ using the Tables 1 and 2, respectively. The first column of the two tables includes all the cases for $a = a_1 \cup a_2$,

Table 1. $s \otimes (1 \oplus r_1)$ and $a = \{s\}$.

$\|s \otimes (1 \oplus r_1)\| (\{a\})$			
$a = a_1 \cup a_2$	$\|s\| (\{a_1\})$	$\|1 \oplus r_1\| (\{a_2\})$	$\cdot$
$a_1 = \emptyset, a_2 = \{s\}$	$\hat{0}$	$\hat{0} + \hat{0}$	$\hat{0}$
$a_1 = \{s\}, a_2 = \emptyset$	k_s	$\hat{1} + \hat{0}$	k_s
$a_1 = \{s\}, a_2 = \{s\}$	k_s	$\hat{0} + \hat{0}$	$\hat{0}$
$+$			k_s

Table 2. $s \otimes (1 \oplus r_1)$ and $a = \{s, r_1\}$.

$\|s \otimes (1 \oplus r_1)\| (\{a\})$			
$a = a_1 \cup a_2$	$\|s\| (\{a_1\})$	$\|1 \oplus r_1\| (\{a_2\})$	$\cdot$
$a_1 = \emptyset, a_2 = \{s, r_1\}$	$\hat{0}$	$\hat{0} + k_{r_1}$	$\hat{0}$
$a_1 = \{s, r_1\}, a_2 = \emptyset$	k_s	$\hat{1} + \hat{0}$	k_s
$a_1 = \{s\}, a_2 = \{r_1\}$	k_s	$\hat{0} + k_{r_1}$	$k_s \cdot k_{r_1}$
$a_1 = \{r_1\}, a_2 = \{s\}$	$\hat{0}$	$\hat{0} + \hat{0}$	$\hat{0}$
$a_1 = \{s\}, a_2 = \{s, r_1\}$	k_s	$\hat{0} + k_{r_1}$	$k_s \cdot k_{r_1}$
$a_1 = \{s, r_1\}, a_2 = \{s\}$	k_s	$\hat{0} + \hat{0}$	$\hat{0}$
$a_1 = \{r_1\}, a_2 = \{s, r_1\}$	$\hat{0}$	$\hat{0} + k_{r_1}$	$\hat{0}$
$a_1 = \{s, r_1\}, a_2 = \{r_1\}$	k_s	$\hat{0} + k_{r_1}$	$k_s \cdot k_{r_1}$
$a_1 = \{s, r_1\}, a_2 = \{s, r_1\}$	k_s	$\hat{0} + k_{r_1}$	$k_s \cdot k_{r_1}$
$+$			$k_s + (k_s \cdot k_{r_1})$

while the second and third one show the semantics of the respective $wAI(P)$ terms of z. In the fourth column, we derive the weight of the respective synchronization. The weight of z on the given $\{a\}$ is shown at the last row of the tables. Then the overall weight of z on γ is computed as follows:

$$
\begin{aligned}
\|z\| (\gamma) &= \|s \otimes (1 \oplus r_1)\| (\gamma) \\
&= \sum_{a \in \gamma} \Big(\sum_{a = a_1 \cup a_2} \big(\|s\| (\{a_1\}) \cdot \|(1 \oplus r_1)\| (\{a_2\}) \big) \Big) \\
&= \sum_{a \in \gamma} \Big(\sum_{a = a_1 \cup a_2} \big(\|s\| (\{a_1\}) \cdot \big(\sum_{a' \in \{a_2\}} (\|1\| (\{a'\}) + \|r_1\| (\{a'\})) \big) \big) \Big) \\
&= \sum_{a \in \gamma} \Big(\sum_{a = a_1 \cup a_2} \big(\|s\| (\{a_1\}) \cdot \big(\|1\| (\{a_2\}) + \|r_1\| (\{a_2\}) \big) \big) \Big) \\
&= k_s + \big(k_s + (k_s \cdot k_{r_1}) \big) \\
&= k_s + (k_s \cdot k_{r_1}).
\end{aligned}
$$

Next we apply $wAI(P)$ for encoding four coordination schemes in the weighted setup. Due to space limitations we omit the computation of the resulting weight and the respective tables, which can be derived as in the previous example.

Example 2. Consider an architecture with a sender and two receivers, each having a single port s and r_1, r_2, for sending and receiving messages, respectively. Hence, $P = \{s, r_1, r_2\}$, and let k_s, k_{r_1}, k_{r_2}, denote the weights of s, r_1, r_2, respectively. We apply the coordination schemes of Rendezvous, Broadcast, Atomic Broadcast and Causality Chain, and we formalize them by $wAI(P)$.

I) Weighted Rendezvous: Rendezvous scheme requires strong synchronization between the sender and each of the receivers. Hence, it allows the single interaction $\{s, r_1, r_2\}$. In order to encode the weight of this scheme, it suffices to apply the weighted synchronization operator among the three involved ports. Therefore, the $wAI(P)$ element describing weighted Rendezvous is $z = s \otimes r_1 \otimes r_2$. Following a methodology similar to the previous example, we can derive that for $\gamma = \{\{s, r_1, r_2\}\} \in \Gamma(P)$ the overall weight of Rendezvous is: $\|z\|(\gamma) = \|s \otimes r_1 \otimes r_2\|(\gamma) = k_s \cdot k_{r_1} \cdot k_{r_2}$. For instance, in Fuzzy semiring, the resulting value represents the minimum of the weights associated with each port in the architecture.

II) Weighted Broadcast: In our example, Broadcast scheme allows the interactions $\{s\}, \{s, r_1\}, \{s, r_2\}$, and $\{s, r_1, r_2\}$. The $wAI(P)$ element encoding weighted Broadcast is $z = s \otimes (1 \oplus r_1) \otimes (1 \oplus r_2)$. Indeed, for $\gamma = \{\{s\}, \{s, r_1\}, \{s, r_2\}, \{s, r_1, r_2\}\} \in \Gamma(P)$ the weight of Broadcast is: $\|z\|(\gamma) = \|s \otimes (1 \oplus r_1) \otimes (1 \oplus r_2)\|(\gamma) = k_s + (k_s \cdot k_{r_1}) + (k_s \cdot k_{r_2}) + (k_s \cdot k_{r_1} \cdot k_{r_2})$. In $\mathbb{R}_{max}$ semiring for instance, this value returns the maximum of the sum of the weights of each interaction on γ.

III) Weighted Atomic Broadcast: In the Atomic Broadcast scheme, a message is either received by all receivers or by none of them. Hence, in our example, the scheme allows the interactions $\{s\}$ and $\{s, r_1, r_2\}$. The $wAI(P)$ element describing weighted Atomic Broadcast is $z = s \otimes (1 \oplus r_1 \otimes r_2)$. We let $\gamma = \{\{s\}, \{s, r_1, r_2\}\} \in \Gamma(P)$. Then the weight of the Atomic Broadcast is $\|z\|(\gamma) = \|s \otimes (1 \oplus r_1 \otimes r_2)\|(\gamma) = k_s + (k_s \cdot k_{r_1} \cdot k_{r_2})$. In Viterbi semiring for instance, this value returns the maximum of the weights of the two interactions when applying the Atomic Broadcast scheme to the architecture.

IV) Weighted Causality Chain: In the Causality Chain scheme, a message is either not received by any of the two receivers, or it is received by the first one of them, or by both of them. Thus, there are three possible interactions, namely $\{s\}, \{s, r_1\}, \{s, r_1, r_2\}$. The $wAI(P)$ element for weighted Causality Chain is $z = s \otimes (1 \oplus r_1 \otimes (1 \oplus r_2))$. We let $\gamma = \{\{s\}, \{s, r_1\}, \{s, r_1, r_2\}\} \in \Gamma(P)$. Then the weight of the scheme on γ is given by $\|z\|(\gamma) = \|s \otimes (1 \oplus r_1 \otimes (1 \oplus r_2))\|(\gamma) = k_s + (k_s \cdot k_{r_1}) + (k_s \cdot k_{r_1} \cdot k_{r_2})$. For instance, in $\mathbb{R}_{min}$ semiring, this value represents for Causality Chain the minimum of the sum of the weights of each interaction on the given γ.

5   The Weighted Algebra of Connectors

In architectures, the components communicate through their ports and their allowed interactions are defined by the imposed coordination scheme. In turn, connectors specify the synchronization constraints among these interactions by relating a set of typed ports. Types extend ports with synchronization modes, and specifically in this work, with Rendezvous and Broadcast mode [3]. Rendezvous requires that all the components interact simultaneously, while in Broadcast, a component initiates the interactions with some of the rest components.

In this section, we are interested in encoding the weight of connectors in architectures. For this, we study the weighted Algebra of Connectors over P and K, that extends $wAI(P)$ with two typing operators, namely triggers and synchrons that correspond to Rendezvous and Broadcast mode, respectively. We prove several properties for $wAC(P)$ and we show that it can encode sufficiently several connectors in the weighted setup.

Definition 3. *Let P be a set of ports such that $0, 1 \notin P$. The syntax of the weighted Algebra of Connectors ($wAC(P)$ for short) over P and K is given by*

$$\sigma ::= [0] \mid [1] \mid [p] \mid [\zeta] \quad (synchron)$$

$$\tau ::= [0]' \mid [1]' \mid [p]' \mid [\zeta]' \quad (trigger)$$

$$\zeta ::= \sigma \mid \tau \mid \zeta \oplus \zeta \mid \zeta \otimes \zeta$$

where $p \in P$, "$\oplus$" denotes the weighted union operator, "$\otimes$" denotes the weighted fusion operator, and "$[\cdot]$", "$[\cdot]'$" are the synchron and trigger unary typing operators, respectively.

Similarly to the Algebra of Connectors introduced in [3], the new operators in $wAC(P)$, specifically, "$[\cdot]$" and "$[\cdot]'$", are assigned the characterization "typing operators" since they encode the type of the synchronization mode applied to the respective ports. Particularly, a trigger is responsible for initiating an interaction, while a synchron requires the simultaneous interaction with other ports. It should be clear that the typing operators in $wAC(P)$ coincide with the ones from the work of [3], since the synchronization mode that they encode is a qualitative feature. The difference is that in this work, the typing operators are applied among connectors in the weighted setup.

We call ζ a $wAC(P)$ connector over P and K. When the latter are understood, we refer to ζ as a $wAC(P)$ connector. Also, we write $[\zeta]^\alpha$ for $\alpha \in \{0, 1\}$ to denote a typed $wAC(P)$ connector. When $\alpha = 0$, ζ is a synchron, otherwise for $\alpha = 1$ it is a trigger. Moreover, we call ζ a *fusion-$wAC(P)$ connector* when $\zeta = [\zeta_1]^{\alpha_1} \otimes \ldots \otimes [\zeta_n]^{\alpha_n}$, where $\zeta_1, \ldots, \zeta_n \in wAC(P)$ and $\alpha_1, \ldots, \alpha_n \in \{0, 1\}$.

Let ζ be a fusion-$wAC(P)$ connector. We denote by $\#_T \zeta$ the number of its trigger elements, which we call the *degree* of ζ. For $\zeta = \bigoplus_{i \in [n]} \zeta_i$, where all ζ_i are fusion-$wAC(P)$ connectors, we let $\#_T \zeta = \max \{\#_T \zeta_i \mid i \in [n]\}$. We say that ζ has a *strictly positive degree* iff $\min \{\#_T \zeta_i \mid i \in [n]\} > 0$. We use the notion of the degree in the semantics of $wAC(P)$ and in Sect. 6, for proving a concept of congruence relation between fusion-$wAC(P)$ connectors.

The intuition behind the $wAC(P)$ semantics, presented in Definition 4, is to encode the weight of a $wAC(P)$ connector according to the coordination scheme imposed on an architecture. For this, we firstly relate a $wAC(P)$ connector ζ with a $wAI(P)$ element and by the semantics of the latter we get the weight of ζ on a given interactions set $\gamma \in \Gamma(P)$. Note that weighted union has the same meaning both in $wAC(P)$ and $wAI(P)$, while weighted fusion is a generalization of weighted synchronization in $wAI(P)$. This is clarified by the semantics of $wAC(P)$ presented below.

Definition 4. *Let ζ be a $wAC(P)$ connector over P and K. The semantics of ζ is a $wAI(P)$ element defined by the function $|\cdot| : wAC(P) \to wAI(P)$ as follows:*

- $\big| [p] \big| = p$, *for* $p \in P \cup \{0,1\}$,
- $\big| [p]' \big| = p$, *for* $p \in P \cup \{0,1\}$,
- $\big| [\zeta] \big| = |\zeta|$,
- $\big| [\zeta]' \big| = |\zeta|$,
- $|\zeta_1 \oplus \zeta_2| = |\zeta_1| \oplus |\zeta_2|$,
- $\big| [\zeta_1] \otimes [\zeta_2] \big| = |\zeta_1| \otimes |\zeta_2|$,
- $\big| [\zeta_1]^{\alpha_1} \otimes \ldots \otimes [\zeta_n]^{\alpha_n} \big| = \displaystyle\bigoplus_{\substack{i \in [n], \\ \alpha_i = 1}} \left(|\zeta_i| \otimes \bigotimes_{\substack{k \neq i, \\ \alpha_k \in \{0,1\}}} (1 \oplus |\zeta_k|) \right)$, *where* $\#_T \big([\zeta_1]^{\alpha_1} \otimes$

  $\ldots \otimes [\zeta_n]^{\alpha_n} \big) > 0$ *and* $\alpha_1, \ldots, \alpha_n \in \{0,1\}$.

Note that the weighted fusion operator in $wAC(P)$ is applied among typed $wAC(P)$ connectors, while this is not the case for the weighted union operator. Also, it holds that $\left| \bigoplus_{i \in [n]} \zeta_i \right| = \bigoplus_{i \in [n]} |\zeta_i|$ and $\left| \bigotimes_{i \in [n]} [\zeta_i] \right| = \bigotimes_{i \in [n]} |\zeta_i|$ for $\zeta_1, \ldots, \zeta_n \in wAC(P)$.

Next we simply write $0, 1, p$, for $[0], [1], [p]$, respectively, and $0', 1', p'$, for $[0]', [1]', [p]'$, respectively. For instance, let $P = \{p, q\}$. Then $[[p]]'$ is written $[p]'$, $[p] \oplus [q]'$ is written $p \oplus q'$, $[[p] \oplus [q]']$ is written $[p \oplus q']$, and $[[p]]' \otimes [[q]]$ is written $[p]' \otimes [q]$.

It should be clear that given a $\zeta \in wAC(P)$, in order to proceed the computations on $|\zeta| \in wAI(P)$ we need to consider the corresponding equivalence class and apply Corollary 1, i.e., that $(wAI(P)/\equiv, \oplus, \otimes, \bar{0}, \bar{1})$ is a commutative and idempotent semiring. Though, for simplicity, in the rest of the paper, we identify $\overline{|\zeta|}$ with the representative $|\zeta|$.

In the next example, we clarify the above conventions in our notations.

Example 3. Let $P = \{p, q, r\}$. We apply the weighted fusion operator to connectors p and $q \oplus r$. For this, we need to specify a typing operator among them. We choose to synchronize the connectors p and $q \oplus r$. Hence, the resulting connector is $[p] \otimes [q \oplus r]$ and its $wAI(P)$ element is computed as follows:
$$\big| [p] \otimes [q \oplus r] \big| = |p| \otimes |q \oplus r| = |p| \otimes (|q| \oplus |r|) = p \otimes (q \oplus r) = (p \otimes q) \oplus (p \otimes r).$$

On the other hand, if we let p serve as a trigger, then we obtain the $wAC(P)$ connector $[p]' \otimes [q \oplus r]$. In turn, its $wAI(P)$ element is computed as follows:
$$\big|\, [p]' \otimes [q \oplus r] \,\big| = |p| \otimes (1 \oplus |q \oplus r|) = |p| \otimes \big(1 \oplus (|q| \oplus |r|)\big) = p \otimes \big(1 \oplus (q \oplus r)\big) = p \otimes (1 \oplus q \oplus r) = p \oplus (p \otimes q) \oplus (p \otimes r).$$

Two connectors $\zeta_1, \zeta_2 \in wAC(P)$ are called equivalent and we write $\zeta_1 \equiv \zeta_2$, when $|\zeta_1| = |\zeta_2|$, i.e., when they return the same $wAI(P)$ elements. Hence, $\|\, |\zeta_1| \,\|(\gamma) = \|\, |\zeta_2| \,\|(\gamma)$ for every interactions set $\gamma \in \Gamma(P)$. Clearly "$\equiv$" is an equivalence relation. We define the quotient set $wAC(P)/\equiv$ of "$\equiv$" on $wAC(P)$ and for every $\zeta \in wAC(P)$ we denote by $\overline{\zeta}$ its equivalence class. We define on $wAC(P)/\equiv$ the operations $\overline{\zeta_1} \oplus \overline{\zeta_2} = \overline{\zeta_1 \oplus \zeta_2}$ and $\overline{[\zeta_1]^\alpha} \otimes \overline{[\zeta_2]^\beta} = \overline{[\zeta_1]^\alpha \otimes [\zeta_2]^\beta}$ for every $\zeta_1, \zeta_2 \in wAC(P)$ and $\alpha, \beta \in \{0,1\}$. The operations are well-defined, since, if $\overline{\zeta_1} = \overline{\zeta_3}$ and $\overline{\zeta_2} = \overline{\zeta_4}$, then $\overline{\zeta_1} \oplus \overline{\zeta_2} = \overline{\zeta_3} \oplus \overline{\zeta_4}$ and $\overline{[\zeta_1]^\alpha} \otimes \overline{[\zeta_2]^\beta} = \overline{[\zeta_3]^\alpha} \otimes \overline{[\zeta_4]^\beta}$ for every $\zeta_1, \zeta_2, \zeta_3, \zeta_4 \in wAC(P)$ and $\alpha, \beta \in \{0,1\}$.

In [3], the authors proved several properties for the Algebra of Connectors using axioms. Next, we show that $wAC(P)$ acknowledges the respective properties in the weighted setup. For this, we use the equivalence classes of $wAC(P)$.

By the following result we derive that connector [1] serves as the neutral element of weighted fusion operator as well as that the weighted union and weighted fusion operators satisfy the commutativity property with respect to the typing operator.

Proposition 2. *Let* $\zeta_1, \zeta_2 \in wAC(P)$ *and* $\alpha, \beta \in \{0,1\}$. *Then*

(i) $\overline{[\zeta_1]^\alpha} \otimes \overline{[1]} = \overline{[\zeta_1]^\alpha} = \overline{[1]} \otimes \overline{[\zeta_1]^\alpha}$

(ii) $\overline{[\zeta_1]^\alpha} \oplus \overline{[\zeta_2]^\beta} = \overline{[\zeta_2]^\beta} \oplus \overline{[\zeta_1]^\alpha}$

(iii) $\overline{[\zeta_1]^\alpha} \otimes \overline{[\zeta_2]^\beta} = \overline{[\zeta_2]^\beta} \otimes \overline{[\zeta_1]^\alpha}$.

Proof. We use the equivalence classes of $wAC(P)$ as well as we apply Definition 4 and Corollary 1.

Next proposition states some nice properties for $wAC(P)/\equiv$ that allow simplifying the semantics of $wAC(P)$. Specifically, (iii) and (iv) state the associativity property of weighted fusion operator when a specific type, synchron or trigger, is applied for simple grouping.

Proposition 3. *Let* $\zeta_1, \zeta_2, \zeta_3 \in wAC(P)$ *and* $\alpha, \beta \in \{0,1\}$. *Then*

(i) $\overline{\big[\,[\zeta_1]^\alpha\,\big]^\beta} = \overline{[\zeta_1]^\beta}$

(ii) $\overline{[\zeta_1 \oplus \zeta_2]^\alpha} = \overline{[\zeta_1]^\alpha} \oplus \overline{[\zeta_2]^\alpha}$

(iii) $\overline{\big[\,[\zeta_1] \otimes [\zeta_2]\,\big] \otimes [\zeta_3]} = \overline{[\zeta_1] \otimes \big[\,[\zeta_2] \otimes [\zeta_3]\,\big]}$

(iv) $\overline{\big[\,[\zeta_1]' \otimes [\zeta_2]'\,\big]' \otimes [\zeta_3]'} = \overline{[\zeta_1]' \otimes \big[\,[\zeta_2]' \otimes [\zeta_3]'\,\big]'}$.

Proof. We use the equivalence classes of $wAC(P)$ as well as we apply Definition 4 and Corollary 1.

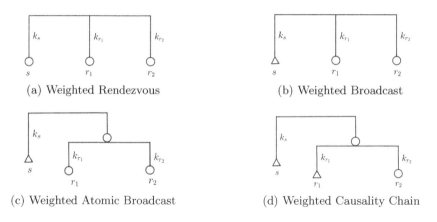

(a) Weighted Rendezvous

(b) Weighted Broadcast

(c) Weighted Atomic Broadcast

(d) Weighted Causality Chain

Fig. 1. Representation of $wAC(P)$ connectors.

In the sequel, we apply $wAC(P)$ for modeling several connectors in the weighted setup. For the resulting $wAC(P)$ connectors we follow the representation considered in [3]. In particular, we use triangles and circles in order to represent triggers and synchrons, respectively. Then we connect the involved $wAC(P)$ connectors with lines which are labeled by the respective weight from K. In turn, we draw the resulting $wAC(P)$ connector incrementally.

Example 4. Consider the architecture of Example 2 with one sender and two receivers. We have that $P = \{s, r_1, r_2\}$ is the set of ports and k_s, k_{r_1}, k_{r_2}, denote the weight of s, r_1, r_2, respectively. We define the $wAC(P)$ connectors for Rendezvous, Broadcast, Atomic Broadcast and Causality Chain scheme.

I) Weighted Rendezvous: In Rendezvous scheme, the involved components should be strongly synchronized. Hence, the respective connector should restrict to the synchron typing without applying any trigger typing. Then a $wAC(P)$ connector should encode the "cost" for this synchronization. Thus, the $wAC(P)$ connector for weighted Rendezvous in our example, is given by $\zeta = [s] \otimes [r_1] \otimes [r_2]$ and is shown in Fig. 1a. In particular, the $wAC(P)$ connectors s, r_1, r_2, are represented with circles, since they are all typed with synchrons. Moreover, the connection is performed simultaneously, and hence the three $wAC(P)$ connectors occur at the same level. Then the $wAI(P)$ element of ζ is computed as follows: $|\zeta| = |[s] \otimes [r_1] \otimes [r_2]| = |s| \otimes |r_1| \otimes |r_2| = s \otimes r_1 \otimes r_2$.

II) Weighted Broadcast: In a Broadcast scheme, a component triggers the interaction with some of the rest components, and hence a connector should assign a trigger typing along with some synchrons, respectively. For our example, s is typed with trigger, while r_1 and r_2 are typed with synchrons. In turn, we apply the weighted fusion operator among these three $wAC(P)$ connectors in order to encode the corresponding weight. Therefore, the resulting $wAC(P)$ connector for the weighted Broadcast scheme is $\zeta = [s]' \otimes [r_1] \otimes [r_2]$. The connector is shown in Fig. 1b, where the trigger for s and the synchrons for

r_1 and r_2 are indicated by a triangle and two circles, respectively, all occurring at the same level. Then we derive the $wAI(P)$ element of ζ as follows: $|\zeta| = \big|[s]' \otimes [r_1] \otimes [r_2]\big| = |s| \otimes (1 \oplus |r_1|) \otimes (1 \oplus |r_2|) = s \otimes (1 \oplus r_1) \otimes (1 \oplus r_2)$.

III) Weighted Atomic Broadcast: In Atomic Broadcast scheme, a component initiates the interactions either with all the other components or with none of them. In turn, for our example, a $wAC(P)$ connector should encode the weight for synchronizing the trigger of the sender $[s]'$ and the synchron of the two receivers $[[r_1] \otimes [r_2]]$. The $wAC(P)$ connector $[r_1] \otimes [r_2]$ encodes the weight for the simultaneous connection of the synchrons $[r_1]$ and $[r_2]$. Hence, the $wAC(P)$ connector for the weighted Atomic Broadcast is $\zeta = [s]' \otimes [[r_1] \otimes [r_2]]$, and is presented in Fig. 1c. The synchrons for r_1 and r_2 are denoted with circles and occur at the same level, the trigger for s is represented with a triangle, and a circle is used for the synchronization of the latter with $[[r_1] \otimes [r_2]]$. The corresponding $wAI(P)$ element is computed as follows: $|\zeta| = \big|[s]' \otimes [[r_1] \otimes [r_2]]\big| = |s| \otimes \big(1 \oplus |[r_1] \otimes [r_2]|\big) = |s| \otimes \big(1 \oplus |r_1| \otimes |r_2|\big) = s \otimes (1 \oplus r_1 \otimes r_2)$.

IV) Weighted Causality Chain: In Causality chain scheme, each component initiates the interaction with the rest of the components but the "last" one. Hence, for our example, a $wAC(P)$ connector should apply a trigger typing to the sender s and the receiver r_1. For this, in Fig. 1d, triangles are used to denote the trigger for s and r_1, while the synchron for r_2 is denoted by a circle. Then we synchronize $[r_1]'$ with $[r_2]$ and in turn, we synchronize $[s]'$ with $[[r_1]' \otimes [r_2]]$. The two synchronizations are depicted with a circle in Fig. 1d, at lower and higher level, respectively. Hence, the $wAC(P)$ connector for the weighted Causality Chain is $\zeta = [s]' \otimes [[r_1]' \otimes [r_2]]$ and its $wAI(P)$ element is computed as follows: $|\zeta| = \big|[s]' \otimes [[r_1]' \otimes [r_2]]\big| = |s| \otimes \big(1 \oplus |[r_1]' \otimes [r_2]|\big) = |s| \otimes \big(1 \oplus |r_1| \otimes (1 \oplus |r_2|)\big) = s \otimes \big(1 \oplus r_1 \otimes (1 \oplus r_2)\big)$.

6   On Congruence Relation for Fusion-$wAC(P)$ Connectors

In this section, we are interested in the congruence problem of weighted connectors. In [3], the authors achieved to introduce a congruence relation for their Algebra of Connectors and provided conditions for proving congruence. Congruence relation is important because in contrast to equivalence relation, it allows to use connectors interchangeably whenever it is required and without causing undesirable alterations in the architecture.

However, it occurs that providing a congruence relation in the weighted setup, it is not an easy task. Congruence relation implies that given two equivalent elements, if we apply any operator from the algebra on them, then we obtain again two equivalent elements [2,6,8]. In our setting, applying the weighted fusion operator on $wAC(P)$ connectors would require specifying a typing. In [3], the authors resolved this issue using the syntactic equality resulting by the several axioms defined for the connectors. In the weighted framework, we can only derive results by semantic equivalence and hence, we cannot follow a similar method. For this, we need to restrict the problem on fusion-$wAC(P)$ connectors which are typed by definition. In turn, we propose a concept of congruence relation that allows replacing a fusion-$wAC(P)$ connector with an alternative one.

Note that our concept of congruence relation is not a "true" congruence in terms that (i) it does not apply to any $wAC(P)$ connector and (ii) it takes equivalent fusion-$wAC(P)$ connectors and returns equivalent $wAC(P)$ connectors. Extending our results for studying a congruence relation for $wAC(P)$ connectors in general, is an interesting open problem that is left as future work.

Next we show that two equivalent fusion-$wAC(P)$ connectors are not in general interchangeable. In turn, we define a concept of congruence relation for fusion-$wAC(P)$ connectors, and we provide two theorems for proving such a congruence by extending the respective results from [3] in the weighted setup.

Example 5. Let $P = \{p, q\}$, where the ports p, q, have weights k_p, k_q, respectively. The fusion-$wAC(P)$ connectors $[p]'$ and $[p]$ are equivalent, i.e., $[p]' \equiv [p]$, since $\left|[p]'\right| = p$ and $\left|[p]\right| = p$, respectively. Though, we show that the $wAC(P)$ connectors $[p]' \otimes [q]$ and $[p] \otimes [q]$ are not equivalent. Indeed, their $wAI(P)$ elements are: $\left|[p]' \otimes [q]\right| = |p| \otimes (1 \oplus |q|) = p \otimes (1 \oplus q) = p \oplus (p \otimes q)$ and $\left|[p] \otimes [q]\right| = |p| \otimes |q| = p \otimes q$, respectively. We let $\gamma = \{\{p\}, \{p, q\}\} \in \Gamma(P)$. For the former $wAI(P)$ element we have that $\|p \oplus (p \otimes q)\|(\gamma) = k_p + (k_p \cdot k_q)$, while for the latter we get that $\|p \otimes q\|(\gamma) = k_p \cdot k_q$.

Now we define a concept of a congruence relation for fusion-$wAC(P)$ connectors.

Definition 5. *We denote by "$\cong$" the largest congruence relation for fusion-$wAC(P)$ connectors, contained in $\equiv$ of $wAC(P)$, i.e., the largest relation satisfying the following: For fusion-$wAC(P)$ connectors ζ_1, ζ_2 and $r \notin P$,*

$$\zeta_1 \cong \zeta_2 \Rightarrow \forall E \in wAC\,(P \cup \{r\})\,, E(\zeta_1/r) \equiv E(\zeta_2/r)$$

where $E(\zeta/r)$ denotes the expression obtained from E by replacing all occurrences of $wAC(P)$ connector r by ζ.

By Example 5, we obtain that the weighted fusion operator does not preserve the equivalence of fusion-$wAC(P)$ connectors. Thus, we need further conditions for proving their congruence. According to the following theorem, it suffices to assign identical typing on equivalent fusion-$wAC(P)$ connectors.

Theorem 1. *Let ζ_1, ζ_2 be fusion-$wAC(P)$ connectors. Then*

$$\zeta_1 \equiv \zeta_2 \Leftrightarrow [\zeta_1]^\alpha \cong [\zeta_2]^\alpha\,, \text{ for any } \alpha \in \{0, 1\}\,.$$

Proof. The right-to-left implication is direct. Assume now that $\zeta_1 \equiv \zeta_2$, i.e., $|\zeta_1| = |\zeta_2|$. We have to prove that $[\zeta_1]^\alpha \cong [\zeta_2]^\alpha$, i.e., for every $E \in wAC\,(P \cup \{r\})$ it holds that $E([\zeta_1]^\alpha/r) \equiv E([\zeta_2]^\alpha/r)$. We assume that r occurs only once in E, otherwise we apply the proof iteratively. It suffices to consider and prove the following equivalences: (1) $[\zeta_1]^\alpha \equiv [\zeta_2]^\alpha$, (2) $[\zeta_1]^\alpha \oplus \zeta \equiv [\zeta_2]^\alpha \oplus \zeta$, (3) $[\zeta_1]^\alpha \otimes$

$[\xi_1]^{\alpha_1} \otimes \ldots \otimes [\xi_n]^{\alpha_n} \equiv [\zeta_2]^\alpha \otimes [\xi_1]^{\alpha_1} \otimes \ldots \otimes [\xi_n]^{\alpha_n}$, and (4) the symmetric case for (2) and the cases for any other position of $[\zeta_1]^\alpha$, $[\zeta_2]^\alpha$ in (3), where $\zeta, \xi_1, \ldots, \xi_n \in wAC(P)$ and $\alpha, \alpha_1, \ldots, \alpha_n \in \{0, 1\}$. The proof of cases (1) and (2) is direct. Next we prove case (3) and we let $\#_T([\xi_1]^{\alpha_1} \otimes \ldots \otimes [\xi_n]^{\alpha_n}) > 0$. For $\alpha = 0$, we have:

$$\left| [\zeta_1] \otimes [\xi_1]^{\alpha_1} \otimes \ldots \otimes [\xi_n]^{\alpha_n} \right|$$

$$= \bigoplus_{\substack{i \in [n], \\ \alpha_i = 1}} \left(|\xi_i| \otimes \bigotimes_{\substack{k \neq i, \\ \alpha_k \in \{0,1\}}} (1 \oplus |\xi_k|) \otimes (1 \oplus |\zeta_1|) \right)$$

$$= \bigoplus_{\substack{i \in [n], \\ \alpha_i = 1}} \left(|\xi_i| \otimes \bigotimes_{\substack{k \neq i, \\ \alpha_k \in \{0,1\}}} (1 \oplus |\xi_k|) \otimes (1 \oplus |\zeta_2|) \right)$$

$$= \left| [\zeta_2] \otimes [\xi_1]^{\alpha_1} \otimes \ldots \otimes [\xi_n]^{\alpha_n} \right|.$$

Now, for $\alpha = 1$, we have:

$$\left| [\zeta_1]' \otimes [\xi_1]^{\alpha_1} \otimes \ldots \otimes [\xi_n]^{\alpha_n} \right|$$

$$= \left(|\zeta_1| \otimes \bigotimes_{l \in [n]} (1 \oplus |\xi_l|) \right) \oplus \bigoplus_{\substack{i \in [n], \\ \alpha_i = 1}} \left(|\xi_i| \otimes \bigotimes_{\substack{k \neq i, \\ \alpha_k \in \{0,1\}}} (1 \oplus |\xi_k|) \otimes (1 \oplus |\zeta_1|) \right)$$

$$= \left(|\zeta_2| \otimes \bigotimes_{l \in [n]} (1 \oplus |\xi_l|) \right) \oplus \bigoplus_{\substack{i \in [n], \\ \alpha_i = 1}} \left(|\xi_i| \otimes \bigotimes_{\substack{k \neq i, \\ \alpha_k \in \{0,1\}}} (1 \oplus |\xi_k|) \otimes (1 \oplus |\zeta_2|) \right)$$

$$= \left| [\zeta_2]' \otimes [\xi_1]^{\alpha_1} \otimes \ldots \otimes [\xi_n]^{\alpha_n} \right|.$$

The omitted cases for (3) as well as case (4) are proved similarly. Finally, for any other form of the expression E we apply the presented cases iteratively, and our proof is completed.

Next we prove a second theorem that provides a method for checking the congruence of two fusion-$wAC(P)$ connectors. According to this result, congruence is ensured for two equivalent fusion-$wAC(P)$ connectors with simultaneously zero or strictly positive degree, when their weighted fusion with $[1]'$ preserves their equivalence. For the proof of the theorem, we need the next proposition.

Proposition 4. *Let* $\zeta = [\zeta_1]^{\alpha_1} \otimes \ldots \otimes [\zeta_n]^{\alpha_n}$, *where* $\zeta_i \in wAC(P)$ *and* $\alpha_i \in \{0, 1\}$ *for* $i \in [n]$. *Then*

$$|\zeta| \oplus \bigotimes_{i \in [n]} (1 \oplus |\zeta_i|) = \bigotimes_{i \in [n]} (1 \oplus |\zeta_i|).$$

Proof. It occurs by induction on n and by applying Corollary 1.

Theorem 2. *Let ζ_1, ζ_2 be fusion-$wAC(P)$ connectors. Then*

$$\zeta_1 \cong \zeta_2 \Leftrightarrow \begin{cases} \zeta_1 \equiv \zeta_2 \\ \zeta_1 \otimes [1]' \equiv \zeta_2 \otimes [1]' \\ \#_T\zeta_1 > 0 \Leftrightarrow \#_T\zeta_2 > 0. \end{cases}$$

Proof. The left-to-right implication is direct. For the other direction we have to prove that for every $E \in wAC(P \cup \{r\})$ it holds that $E(\zeta_1/r) \equiv E(\zeta_2/r)$. Since ζ_1 and ζ_2 are fusion-$wAC(P)$ connectors, we let $\zeta_1 = [\zeta_{1,1}]^{\alpha_1} \otimes \ldots \otimes [\zeta_{1,n}]^{\alpha_n}$ and $\zeta_2 = [\zeta_{2,1}]^{\beta_1} \otimes \ldots \otimes [\zeta_{2,m}]^{\beta_m}$, where $\zeta_{1,i}, \zeta_{2,j} \in wAC(P)$ and $\alpha_i, \beta_j \in \{0,1\}$ for $i \in [n]$ and $j \in [m]$. Let r occur only once in E, and otherwise apply the proof iteratively. It suffices to consider and prove the equivalences: (1) $\zeta_1 \equiv \zeta_2$, (2) $\zeta_1 \oplus \zeta \equiv \zeta_2 \oplus \zeta$, (3) $\zeta_1 \otimes [\xi_1]^{\delta_1} \otimes \ldots \otimes [\xi_r]^{\delta_r} \equiv \zeta_2 \otimes [\xi_1]^{\delta_1} \otimes \ldots \otimes [\xi_r]^{\delta_r}$, and (4) the symmetric case for (2) and the cases for any other position of ζ_1, ζ_2 in (3), where $\zeta, \xi_1, \ldots, \xi_n \in wAC(P)$ and $\delta_1, \ldots, \delta_r \in \{0,1\}$. The proof of cases (1) and (2) is direct. We prove case (3) and we assume that $\#_T\zeta_1 > 0, \#_T\zeta_2 > 0$. By $\zeta_1 \equiv \zeta_2$ we have that $|\zeta_1| = |\zeta_2|$. In turn, it holds that:

$$|\zeta_1| = |\zeta_2| \Rightarrow \bigoplus_{\substack{i_1 \in [n], \\ \alpha_{i_1} = 1}} \left(|\zeta_{1,i_1}| \otimes \bigotimes_{\substack{k_1 \neq i_1, \\ \alpha_{k_1} \in \{0,1\}}} (1 \oplus |\zeta_{1,k_1}|) \right)$$

$$= \bigoplus_{\substack{i_2 \in [m], \\ \beta_{i_2} = 1}} \left(|\zeta_{2,i_2}| \otimes \bigotimes_{\substack{k_2 \neq i_2, \\ \alpha_{k_2} \in \{0,1\}}} (1 \oplus |\zeta_{2,k_2}|) \right). \qquad (\Sigma_1)$$

Also, by $\zeta_1 \otimes [1]' \equiv \zeta_2 \otimes [1]'$ we get $|\zeta_1 \otimes [1]'| = |\zeta_2 \otimes [1]'|$. In turn, it holds that:

$$|\zeta_1 \otimes [1]'| = |\zeta_2 \otimes [1]'|$$

$$\Rightarrow \bigoplus_{\substack{i_1 \in [n], \\ \alpha_{i_1} = 1}} \left(|\zeta_{1,i_1}| \otimes \bigotimes_{\substack{k_1 \neq i_1, \\ \alpha_{k_1} \in \{0,1\}}} (1 \oplus |\zeta_{1,k_1}|) \right) \oplus \bigotimes_{i \in [n]} (1 \oplus |\zeta_{1,i}|)$$

$$= \bigoplus_{\substack{i_2 \in [m], \\ \beta_{i_2} = 1}} \left(|\zeta_{2,i_2}| \otimes \bigotimes_{\substack{k_2 \neq i_2, \\ \beta_{k_2} \in \{0,1\}}} (1 \oplus |\zeta_{2,k_2}|) \right) \oplus \bigotimes_{j \in [m]} (1 \oplus |\zeta_{2,j}|)$$

$$\Rightarrow \bigotimes_{i \in [n]} (1 \oplus |\zeta_{1,i}|) = \bigotimes_{j \in [m]} (1 \oplus |\zeta_{2,j}|), \qquad (\Sigma_2)$$

where the last step holds by Proposition 4. We assume that $\#_T\big([\xi_1]^{\delta_1} \otimes \ldots \otimes [\xi_r]^{\delta_r}\big) > 0$. Then

$$
\begin{aligned}
|\zeta_1 \otimes \xi| &= \bigoplus_{\substack{i_1 \in [n], \\ \alpha_{i_1}=1}} \bigg(|\zeta_{1,i_1}| \otimes \bigotimes_{\substack{k_1 \neq i_1, \\ \alpha_{k_1} \in \{0,1\}}} (1 \oplus |\zeta_{1,k_1}|) \otimes \bigotimes_{l \in [r]}(1 \oplus |\xi_l|)\bigg) \oplus \\
&\quad \bigoplus_{\substack{\lambda \in [r], \\ \delta_\lambda=1}} \bigg(|\xi_\lambda| \otimes \bigotimes_{\substack{\mu \neq \lambda, \\ \delta_\mu \in \{0,1\}}} (1 \oplus |\xi_\mu|) \otimes \bigotimes_{i \in [n]}(1 \oplus |\zeta_{1,i}|)\bigg) \\
&= \bigg(\bigoplus_{\substack{i_1 \in [n], \\ \alpha_{i_1}=1}} \Big(|\zeta_{1,i_1}| \otimes \bigotimes_{\substack{k_1 \neq i_1, \\ \alpha_{k_1} \in \{0,1\}}} (1 \oplus |\zeta_{1,k_1}|)\Big)\bigg) \otimes \bigotimes_{l \in [r]}(1 \oplus |\xi_l|) \oplus \\
&\quad \bigg(\bigoplus_{\substack{\lambda \in [r], \\ \delta_\lambda=1}} \Big(|\xi_\lambda| \otimes \bigotimes_{\substack{\mu \neq \lambda, \\ \delta_\mu \in \{0,1\}}} (1 \oplus |\xi_\mu|)\Big)\bigg) \otimes \bigotimes_{i \in [n]}(1 \oplus |\zeta_{1,i}|) \\
&\overset{\Sigma_1}{\underset{\Sigma_2}{=}} \bigg(\bigoplus_{\substack{i_2 \in [m], \\ \beta_{i_2}=1}} \Big(|\zeta_{2,i_2}| \otimes \bigotimes_{\substack{k_2 \neq i_2, \\ \beta_{k_2} \in \{0,1\}}} (1 \oplus |\zeta_{2,k_2}|)\Big)\bigg) \otimes \bigotimes_{l \in [r]}(1 \oplus |\xi_l|) \oplus \\
&\quad \bigg(\bigoplus_{\substack{\lambda \in [r], \\ \delta_\lambda=1}} \Big(|\xi_\lambda| \otimes \bigotimes_{\substack{\mu \neq \lambda, \\ \delta_\mu \in \{0,1\}}} (1 \oplus |\xi_\mu|)\Big)\bigg) \otimes \bigotimes_{j \in [m]}(1 \oplus |\zeta_{2,j}|) \\
&= |\zeta_2 \otimes \xi|.
\end{aligned}
$$

The omitted cases for (3) as well as case (4) are proved similarly. Finally, for any other form of the expression E we apply the presented cases iteratively, and our proof is completed.

The congruences in the next proposition, are derived by Theorem 2.

Proposition 5. *Let $\zeta, \zeta_1, \zeta_2, \zeta_3$ be fusion-$wAC(P)$ connectors where $\#_T\zeta > 0$. Then*

(i) $\zeta \otimes [0']' \cong \zeta$
(ii) $[\zeta_1]' \otimes [\zeta_2] \otimes [\zeta_3] \cong [\zeta_1]' \otimes [[\zeta_2]' \otimes [\zeta_3]']$
(iii) $[\zeta_1]' \otimes [\zeta_2]' \cong [[\zeta_1]' \otimes [\zeta_2]']'$.

Proof. We apply Theorem 2 and Corollary 1.

7   Discussion

In [3], the authors proved the soundness of their algebras and investigated the conditions under which completeness also holds. Proving such results in the weighted setting, is in general, much harder. According to [12], soundness has been only defined for multi-valued logics with values in the bounded distributive lattice $[0, 1]$ with the usual max and min operations (cf. [10]). In turn, in [12], the authors introduced a notion of soundness in the context of weighted propositional configuration logic formulas, with weights ranging over a commutative semiring. That formulas served for encoding the quantitative features of architectures styles.

Following the work of [12], we could provide an analogous definition of soundness for our weighted algebras. In this case, it occurs that proving soundness would require semiring K to be idempotent with respect to its first and second operation. Idempotency for the second operation of K is required by the weighted synchronization and fusion operators in $wAI(P)$ and $wAC(P)$, respectively. However, in this paper, K is idempotent only with respect to its first operation. A further investigation of soundness for our algebras along with the consideration of other algebraic structures is left as future work.

On the other hand, the notion of completeness does not comply in general, in the weighted setup. Indeed, due to the presence of weights we cannot ensure that two arbitrary constructs with the same weight have also the same syntax. In our setting, let for instance $z_1, z_2 \in wAI(P)$. Then z_1 and z_2 can return the same weight, while they encode different coordination schemes.

8 Conclusion

In this paper, we extended the results of [3] in the weighted setup. Specifically, we studied the weighted Algebra of Interactions over a set of ports P and a commutative and idempotent semiring K, $wAI(P)$, that was interpreted by polynomials in $K \langle \Gamma(P) \rangle$. We proved that the structure $(wAI(P)/ \equiv, \oplus, \otimes, \bar{0}, \bar{1})$ is a commutative and idempotent semiring, and we applied $wAI(P)$ for encoding the weight of well-known coordination schemes. In turn, we studied the weighted Algebra of Connectors over P and K, $wAC(P)$, whose semantics were defined as $wAI(P)$ elements. In turn, from the semantics of the latter we can derive the weight of $wAC(P)$ connectors for a given interactions set over P. We proved several properties for $wAC(P)$ and we showed the expressiveness of the algebra by modeling several connectors in the weighted setup. Finally, we defined a concept of congruence relation for fusion-$wAC(P)$ connectors and we derived two theorems for proving such a congruence.

There are several directions for future work. An important open problem is providing a congruence relation for $wAC(P)$ connectors in general, as well as investigating a different weighted framework for connectors in order to solve their congruence problem. Future work is also studying our weighted algebras over alternative structures than K, in order to prove their soundness. Moreover, in [4,5], the authors used glue operators as composition operators and formalized the behavior of the interacting components. Therefore, future research includes modeling both the behavior and the coordination of component-based systems in the weighted setup. On the other hand, several theories have formalized connectors as stateful entities whose interactions may be modified during a system's operation [2,13]. In other words, it would be interesting to extend our results for connectors with dynamic interactions. In addition to these theoretical directions, future work includes implementing the presented formal framework.

Acknowledgement. We are deeply grateful to the anonymous referees for their constructive comments and suggestions that brought the paper in its current form, especially regarding the results of Sect. 6.

References

1. Arbab, F.: Reo: a channel-based coordination model for component composition. Math. Struct. Comput. Sci. **14**(3), 329–366 (2004). https://doi.org/10.1017/S0960129504004153
2. Bernardo, M., Ciancarini, P., Donatiello, L.: On the formalization of architectural types with process algebras. In: Knight, J.C., Rosenblum, D.S. (eds.) ACM SIGSOFT, pp. 140–148. ACM (2000). https://doi.org/10.1145/355045.355064
3. Bliudze, S., Sifakis, J.: The algebra of connectors - structuring interaction in BIP. IEEE Trans. Computers **57**(10), 1315–1330 (2008). https://doi.org/10.1109/TC.2008.26
4. Bliudze, S., Sifakis, J.: A notion of glue expressiveness for component-based systems. In: van Breugel, F., Chechik, M. (eds.) CONCUR 2008. LNCS, vol. 5201, pp. 508–522. Springer, Heidelberg (2008). https://doi.org/10.1007/978-3-540-85361-9_39
5. Bliudze, S., Sifakis, J.: Synthesizing glue operators from glue constraints for the construction of component-based systems. In: Apel, S., Jackson, E. (eds.) SC 2011. LNCS, vol. 6708, pp. 51–67. Springer, Heidelberg (2011). https://doi.org/10.1007/978-3-642-22045-6_4
6. Bruni, R., Lanese, I., Montanari, U.: A basic algebra of stateless connectors. Theor. Comput. Sci. **366**(1–2), 98–120 (2006). https://doi.org/10.1016/j.tcs.2006.07.005
7. Bruni, R., Melgratti, H., Montanari, U.: A survey on basic connectors and buffers. In: Beckert, B., Damiani, F., de Boer, F.S., Bonsangue, M.M. (eds.) FMCO 2011. LNCS, vol. 7542, pp. 49–68. Springer, Heidelberg (2013). https://doi.org/10.1007/978-3-642-35887-6_3
8. Droste, M., Kuich, W., Vogler, H. (eds.): Handbook of Weighted Automata. Springer, Heidelberg (2009). https://doi.org/10.1007/978-3-642-01492-5
9. Fountoukidou, C.C., Pittou, M.: A formal algebraic approach for the quantitative modeling of connectors in architectures (2022). arXiv:2202.06594
10. Hájek, P. (ed.): Metamathematics of Fuzzy Logic. Kluwer Academic Publishers (1998). https://doi.org/10.1007/978-94-011-5300-3
11. Nawaz, M.S., Sun, M.: Using PVS for modeling and verification of probabilistic connectors. In: Hojjat, H., Massink, M. (eds.) FSEN 2019. LNCS, vol. 11761, pp. 61–76. Springer, Cham (2019). https://doi.org/10.1007/978-3-030-31517-7_5
12. Paraponiari, P., Rahonis, G.: Weighted propositional configuration logics: A specification language for architectures with quantitative features. Inform. Comput. **282** (2022). https://doi.org/10.1016/j.ic.2020.104647
13. Rouland, Q., Hamid, B., Jaskolka, J.: Formalizing reusable communication models for distributed systems architecture. In: Abdelwahed, E.H., Bellatreche, L., Golfarelli, M., Méry, D., Ordonez, C. (eds.) MEDI 2018. LNCS, vol. 11163, pp. 198–216. Springer, Cham (2018). https://doi.org/10.1007/978-3-030-00856-7_13

14. Sifakis, J.: Rigorous systems design. Found. Trends Signal Process. **6**(4), 293–362 (2013). https://doi.org/10.1561/1000000034
15. Sun, M., Zhang, X.: A relational model for probabilistic connectors based on timed data distribution streams. In: Jansen, D.N., Prabhakar, P. (eds.) FORMATS 2018. LNCS, vol. 11022, pp. 125–141. Springer, Cham (2018). https://doi.org/10.1007/978-3-030-00151-3_8

Watson-Crick Powers of a Word

Lila Kari[1]($\boxtimes$) and Kalpana Mahalingam[2] 🆔

[1] School of Computer Science, University of Waterloo, Waterloo, Canada
lila@uwaterloo.ca
[2] Department of Mathematics, Indian Institute of Technology Madras,
Chennai, India
kmahalingam@iitm.ac.in

Abstract. In this paper we define and investigate the binary word operation of strong-θ-catenation (denoted by $\otimes$) where θ is an antimorphic involution modelling the Watson-Crick complementarity of DNA single strands. When iteratively applied to a word u, this operation generates all the strong-θ-powers of u (defined as any word in $\{u, \theta(u)\}^{+}$), which amount to all the Watson-Crick powers of u when $\theta = \theta_{DNA}$ (the antimorphic involution on the DNA alphabet $\Delta = \{A, C, G, T\}$ that maps A to T and C to G). In turn, the Watson-Crick powers of u represent DNA strands usually undesirable in DNA computing, since they attach to themselves via intramolecular Watson-Crick complementarity that binds u to $\theta_{DNA}(u)$, and thus become unavailable for other computational interactions. We find necessary and sufficient conditions for two words u and v to commute with respect to the operation of strong-θ-catenation. We also define the concept of $\otimes$-primitive root pair of a word, and prove that it always exists and is unique.

Keywords: DNA computing · Molecular computing · Binary word operations · Algebraic properties

1 Introduction

Periodicity and primitivity of words are fundamental properties in combinatorics on words and formal language theory. Motivated by DNA computing, and the properties of information encoded as DNA strands, Czeizler, Kari, and Seki proposed and investigated the notion of pseudo-primitivity (and pseudo-periodicity) of words in [1,7]. The motivation was that one of the particularities of information-encoding DNA strands is that a word u over the DNA alphabet $\{A, C, G, T\}$ contains basically the same information as its Watson-Crick complement. Thus, in a sense, a DNA word and its Watson-Crick complement are "identical," and notions such as periodicity, power of a word, and primitivity can be generalized by replacing the identity function (producing powers of a word), by a function that models Watson-Crick complementarity (producing

This work was partially supported by Natural Sciences and Engineering Research Council of Canada (NSERC) Discovery Grant R2824A01 to L.K.

© The Author(s), under exclusive license to Springer Nature Switzerland AG 2022
D. Poulakis and G. Rahonis (Eds.): CAI 2022, LNCS 13706, pp. 136–148, 2022.
https://doi.org/10.1007/978-3-031-19685-0_10

pseudo-powers). Traditionally, Watson-Crick complementarity has been modelled mathematically by the antimorphic involution θ_{DNA} over the DNA alphabet $\Sigma = \{A, C, G, T\}$, that maps A to T, C to G and viceversa. Recall that a function θ is an antimorphism on Σ^* if $\theta(uv) = \theta(v)\theta(u)$, for all $u, v \in \Sigma^*$, and is an involution on Σ if $\theta(\theta(a)) = a$, for all $a \in \Sigma$. In [1], a word w was called a θ-power or pseudo-power of u if $w \in u\{u, \theta(u)\}^*$ for some $u \in \Sigma^+$, and θ-primitive or pseudo-primitive if it was not a pseudo-power of any such word, [1]. Pseudo-powers of words over the DNA alphabet have been extensively investigated as a model of DNA strands that can bind to themselves via Watson-Crick complementarity, rendering them unavailable for programmed computational interactions in most types of DNA computing algorithms, [3,10,11,13]. However, given that biologically there is no distinction between a DNA strand and its Watson-Crick complement, the issue remains that there is no biologically-motivated rationale for excluding from the definition of pseudo-power strings that are repetitions of u or $\theta(u)$ but start with $\theta(u)$.

This paper fills the gap by introducing the notion of *strong-θ-power of a word* u, defined as *any* word belonging to the set $\{u, \theta(u)\}^+$. In the particular case when $\theta = \theta_{DNA}$, the Watson-Crick complementarity involution, this will be called a *Watson-Crick power of the word*. Similar with the operation of θ-catenation which was defined and studied in [4] as generating all pseudo-powers of a word, here we define and study a binary operation called strong-θ-catenation (denoted by $\otimes$) which, when iteratedly applied to a single word u, generates all its strong-θ-powers. We find, for example, necessary and sufficient conditions for two words u and v to commute with respect to $\otimes$ (Corollary 17). We also define the concept of $\otimes$-primitive root pair of a word, and prove that it always exists and it is unique (Proposition 24).

The paper is organized as follows. Section 2 introduces definitions and notations and recalls some necessary results. Section 3 defines the operation $\otimes$ (strong-θ-catenation), and lists some of its basic properties. Section 4 studies some word equations that involve both words and their Watson-Crick complements, Sect. 5 investigates conjugacy and commutativity with respect to $\otimes$, and Sect. 6 explores the concept of $\otimes$-primitivity, and that of $\otimes$-primitive root pair of a word.

2 Preliminaries

An alphabet Σ is a finite non-empty set of symbols. The set of all words over Σ, including the empty word λ is denoted by Σ^* and $\Sigma^+ = \Sigma^* \setminus \{\lambda\}$ is the set of all non-empty words over Σ. The length of a word $w \in \Sigma^*$ is the number of symbols in the word and is denoted by $|w|$. We denote by $|u|_a$, the number of occurrences of the letter a in u and by $\mathrm{Alph}(u)$, the set of all symbols occurring in u. A word $w \in \Sigma^+$ is said to be *primitive* if $w = u^i$ implies $w = u$ and $i = 1$. Let Q denote the set of all primitive words.

For every word $w \in \Sigma^+$, there exists a unique word $\rho(w) \in \Sigma^+$, called the *primitive root* of w, such that $\rho(w) \in Q$ and $w = \rho(w)^n$ for some $n \geq 1$.

A function $\phi : \Sigma^* \to \Sigma^*$ is called a *morphism* on Σ^* if for all words $u, v \in \Sigma^*$ we have that $\phi(uv) = \phi(u)\phi(v)$, an *antimorphism* on Σ^* if $\phi(uv) = \phi(v)\phi(u)$ and an *involution* if $\phi(\phi(x)) = x$ for all $x \in \Sigma^*$.

A function $\phi : \Sigma^* \to \Sigma^*$ is called a *morphic involution on Σ^** (respectively, an *antimorphic involution on Σ^**) if it is an involution on Σ extended to a morphism (respectively, to an antimorphism) on Σ^*. For convenience, in the remainder of this paper we use the convention that the letter ϕ denotes an involution that is either morphic or antimorphic (such a function will be termed *(anti)morphic involution*), that the letter θ denotes an antimorphic involution, and that the letter μ denotes a morphic involution.

Definition 1. *For a given $u \in \Sigma^*$, and an (anti)morphic involution ϕ, the set $\{u, \phi(u)\}$ is denoted by u_ϕ, and is called a ϕ-complementary pair, or ϕ-pair for short. The length of a ϕ-pair u_ϕ is defined as $|u_\phi| = |u| = |\phi(u)|$.*

Note that if θ_{DNA} is the Watson-Crick complementarity function over the DNA alphabet $\{A, C, G, T\}$, that is, the antimorphic involution that maps A to T, C to G, and viceversa, then a θ_{DNA}-complementary pair $\{u, \theta_{DNA}(u)\}$ models a pair of Watson-Crick complementary DNA strands.

A ϕ-power of u (also called pseudo-power in [1]) is a word of the form $u_1 u_2 \cdots u_n$ for some $n \geq 1$, where $u_1 = u$ and for any $2 \leq i \leq n$, $u_i \in \{u, \phi(u)\}$. A word $w \in \Sigma^*$ is called a palindrome if $w = w^R$, where the reverse, or mirror image operator is defined as $\lambda = \lambda^R$ and $(a_1 a_2 \ldots a_n)^R = a_n \ldots a_2 a_1$, where $a_i \in \Sigma$ for all $1 \leq i \leq n$. A word $w \in \Sigma^*$ is called a ϕ-palindrome if $w = \phi(w)$, and the set of all ϕ-palindromes is denoted by P_ϕ. If $\phi = \mu$ is a morphic involution on Σ^* then the only μ-palindromes are the words over Σ', where $\Sigma' \subseteq \Sigma$, and μ is the identity on Σ'. Lastly, if $\phi = \theta$ is the identity function on Σ extended to an antimorphism on Σ^*, then a θ-palindrome is a classical palindrome, while if $\phi = \mu$ is the identity function on Σ extended to a morphism on Σ^*, then every word is a μ-palindrome. For more definitions and notions regarding words and languages, the reader is referred to [8]. We recall some results from [9].

Lemma 2. *[9] Let $u, v, w \in \Sigma^+$ be such that, $uv = vw$, then for $k \geq 0$, $x \in \Sigma^+$ and $y \in \Sigma^*$, $u = xy$, $v = (xy)^k x$, $w = yx$.*

Two words u and v are said to commute if $uv = vu$. We recall the following result from [5] characterizing θ-conjugacy and θ-commutativity for an antimorphic involution θ (if $\theta = \theta_{DNA}$, these are called Watson-Crick conjugacy, respectively Watson-Crick commutativity). Recall that u is said to be a θ-conjugate of w if $uv = \theta(v)w$ for some $v \in \Sigma^+$, and u is said to θ-commute with v if $uv = \theta(v)u$.

Proposition 3. *[5] For $u, v, w \in \Sigma^+$ and θ an antimorphic involution,*

1. *If $uv = \theta(v)w$, then either there exists $x \in \Sigma^+$ and $y \in \Sigma^*$ such that $u = xy$ and $w = y\theta(x)$, or $u = \theta(w)$.*
2. *If $uv = \theta(v)u$, then $u = x(yx)^i$, $v = yx$, for some $i \geq 0$ and θ-palindromes $x \in \Sigma^*, y \in \Sigma^+$.*

We recall the following from [6].

Proposition 4. *[6] Let $x, y \in \Sigma^+$ and θ an antimorphic involution, such that $xy = \theta(y)\theta(x)$ and $yx = \theta(x)\theta(y)$. Then, one of the following holds:*

1. $x = \alpha^i$, $y = \alpha^k$ for some $\alpha \in P_\theta$
2. $x = [\theta(s)s]^i \theta(s)$, $y = [s\theta(s)]^k s$ for some $s \in \Sigma^+$, $i, k \geq 0$.

3 A Binary Operation Generating Watson-Crick Powers

A binary operation $\circ$ is mapping $\circ : \Sigma^* \times \Sigma^* \to 2^{\Sigma^*}$. A binary word (bw, in short) operation with right identity, called ϕ-catenation, and which generates pseudo-powers of a word u (ϕ-powers, where ϕ is either a morphic or an antimorphic involution) when iteratively applied to it, was defined and studied in [4]. However, one can observe (See Remark 1) that ϕ-catenation does not generate all the words in $\{u, \phi(u)\}^+$. After exploring several binary word operations that each generates a certain subset of $\{u, \phi(u)\}^+$, we select the binary word operation, called strong-ϕ-catenation, which generates the entire set, and discuss some of its properties.

For a given binary operation $\circ$, the *i-th $\circ$-power of a word* is defined by :

$$u^{\circ(0)} = \{\lambda\}, \quad u^{\circ(1)} = u \circ \lambda, \quad u^{\circ(n)} = u^{\circ(n-1)} \circ u, \quad n \geq 2$$

Note that, depending on the operation $\circ$, the i-th power of a word can be a singleton word, or a set of words.

Remark 1. *Let $u, v \in \Sigma^+$ and θ be an antimorphic involution. The following are possible binary operations that, when $\theta = \theta_{DNA}$ is iteratively applied to a word u, generate various sets of Watson-Crick powers of u (these operations include the θ-catenation operation $\odot$ defined in [4]).*

1. *The operation $\odot$ and $\odot'$ and their corresponding n-th power, $n \geq 1$:*

$$u \odot v = \{uv, u\theta(v)\} \ , \ u \odot' v = \{uv, \theta(u)v\}$$

$$u^{\odot(n)} = u\{x_1 x_2 \cdots x_{n-1} \ : \ x_i = u \text{ or } x_i = \theta(u)\}$$

$$u^{\odot'(n)} = \{u^n\} \cup \{x_i y_i \ : \ x_i = [\theta(u)]^i, y_i = u^{n-i}, 1 \leq i \leq n-1\}$$

2. *The operation $\ominus$ and $\ominus'$ and their corresponding n-th power, $n \geq 1$:*

$$u \ominus v = u\theta(v) \ , \ u \ominus' v = \theta(u)v$$

$$u^{\ominus(n)} = u[\theta(u)]^{n-1} \ , \ u^{\ominus'(n)} = \theta(u)u^{n-1}$$

3. *The operation $\oplus$ and its corresponding n-th power, $n \geq 1$:*

$$u \oplus v = \{uv, u\theta(v), \theta(u)v\}$$

$$u^{\oplus(n)} = \{x_1 x_2 \cdots x_n \ : \ x_i = u \text{ or } x_i = \theta(u)\} \setminus \{[\theta(u)]^n\}$$

4. *The operation $\otimes$ and its corresponding n-th power, $n \geq 1$:*

$$u \otimes v = \{uv, u\theta(v), \theta(u)v, \theta(u)\theta(v)\}$$

$$u^{\otimes(n)} = \{x_1 x_2 \cdots x_n \ : \ x_i = u \ or \ x_i = \theta(u)\}$$

Note that, when $\theta = \theta_{DNA}$ is iteratively applied to a word u, the operations $\ominus, \ominus', \odot, \odot', \oplus$ generate some, but not all, Watson-Crick powers of u. The only operation that generates all the Watson-Crick powers of u is $\otimes$. Thus, in the remainder of this paper, we will restrict our discussion to the study of the operation $\otimes$, which we call *strong-ϕ-catenation*. We now give the formal definition.

Definition 5. *Given a morphic or an antimorphic involution ϕ on Σ^* and two words $u, v \in \Sigma^*$, we define the strong-ϕ-catenation operation with respect to ϕ as*

$$u \otimes v = \{uv, u\phi(v), \phi(u)v, \phi(u)\phi(v)\}.$$

Observe now that $u^{\otimes(n)} = \{u, \phi(u)\}^n$ is the set comprising all the nth strong-ϕ-powers of u with respect to ϕ. When $\phi = \theta_{DNA}$ is the Watson-Crick complementarity involution, this set comprises all the Watson-Crick powers of u.

Even though, for simplicity of notation, the notation for strong-ϕ-catenation and strong-ϕ-power does not explicitly include the function ϕ, these two notions are always defined with respect to a given fixed (anti)morphic involution ϕ. For a given (anti)morphic involution ϕ, and a given $n \geq 1$, the following equality relates the set of all strong-ϕ-powers of u with respect to ϕ (generated by strong-ϕ-catenation), to the set of all ϕ-powers of u with respect to ϕ (generated by ϕ-catenation):

$$u^{\otimes(n)} = u^{\odot(n)} \cup \phi(u)^{\odot(n)}$$

As an example, consider the case of θ_{DNA}, the Watson-Crick complementary function, and the words $u = ATC$, $v = GCTA$. Then,

$$u \otimes v = \{ATC \ GCTA, ATC \ TAGC, GAT \ GCTA, GAT \ TAGC\},$$

which is the set of all catenations that involve the word u and the word v (in this order) and their images under θ_{DNA} and

$$u^{\otimes(n)} = \{u_1 u_2 \cdots u_n \ : \ u_i = ATC \ or \ u_i = GAT, \ 1 \leq i \leq n\}$$

Note that $|u \otimes v| = 4$ iff $u, v \notin P_\phi$. It is clear from the above definition that for $u, v \in \Sigma^+$, $u \otimes v = u_\phi v_\phi$.

Remark 2. *For any $u \in \Sigma^+$, we have that $u^{\odot(n)} \subset u^{\otimes(n)}$ and $\theta(u)^{\odot(n)} \subset u^{\otimes(n)}$. However, if $u \in P_\phi$, then $u^{\odot(n)} = \theta(u)^{\odot(n)} = u^{\otimes(n)}$.*

Note that for the operation $\otimes$, $\lambda \otimes u = u \otimes \lambda \neq u$. Hence, the operation $\otimes$ does not have an identity. We have the following observation.

Lemma 6. *For $u \in \Sigma^+$, and ϕ (anti)morphic involution, the following statements hold.*

1. *For all $n \geq 1$, we have that $\alpha \in \{u, \phi(u)\}^n$ iff $\alpha \in u^{\otimes(n)}$.*
2. *For all $n \geq 1$, we have that $u^{\otimes(n)} = \phi(u^{\otimes(n)}) = \phi(u)^{\otimes(n)}$.*
3. *For all $m, n \geq 1$, we have that $(u^{\otimes(m)})^{\otimes(n)} = (u^{\otimes(n)})^{\otimes(m)} = u^{\otimes(mn)}$.*

A bw-operation $\circ$ is called length-increasing if for any $u, v \in \Sigma^+$ and $w \in u \circ v$, $|w| > max\{|u|, |v|\}$. A bw-operation $\circ$ is called propagating if for any $u, v \in \Sigma^*$, $a \in \Sigma$ and $w \in u \circ v$, $|w|_a = |u|_a + |v|_a$. In [4], these notions were generalized to incorporate an (anti)morphic involution ϕ, as follows. A bw-operation $\circ$ is called ϕ-propagating if for any $u, v \in \Sigma^*$, $a \in \Sigma$ and $w \in u \circ v$, $|w|_{a,\phi(a)} = |u|_{a,\phi(a)} + |v|_{a,\phi(a)}$. It was shown in [4] that the operation ϕ-catenation is not propagating but is ϕ-propagating.

A bw-operation $\circ$ is called left-inclusive if for any three words $u, v, w \in \Sigma^*$ we have

$$(u \circ v) \circ w \supseteq u \circ (v \circ w)$$

and is called right-inclusive if

$$(u \circ v) \circ w \subseteq u \circ (v \circ w).$$

Similar to the properties of the operation ϕ-catenation investigated in [4], one can easily observe that the strong-ϕ-catenation operation is length increasing, not propagating and ϕ-propagating. In [4] it was shown that for a morphic involution the ϕ-catenation operation is trivially associative, whereas for an antimorphic involution the ϕ-catenation operation is not associative. In contrast, the strong-ϕ-catenation operation is right inclusive, left inclusive, as well as associative, when ϕ is a morphic as well as an antimorphic involution.

Since the Watson-Crick complementarity function θ_{DNA} is an antimorphic involution, in the remainder of this paper we only investigate antimorphic involution mappings $\phi = \theta$.

4   Watson-Crick Conjugate Equations

In this section we discuss properties of words that satisfy some Watson-Crick conjugate equations, that is, word equations that involve both words and their Watson-Crick complements. It is well known that any two distinct words satisfying a non-trivial equation are powers of a common word. We discuss several examples of word equations over two distinct words x and y that are either power of a θ-palindrome, or a product of θ-palindromes, where θ is an antimorphic involution on Σ^*. We observe that, in most cases, words satisfying a non-trivial conjugacy relation are powers of a common θ-palindromic word. We have the following lemmas which we use later.

Lemma 7. *Let θ be an antimorphic involution and let $x, y \in \Sigma^+$ be such that x and y satisfy one of the following :*

1. $xxy = y\theta(x)x$
2. $\theta(x)xy = yxx$
3. $xxy = yx\theta(x)$
4. $x\theta(x)y = yxx$
5. $xy = \theta(y)x$ and $yx = x\theta(y)$

Then, $x = \alpha^m$ and $y = \alpha^n$ for some $m, n \geq 1$ and $\alpha \in P_\theta$.

Proof. We only prove for the case $xxy = y\theta(x)x$ and omit the rest as they are similar. Let, $xxy = y\theta(x)x$ then, by Lemma 2 we have $xx = pq$, $y = (pq)^i p$ and $\theta(x)x = qp$ where $i \geq 0$. We now have the following cases.

1. If $|p| = |q|$, then $x = p = q = \theta(x)$ and hence, $x = \alpha^m$, $y = \alpha^n$ such that $m, n \geq 1$ and $\alpha \in P_\theta$.
2. If $|p| > |q|$, then $x = p_1 = p_2 q$ and $\theta(x)x = \theta(q)\theta(p_2)p_1 = qp$ which implies $q \in P_\theta$ and $p_1 p_2 = \theta(p_2)p_1$. Thus by Lemma 2 there exist words s, t such that $p_2 = ts$, $p_1 = (st)^j s$ and $\theta(p_2) = st$ which implies that $s, t \in P_\theta$ and $p = (st)^{j+1}s \in P_\theta$. Also, $qp = \theta(x)x = \theta(p_1)p_2 q = \theta(p)q = pq$. Thus p and q are powers of a common θ-palindromic word. Hence, the result.
3. The case when $|p| < |q|$ is similar to the case $|p| > |q|$ and we omit its proof. □

Using a proof technique similar to that of Lemma 7 one can prove the following.

Lemma 8. *Let θ be an antimorphic involution and let $x, y \in \Sigma^+$. If $yx\theta(x) = \theta(x)xy$ then one of the following hold:*

1. $x = \alpha^m$ and $y = \alpha^n$ for some $m, n \geq 1$ and $\alpha \in P_\theta$.
2. $x = [s\theta(s)]^m s$ and $y = [\theta(s)s]^n \theta(s)$ for some $s \in \Sigma^+$.

It is well known that if two words x and y commute (i.e.) $xy = yx$, both x and y are powers of a common word, and the next result follows directly.

Lemma 9. *For $x, y \in \Sigma^+$, if $yxx = xxy$, then $x = \alpha^m$ and $y = \alpha^n$ for some $m, n \geq 1$ and $\alpha \in \Sigma^+$.*

It was shown in [5] that if x θ-commutes with y (i.e.), $xy = \theta(y)x$, then x is a θ-palindrome and y can be expressed as a catenation of two θ-palindromes. Similarly, we now show in Lemma 10 that if xx θ-commutes with y (i.e.), $xxy = \theta(y)xx$ then x is a θ-palindrome and y can be expressed as a product of palindromes. The proofs of the following results are similar to that of the proof of Lemma 9 and hence we omit them.

Lemma 10. *Let θ be an antimorphic involution and let $x, y \in \Sigma^+$. If $xxy = \theta(y)xx$ then, one of the following hold :*

1. $x = \alpha^m$ and $y = \alpha^n$ for some $m, n \geq 1$ and $\alpha \in P_\theta$
2. $y = qx^2$ for $q, x \in P_\theta$.
3. $x = (st)^k s$, $y = ts(st)^k s$ for $k \geq 1$ and $s, t \in P_\theta$.

In the following we find the structure of x that results from xx being a conjugate of $\theta(x)\theta(x)$. We show that such words are either power of a θ-palindrome or a catenation of two θ-palindromes.

Lemma 11. *Let θ be an antimorphic involution and let $x, y \in \Sigma^+$. If $xxy = y\theta(x)\theta(x)$ then one of the following is true:*

1. $x = \alpha^m$ and $y = \alpha^n$ for some $m, n \geq 1$ and $\alpha \in P_\theta$.
2. $x = st$ and $y = [st]^n s$ for some $n \geq 0$ and $s, t \in P_\theta$.

5   Conjugacy and Commutativity with Respect to $\otimes$

In this section we discuss conditions on words $u, w \in \Sigma^+$, such that u is a $\otimes$-conjugate of w, i.e., $u \otimes v = v \otimes w$ for some $v \in \Sigma^+$. We then discuss the special case when $u = w$, i.e., $u \otimes$-commutes with v, and prove a necessary and sufficient condition for $\otimes$-commutativity (Corollary 17).

Proposition 12. *Let $u, v, w \in \Sigma^+$ be such that $uv = vw$ and $u \otimes v = v \otimes w$. Then, either $u = v = w$ or $u = s^m = w$ and $v = s^n$, for $s \in P_\theta$.*

Proof. By definition, for $u, v, w \in \Sigma^+$,

$$u \otimes v = \{uv, u\theta(v), \theta(u)v, \theta(u)\theta(v)\}$$

and similarly,

$$v \otimes w = \{vw, v\theta(w), \theta(v)w, \theta(v)\theta(w)\}$$

Given that $uv = vw$ and $u \otimes v = v \otimes w$. Then, by Lemma 2, we have $u = xy$, $v = (xy)^i x$ and $w = yx$. We now have the following cases.

1. If $u\theta(v) = v\theta(w)$ then, $u\theta(v) = (xy)(\theta(x)\theta(y))^i \theta(x) = (xy)^i x\theta(x)\theta(y)$. If $i \neq 0$, then $x, y \in P_\theta$ and $xy = yx$ and hence, u, v and w are powers of a common θ-palindrome. If $i = 0$ then, $xy\theta(x) = x\theta(x)\theta(y)$ and by Proposition 3, $y = st$ and $\theta(x) = (st)^j s$ where $s, t \in P_\theta$ and hence, $x \in P_\theta$. Thus, $u \otimes v = \{xyx, \theta(y)xx\}$ and $v \otimes w = \{xyx, xx\theta(y)\}$. Since, $u \otimes v = v \otimes w$, $\theta(y)xx = xx\theta(y)$ and by Lemma 9, $x = p^{m_1}$, $y = p^{m_2}$ for $p \in P_\theta$. Thus, $u = p^m = w$, $v = p^n$ for $p \in P_\theta$.
2. The case when $u\theta(v) = \theta(v)w$ is similar to case (1) and we omit it.
3. If $u\theta(v) = \theta(v)\theta(w)$ then, $u\theta(v) = xy(\theta(x)\theta(y))^i \theta(x) = (\theta(x)\theta(y))^i \theta(x)\theta(x)$ $\theta(y) = \theta(v)\theta(w)$. If $i = 0$ then, $x \in P_\theta$ and the case is similar to the previous one. If $i \neq 0$ then $x, y \in P_\theta$ and $yx = xy$ and hence, $y = p^{j_1}$, $x = p^{j_2}$. Thus, $u = w = p^m$ and $v = p^n$ for $p \in P_\theta$.

Hence, the result.   □

A similar proof works for the next result and hence, we omit it.

Proposition 13. *Let $u, v, w \in \Sigma^+$ be such that $uv = v\theta(w)$ and $u \otimes v = v \otimes w$. Then, either $u = v = \theta(w)$ or $u = \alpha^m = w$ and $v = \alpha^n$, for $\alpha \in P_\theta$.*

The following proposition uses Lemma 7, 9 and 11.

Proposition 14. *Let $u, v, w \in \Sigma^+$ be such that $uv = \theta(v)w$ and $u \otimes v = v \otimes w$. Then, either $u = \theta(v) = \theta(w)$ or $u = \alpha^m = w$ and $v = \alpha^n$, for $\alpha \in P_\theta$.*

Proof. Given that $uv = \theta(v)w$ and $u \otimes v = v \otimes w$. Then by Proposition 3, we have either $u = \theta(w)$ and $v = \gamma w$ for some $\gamma \in P_\theta$ or $u = xy$, $v = \theta(x)$, $w = y\theta(x)$ for some $x, y \in \Sigma^*$.

1. If $u = \theta(w)$ and $v = \gamma w$ for $\gamma \in P_\theta$, then

$$u \otimes v = \{\theta(w)\gamma w, \theta(w)\theta(w)\gamma, w\gamma w, w\theta(w)\gamma\}$$

$$= \{\gamma ww, \gamma w\theta(w), \theta(w)\gamma w, \theta(w)\gamma\theta(w)\} = v \otimes w$$

   If $\gamma = \lambda$ then, $u = \theta(v) = \theta(w)$. If not, then we have the following cases.
   - If $\theta(ww)\gamma = \gamma ww$, then by Lemma 11 either $\theta(w) = \alpha^m$ and $\gamma = \alpha^n$ for some $m, n \geq 1$ and $\alpha \in \Sigma^+$ or $\theta(w) = st$ and $\gamma = [st]^n s$ for some $n \geq 0$ and $s, t \in P_\theta$. In the case when $\theta(w) = \alpha^m$ and $\gamma = \alpha^n$ for some $m, n \geq 1$ and $\alpha \in P_\theta$, u, v and w are powers of a common θ palindrome α. If $\theta(w) = st$ and $\gamma = [st]^n s$ for some $n \geq 0$ and $s, t \in P_\theta$, then $u \otimes v = \{w\gamma w, w\theta(w)\gamma\} = \{ts(st)^{n+1}s, ts(st)^{n+1}s\} = \{\gamma w\theta(w), \theta(w)\gamma\theta(w)\} = v \otimes w = \{(st)^{n+1}sst, (st)^{n+1}sst\}$. This implies that s and t are powers of a common word and since, $s, t \in P_\theta$, u, v and w are powers of a common θ palindrome.
   - If $\theta(ww)\gamma = \gamma w\theta(w)$ then by Lemma 7 we have $w = \alpha^m$ and $\gamma = \alpha^n$ for some $m, n \geq 1$ and $\alpha \in P_\theta$ and hence, u, v and w are powers of $\alpha \in P_\theta$.
   - If $\theta(ww)\gamma = \theta(w)\gamma\theta(w)$, then γ and $\theta(w)$ are powers of a common word and since $\gamma \in P_\theta$, u, v and w are powers of a common θ palindrome.
2. If $u = xy$, $v = \theta(x)$, $w = y\theta(x)$ for some $x, y \in \Sigma^*$ then,

$$u \otimes v = \{xy\theta(x), xyx, \theta(y)\theta(x)\theta(x), \theta(y)\theta(x)x\}$$

$$= \{\theta(x)y\theta(x), \theta(x)x\theta(y), xy\theta(x), xx\theta(y)\} = v \otimes w$$

If $xyx = \theta(x)y\theta(x)$, then $x \in P_\theta$ and hence $u \otimes v = \{xyx, \theta(y)xx\} = \{xyx, xx\theta(y)\} = v \otimes w$, which implies $xx\theta(y) = \theta(y)xx$. Then by Lemma 9, x and $\theta(y)$ are powers of a common word α. Since, $x \in P_\theta$, $\alpha \in P_\theta$.
The cases when $xyx = \theta(x)x\theta(y)$ and $xyx = xy\theta(x)$ are similar. If $xyx = xx\theta(y)$, then $yx = x\theta(y)$ and by Proposition 3, we have that $y = st$, $x = (st)^i s$ for some $s, t \in P_\theta$ and hence, $x \in P_\theta$ and the case is similar to the above. Thus, in all cases x and y and hence u, v and w are powers of a common θ-palindrome.   □

The proof of the following is similar to that of Proposition 14 and hence, we omit it.

Proposition 15. *Let $u, v, w \in \Sigma^+$ be such that $uv = \theta(v)\theta(w)$ and $u \otimes v = v \otimes w$. Then, either $u = \theta(v) = w$ or $u = \alpha^m = w$ and $v = \alpha^n$, for $\alpha \in P_\theta$.*

Based on the above results (Propositions 12, 13, 14 and 15), we give a neccessary and sufficient condition on words u, v and w such that $u \otimes v = v \otimes w$.

Theorem 16. *Let $u, v, w \in \Sigma^+$. Then, $u \otimes v = v \otimes w$ iff one of the following holds:*

1. $u = \theta(v) = w$
2. $u = v = w$
3. $u = \theta(v) = \theta(w)$
4. $u = v = \theta(w)$
5. $u = s^m = w$ and $v = s^n$, for $s \in P_\theta$.

Based on the above theorem one can deduce conditions on u and v such that u and $v \otimes$ commute with each other. We have the following corollary.

Corollary 17. *For an antimorphic involution θ and $u, v \in \Sigma^+$, $u \otimes v = v \otimes u$ iff (i) $u = v$, or (ii) $u = \theta(v)$, or (iii) u and v are powers of a common θ-palindrome.*

6 $\otimes$-Primitive Words, and a Word's $\otimes$-Primitive Root Pair

In this section we introduce a special class of primitive words, using the binary word operation $\otimes$. More precisely, similar to the primitive words defined in [8,9] based on the catenation operation, given an antimorphic involution θ we define $\otimes$-primitive words with respect to θ, based on the binary word operation $\otimes$. We study several properties of $\otimes$-primitive words. We also define the notion of $\otimes$-primitive root pair of a word w, and show that every word has a unique $\otimes$-primitive root pair, which is a θ-pair of $\otimes$-primitive words (Proposition 24).

Analogous to the definitions given in [2], we define the following.

Definition 18. *Let θ be an antimorphic involution. A non-empty word w is called $\otimes$-primitive with respect to θ if it cannot be expressed as a non-trivial strong-θ-power of another word.*

By Definition 18, a word $w \in \Sigma^+$ is $\otimes$-primitive if the condition $w \in u^{\otimes(n)}$ for some word u and $i \geq 1$ implies $i = 1$ and $w \in u_\theta$.

Example 19. Consider the Watson-Crick complementarity function θ_{DNA} and the word $w = ACTAGTAGTACTACTAGT$. The word w is not $\otimes$-primitive with respect to θ_{DNA} since $w \in (ACT)^{\otimes(6)}$, whereas the word $x = ACTAAG$ is $\otimes$-primitive with respect to θ_{DNA}.

We now relate the notion of $\otimes$-primitive word with respect to an (anti)morphic involution θ, to that of θ-primitive words introduced in [1], whereby a word w is called θ-primitive if it cannot be expressed as a non-trivial θ-power (pseudo-power) of another word. One can observe that the word w in Example 19 is not θ-primitive and the word $x = ACTAAG$ is θ-primitive. The following holds.

Remark 3. *Given an antimorphic involution θ and a word u in Σ^+, the following are equivalent: (i) u is θ-primitive, (ii) u is $\otimes$-primitive with respect to θ, and (iii) ([4]) u is $\odot$-primitive with respect to θ.*

Thus, if $Q_\otimes$, $Q_\odot$ and Q_θ denote the classes of all $\otimes$-primitive, $\odot$-primitive, and θ-primitive words over Σ^* respectively, then $Q_\otimes = Q_\odot = Q_\theta$. It was shown in [1] that all θ-primitive words are primitive but the converse is not true in general. It then follows that all $\otimes$-primitive words with respect to a given antimorphic involution θ are primitive, but the converse does not generally hold. Thus, $Q_\otimes$ is a strict subset of the class of primitive words. We now recall the following result from [4].

Lemma 20. *Let $\circ$ be a binary word operation that is plus-closed and ϕ-propagating. Then, for every word $w \in \Sigma^+$ there exists a $\circ$-primitive word u and a unique integer $n \geq 1$ such that $w \in u^{\circ(n)}$.*

Since the binary operation $\otimes$ is plus-closed and θ-propagating, by Lemma 20 we conclude the following.

Lemma 21. *Let θ be an antimorphic involution on Σ^*. For all $w \in \Sigma^+$, there exists a word u which is $\otimes$-primitive with respect to θ, such that $w \in u^{\otimes(n)}$ for some $n \geq 1$.*

By Lemma 21, given a non-empty word w, there always exists a $\otimes$-primitive word u such that w is a strong θ-power of u. In general, for a binary operation $\circ$, the authors in [2] call a $\circ$-primitive word u a "$\circ$-root of w," if $w \in u^{\circ(n)}$ for some $n \geq 1$. Note that a word w may have several $\circ$-roots. For example, for the word w in Example 19, we have that $w \in x^{\otimes(6)} = (ACT)^{\otimes(6)} = (\theta(x))^{\otimes(6)} = (AGT)^{\otimes(6)}$, that is, there are two $\otimes$-primitive words, x and $\theta(x)$, which are $\otimes$-roots of w. However, uniqueness can still be ensured if we select the θ-pair $x_\theta = \{x, \theta(x)\}$, such that x is $\otimes$-primitive and $w \in x_\theta^+$. We give a formal definition in the following.

Definition 22. *Given an antimorphic involution θ, the $\otimes$-primitive root pair of a word $w \in \Sigma^+$ relative to θ (or simply the $\otimes$-primitive root pair of w) is the θ-pair $u_\theta = \{u, \theta(u)\}$ which satisfies the property that u is $\otimes$-primitive and $w \in u^{\otimes(n)}$ for some $n \geq 1$.*

For example, in Example 19 the $\otimes$-primitive root pair of w is $x_\theta = \{ACT, AGT\}$. In the following we will prove that, for a given antimorphic involution θ, the $\otimes$-primitive root pair of a word $w \in \Sigma^+$ always exists and it is unique. Indeed, by Lemma 21 and Lemma 6, it follows that a $\otimes$-primitive root pair of a word $w \in \Sigma^+$ always exists. We now prove that every word $w \in \Sigma^+$ has a unique $\otimes$-primitive root pair relative to θ, which we will denote by $\rho_\theta^\otimes(w)$. We use the following result from [1].

Theorem 23. *[1] Let $u, v, w \in \Sigma^+$ such that $w \in u\{u, \theta(u)\}^* \cap v\{v, \theta(v)\}^*$. Then u and v have a common θ-primitive root.*

Proposition 24. *Given an antimorphic involution θ and a word $w \in \Sigma^+$, its $\otimes$-primitive root pair $\rho_\theta^\otimes(w)$ is unique.*

Proof. For $w \in \Sigma^+$, by Lemma 21 there exists a $\otimes$-primitive word u such that $w \in u^{\otimes(n)}$, for some $n \geq 1$. Suppose there exists another $\otimes$-primitive word v such that $w \in v^{\otimes(m)}$ for some $m \geq 1$, i.e., $w \in \{u, \theta(u)\}^n$ and $w \in \{v, \theta(v)\}^m$. We then have the following cases:

1. If $w \in u\{u, \theta(u)\}^{n-1}$ and $w \in v\{v, \theta(v)\}^{m-1}$ then, by Theorem 23, u and v have a common θ-primitive root t. That is $u \in t^{\odot(k_1)}$ and $v \in t^{\odot(k_2)}$ for some θ-primitive t and $k_1, k_2 \geq 1$. Hence, by Remark 2, $u \in t^{\otimes(k_1)}$ and $v \in t^{\otimes(k_2)}$.
2. If $w \in u\{u, \theta(u)\}^{n-1}$ and $w \in \theta(v)\{v, \theta(v)\}^{m-1}$ then, by Theorem 23, u and $\theta(v)$ have a common θ-primitive root t. That is $u \in t^{\odot(k_1)}$ and $\theta(v) \in t^{\odot(k_2)}$ for some θ-primitive t and $k_1, k_2 \geq 1$. Hence, by Remark 2, $u \in t^{\otimes(k_1)}$ and $\theta(v) \in t^{\otimes(k_2)}$ which implies $v \in t^{\otimes(k_2)}$.
3. The case when $w \in \theta(u)\{u, \theta(u)^{n-1}\} \cap \theta(v)\{v, \theta(v)\}^{m-1}$ and the case when $w \in \theta(u)\{u, \theta(u)\}^{n-1} \cap v\{v, \theta(v)\}^{m-1}$ are similar to the previous cases.

By Remark 3, we have that t is also $\otimes$-primitive and thus, in all three situations above, both u and v are strong-ϕ-powers of t. Since both u and v are $\otimes$-primitive, it follows that $u, v \in \{t, \theta(t)\}$, which further implies that $u_\theta = v_\theta$. Thus, the $\otimes$-primitive root pair of w, denoted by $\rho_\theta^\otimes(w)$, is unique. □

We now try to find conditions on u and v such that $u^{\otimes m} = v^{\otimes n}$ for $m, n \geq 1$ and $m \neq n$. Without loss of generality, we assume that $m < n$ and $|u| > |v|$.

Lemma 25. *If $u^{\otimes m} = v^{\otimes n}$ for some $m, n \geq 1$ and $m \neq n$ then, $u = s^{k_1}$, $v = s^{k_2}$ for some $s \in P_\theta$.*

Proof. Let $\alpha_1, \alpha_2 \in u^{\otimes m}$ such that $\alpha_1 \in u\{u, \theta(u)\}^*$ and $\alpha_2 \in \theta(u)\{u, \theta(u)\}^*$. Since $m \neq n$, there exists $\beta_1, \beta_2 \in v\{v, \theta(v)\}^*$ such that $\alpha_1 = \beta_1$ and $\alpha_2 = \beta_2$. Then by Theorem 23, u, $\theta(u)$ and v have a common θ-primitive root. Hence, u and v are powers of a common θ-palindrome. □

7 Conclusions

This paper defines and investigates the binary word operation strong-θ-catenation which, when iteratively applied to a word u, generates all the strong-θ-powers of u (if $\theta = \theta_{DNA}$ these become all the Watson-Crick powers of u). Future topics of research include extending the strong-θ-catenation to languages and investigating its properties, as well as exploring a commutative version of strong-θ-catenation, similarly to the bi-catenation of words which extends the catenation operation, and was defined in [12] as $u \star v = \{uv, vu\}$.

References

1. Czeizler, E., Kari, L., Seki, S.: On a special class of primitive words. Theoret. Comput. Sci. **411**(3), 617–630 (2010)
2. Hsiao, H., Huang, C., Yu, S.S.: Word operation closure and primitivity of languages. J. Autom. Lang. Comb. **19**(1), 157–171 (2014)
3. Jonoska, N., Mahalingam, K.: Languages of DNA based code words. In: Chen, J., Reif, J. (eds.) DNA 2003. LNCS, vol. 2943, pp. 61–73. Springer, Heidelberg (2004). https://doi.org/10.1007/978-3-540-24628-2_8
4. Kari, L., Kulkarni, M.: Generating the pseudo-powers of a word. J. Univ. Comput. Sci. **8**(2), 243–256 (2002)
5. Kari, L., Mahalingam, K.: Watson-crick conjugate and commutative words. In: Garzon, M.H., Yan, H. (eds.) DNA 2007. LNCS, vol. 4848, pp. 273–283. Springer, Heidelberg (2008). https://doi.org/10.1007/978-3-540-77962-9_29
6. Kari, L., Mahalingam, K.: Watson-Crick palindromes in DNA computing. Nat. Comput. **9**, 297–316 (2010). https://doi.org/10.1007/s11047-009-9131-2
7. Kari, L., Seki, S.: An improved bound for an extension of Fine and Wilf's theorem and its optimality. Fund. Inform. **101**, 215–236 (2010)
8. Lothaire, M.: Combinatorics on Words. Cambridge University Press, Cambridge (1997)
9. Lyndon, R.C., Schützenberger, M.P.: The equation $a^M = b^N c^P$ in a free group. Mich. Math. J. **9**, 289–298 (1962)
10. Mauri, G., Ferretti, C.: Word design for molecular computing: a survey. In: Chen, J., Reif, J. (eds.) DNA 2003. LNCS, vol. 2943, pp. 37–47. Springer, Heidelberg (2004). https://doi.org/10.1007/978-3-540-24628-2_5
11. Păun, G., Rozenberg, G., Salomaa, A.: DNA Computing: New Computing Paradigms. Springer, Berlin, Heidelberg (1998). https://doi.org/10.1007/978-3-662-03563-4
12. Shyr, H., Yu, S.: Bi-catenation and shuffle product of languages. Acta Informatica **35**, 689–707 (1998). https://doi.org/10.1007/s002360050139
13. Tulpan, D.C., Hoos, H.H., Condon, A.E.: Stochastic local search algorithms for DNA word design. In: Hagiya, M., Ohuchi, A. (eds.) DNA 2002. LNCS, vol. 2568, pp. 229–241. Springer, Heidelberg (2003). https://doi.org/10.1007/3-540-36440-4_20

A Normal Form for Matrix Multiplication Schemes

Manuel Kauers [ID] and Jakob Moosbauer[(✉)] [ID]

Institute for Algebra, Johannes Kepler University, Linz, Austria
{manuel.kauers,jakob.moosbauer}@jku.at

Abstract. Schemes for exact multiplication of small matrices have a large symmetry group. This group defines an equivalence relation on the set of multiplication schemes. There are algorithms to decide whether two schemes are equivalent. However, for a large number of schemes a pairwise equivalence check becomes cumbersome. In this paper we propose an algorithm to compute a normal form of matrix multiplication schemes. This allows us to decide pairwise equivalence of a larger number of schemes efficiently.

1 Introduction

Computing the product of two $n \times n$ matrices using the straightforward algorithm costs $O(n^3)$ operations. Strassen found a multiplication scheme that allows to multiply two 2×2 matrices using only 7 multiplications instead of 8 [13]. This scheme can be applied recursively to compute the product of $n \times n$ matrices in $O(n^{\log_2 7})$ operations. This discovery lead to a large amount of research on finding the smallest ω such that two $n \times n$ matrices can be multiplied using at most $O(n^\omega)$ operations. The currently best known bound is $\omega < 2.37286$ and is due to Alman and Williams [1].

Another interesting question is to find the exact number of multiplications needed to multiply two $n \times n$ matrices for small numbers n. For $n = 2$ Strassen provided the upper bound of 7. Winograd showed that we also need at least 7 multiplications [14]. De Groote proved that Strassen's algorithm is unique [6] modulo a group of equivalence transformations.

For the case $n = 3$ Laderman was the first to present a scheme that uses 23 multiplications [9], which remains the best known upper bound, unless the coefficient domain is commutative [11]. The currently best lower bound is 19 and was proved by Bläser [3]. There are many ways to multiply two 3×3 matrices using 23 multiplications [2,5,7,8,10,12].

For every newly found algorithm the question arises whether it is really new or it can be mapped to a known solution by one of the transformations described by de Groote. These transformations define an equivalence relation on the set of

M.K. was supported by the Austrian Science Fund (FWF) grant P31571-N32.
J.M. was supported by the Land Oberösterreich through the LIT-AI Lab.

ⓒ The Author(s), under exclusive license to Springer Nature Switzerland AG 2022

D. Poulakis and G. Rahonis (Eds.): CAI 2022, LNCS 13706, pp. 149–160, 2022.
https://doi.org/10.1007/978-3-031-19685-0_11

matrix multiplication algorithms. Some authors used invariants of the action of the transformation group to prove that their newly found schemes are inequivalent to the known algorithms. The works of Berger et al. [2] and Heule et al. [7] provide algorithms to check if two given schemes are equivalent. Berger et al. give an algorithm that can check equivalence over the ground field $\mathbb{R}$ if the schemes fulfill a certain assumption. Heule et al. provide an algorithm to check equivalence over finite fields.

Heule et al. presented more than 17,000 schemes for multiplying 3×3 matrices and showed that they are pairwise nonequivalent, at least when viewed over the ground field $\mathbb{Z}_2$. Their collection has since been extended to more than 64,000 pairwise inequivalent schemes. For testing whether a newly found scheme is really new, we would need to do an equivalence test for each of these schemes. Due to the large number of schemes this becomes expensive.

In this paper we propose an algorithm that computes a normal form for the equivalence class of a given scheme over a finite field. If all known schemes already are in normal form, then deciding whether a newly found scheme is equivalent to any of them is reduced to a normal form computation for the new scheme and a cheap syntactic comparison to every old scheme. Although the transformation group over a finite field is finite, it is so large that checking equivalence by computing every transformation is not feasible. Thus, Heule et al. use a strategy that iteratively maps one scheme to another part by part. We use a similar strategy to find a minimal element of an equivalence class.

2   Matrix Multiplication Schemes

Let K be a field and let $\mathbf{A}, \mathbf{B} \in K^{n \times n}$. The computation of the matrix product $\mathbf{C} = \mathbf{AB}$ by a Strassen-like algorithm proceeds in two stages. In the first stage we compute some intermediate products $M_1, \ldots, M_r$ of linear combinations of entries of $\mathbf{A}$ and linear combinations of entries of $\mathbf{B}$. In the second stage we compute the entries of $\mathbf{C}$ as linear combinations of the M_i.

For example if $n = 2$, we can write

$$\mathbf{A} = \begin{pmatrix} a_{1,1} & a_{1,2} \\ a_{2,1} & a_{2,2} \end{pmatrix} \quad \mathbf{B} = \begin{pmatrix} b_{1,1} & b_{1,2} \\ b_{2,1} & b_{2,2} \end{pmatrix} \quad \text{and} \quad \mathbf{C} = \begin{pmatrix} c_{1,1} & c_{1,2} \\ c_{2,1} & c_{2,2} \end{pmatrix}.$$

Strassen's algorithm computes $\mathbf{C}$ in the following way:

$$M_1 = (a_{1,1} + a_{2,2})(b_{1,1} + b_{2,2})$$
$$M_2 = (a_{2,1} + a_{2,2})(b_{1,1})$$
$$M_3 = (a_{1,1})(b_{1,2} - b_{2,2})$$
$$M_4 = (a_{2,2})(b_{2,1} - b_{1,1})$$
$$M_5 = (a_{1,1} + a_{1,2})(b_{2,2})$$
$$M_6 = (a_{2,1} - a_{1,1})(b_{1,1} + b_{1,2})$$
$$M_7 = (a_{1,2} - a_{2,2})(b_{2,1} + b_{2,2})$$

$$c_{1,1} = M_1 + M_4 - M_5 + M_7$$
$$c_{1,2} = M_3 + M_5$$
$$c_{2,1} = M_2 + M_4$$
$$c_{2,2} = M_1 - M_2 + M_3 + M_6.$$

A Strassen-like multiplication algorithm that computes the product of two $n \times n$ matrices using r multiplications has the form

$$M_1 = (\alpha_{1,1}^{(1)} a_{1,1} + \alpha_{1,2}^{(1)} a_{1,2} + \cdots)(\beta_{1,1}^{(1)} b_{1,1} + \beta_{1,2}^{(1)} b_{1,2} + \cdots)$$

$$\vdots$$

$$M_r = (\alpha_{1,1}^{(r)} a_{1,1} + \alpha_{1,2}^{(r)} a_{1,2} + \cdots)(\beta_{1,1}^{(r)} b_{1,1} + \beta_{1,2}^{(r)} b_{1,2} + \cdots)$$
$$c_{1,1} = \gamma_{1,1}^{(1)} M_1 + \gamma_{1,1}^{(2)} M_2 + \cdots + \gamma_{1,1}^{(r)} M_r$$

$$\vdots$$

$$c_{n,n} = \gamma_{n,n}^{(1)} M_1 + \gamma_{n,n}^{(2)} M_2 + \cdots + \gamma_{n,n}^{(r)} M_r.$$

All the information about such a multiplication scheme is contained in the coefficients $\alpha_{i,j}, \beta_{i,j}$ and $\gamma_{i,j}$. We can write these coefficients as a tensor in $K^{n \times n} \otimes K^{n \times n} \otimes K^{n \times n}$:

$$\sum_{l=1}^{r} ((\alpha_{i,j}^{(l)}))_{i=1,j=1}^{n,n} \otimes ((\beta_{i,j}^{(l)}))_{i=1,j=1}^{n,n} \otimes ((\gamma_{i,j}^{(l)}))_{i=1,j=1}^{n,n}. \qquad (1)$$

A multiplication scheme, seen as an element of $K^{n \times n} \otimes K^{n \times n} \otimes K^{n \times n}$ is equal to the matrix multiplication tensor defined by $\sum_{i,j,k=1}^{n} E_{i,k} \otimes E_{k,j} \otimes E_{i,j}$ where $E_{u,v}$ is the matrix with 1 at position (u, v) and zeros everywhere else [4]. Formulas become a bit more symmetric if we look at the tensor $\sum_{i,j,k=1}^{n} E_{i,k} \otimes E_{k,j} \otimes E_{j,i}$ corresponding to the product $\mathbf{C}^T = \mathbf{AB}$, so we will consider this tensor instead.

We represent a scheme as a table containing the matrices in this tensor. We will refer to the rows and columns of this table as the rows and columns of a scheme. For example Strassen's algorithm is represented as shown in Table 1.

3   The Symmetry Group

There are several transformations that map one matrix multiplication scheme to another one. We call two schemes equivalent if they can be mapped to each other by one of these transformations. De Groote [6] first described the transformations and showed that Strassen's algorithm is unique modulo this equivalence.

The first transformation is permuting the rows of a scheme. This corresponds to just changing the order of the M_i's in the algorithm. Another transformation comes from the fact that $\mathbf{AB} = \mathbf{C}^T \Leftrightarrow \mathbf{B}^T \mathbf{A}^T = \mathbf{C}$. It acts on a tensor by transforming a summand $A \otimes B \otimes C$ to $B^T \otimes A^T \otimes C^T$. Moreover, it follows from the condition that the sum (1) is equal to the matrix multiplication tensor, that

Table 1. Strassen's algorithm.

	α	β	γ
1	$\begin{pmatrix} 1 & 0 \\ 0 & 1 \end{pmatrix}$	$\begin{pmatrix} 1 & 0 \\ 0 & 1 \end{pmatrix}$	$\begin{pmatrix} 1 & 0 \\ 0 & 1 \end{pmatrix}$
2	$\begin{pmatrix} 0 & 0 \\ 1 & 1 \end{pmatrix}$	$\begin{pmatrix} 1 & 0 \\ 0 & 0 \end{pmatrix}$	$\begin{pmatrix} 0 & 0 \\ 1 & -1 \end{pmatrix}$
3	$\begin{pmatrix} 1 & 0 \\ 0 & 0 \end{pmatrix}$	$\begin{pmatrix} 0 & 1 \\ 0 & -1 \end{pmatrix}$	$\begin{pmatrix} 0 & 1 \\ 0 & 1 \end{pmatrix}$
4	$\begin{pmatrix} 0 & 0 \\ 0 & 1 \end{pmatrix}$	$\begin{pmatrix} -1 & 0 \\ 1 & 0 \end{pmatrix}$	$\begin{pmatrix} 1 & 0 \\ 1 & 0 \end{pmatrix}$
5	$\begin{pmatrix} 1 & 1 \\ 0 & 0 \end{pmatrix}$	$\begin{pmatrix} 0 & 0 \\ 0 & 1 \end{pmatrix}$	$\begin{pmatrix} -1 & 1 \\ 0 & 0 \end{pmatrix}$
6	$\begin{pmatrix} -1 & 0 \\ 1 & 0 \end{pmatrix}$	$\begin{pmatrix} 1 & 1 \\ 0 & 0 \end{pmatrix}$	$\begin{pmatrix} 0 & 0 \\ 0 & 1 \end{pmatrix}$
7	$\begin{pmatrix} 0 & 1 \\ 0 & -1 \end{pmatrix}$	$\begin{pmatrix} 0 & 0 \\ 1 & 1 \end{pmatrix}$	$\begin{pmatrix} 1 & 0 \\ 0 & 0 \end{pmatrix}$

also a cyclic permutation of the coefficients α, β and γ is a symmetry transformation. Taking those together we get an action that is composed by an arbitrary permutation of the columns of a scheme and transposing all the matrices if the permutation is odd.

Finally, we can use that for any invertible matrix V we have $\mathbf{AB} = \mathbf{A}VV^{-1}\mathbf{B}$. The corresponding action on a tensor $A \otimes B \otimes C$ maps it to $AV \otimes V^{-1}B \otimes C$. Since we can permute A, B and C we also can insert invertible matrices U and W which results in the action

$$(U, V, W) * A \otimes B \otimes C = UAV^{-1} \otimes VBW^{-1} \otimes WCU^{-1}. \tag{2}$$

This transformation is called the sandwiching action.

If we combine all these transformations we get the group $G = S_r \times S_3 \ltimes \mathrm{GL}(K, n)^3$ of symmetries of $n \times n$ matrix multiplication schemes with r rows. By $\mathrm{Aut}(G)$ we denote the group of automorphisms of a group G.

Definition 1. *Let* $\varphi \colon S_3 \to \mathrm{Aut}(\mathrm{GL}(K, n)^3)$ *be defined by*

$$\varphi(\pi) = \begin{cases} (U, V, W) \mapsto \pi((U, V, W)) & \text{if } \mathrm{sgn}(\pi) = 1 \\ (U, V, W) \mapsto \pi((V^{-T}, W^{-T}, U^{-T})) & \text{if } \mathrm{sgn}(\pi) = -1 \end{cases}$$

The symmetry group of $n \times n$ matrix multiplication schemes with r rows is defined over the set $G = S_r \times S_3 \times \mathrm{GL}(K, n)^3$ with the multiplication given by

$$(\sigma_1, \pi_1, (U_1, V_1, W_1)) \cdot (\sigma_2, \pi_2, (U_2, V_2, W_2)) =$$
$$(\sigma_1\sigma_2, \pi_1\pi_2, (U_1, V_1, W_1)\varphi(\pi_1)((U_2, V_2, W_2))).$$

The action $g * s$ of a group element $g = (\sigma, \pi, (U, V, W)) \in G$ on a multiplication scheme $s \in (K^{n \times n})^{r \times 3}$ is defined by first letting σ permute the rows of s then letting π permute the columns of s and transposing every matrix if $\text{sgn}(\pi) = -1$ and finally letting U, V and W act on every row as defined in Eq. (2).

One can show that this action fulfills the criteria of a group action.

4   Minimal Orbit Elements

Two schemes are equivalent if they belong to the same orbit under the action of the group G. Our goal in this section is to define a normal form for every orbit. The particular choice of the normal form is partly motivated by implementation convenience and not by any special properties. From now on we assume that K is a finite field. Since over a finite field the symmetry group is finite we could decide equivalence or compute a normal form by exhaustive search. However, already for $n = 3$ the symmetry group over $\mathbb{Z}_2$ has a size of $23! \cdot 6 \cdot 4741632 \approx 7 \cdot 10^{29}$.

Definition 2. *Let $s \in (K^{n \times n})^{r \times 3}$ be a matrix multiplication scheme. The rank pattern of the scheme is defined as the table*

$$((\text{ranks}_{i,1}, \text{ranks}_{i,2}, \text{ranks}_{i,3}))_{i=1}^{r}.$$

The rank vector of a row (A, B, C) is $(\text{rank}(A), \text{rank}(B), \text{rank}(C))$.

Since the matrices U, V and W are invertible, the sandwiching action leaves the rank pattern invariant. Transposing the matrices does not change their rank either. Therefore the only way a group element changes the rank pattern of a scheme is by permuting it accordingly. So for two equivalent schemes their rank patterns only differ by a permutation of rows and columns. This allows us to permute the rows and columns of the scheme such that the rank pattern becomes maximal under lexicographic order.

This maximal rank pattern is a well-known invariant of the symmetry group that has been used to show that two schemes are not equivalent. For example Courtois et al. [5] and Oh et al. [10] used this test to prove that their schemes were indeed new. However, this method only provides a sufficient condition for the inequivalence of schemes and can not decide equivalence of schemes. In Heule et al.'s data for certain rank patterns there are almost 1000 inequivalent schemes having this rank pattern.

We choose the normal form to be an orbit element which has a maximal rank pattern and is minimal under a certain lexicographic order. For doing so fix a total order on K such that $0 < 1 < x$ for all $x \in K \setminus \{0, 1\}$. The order need not be compatible with $+$ or $\cdot$ in any sense. For the matrices in the schemes we use colexicographic order by columns, with columns compared by lexicographic order. This means for two column vectors $v = (x_1, \ldots, x_n)^T$ and $v' = (x'_1, \ldots, x'_n)^T$ we define recursively

$$v < v' :\Leftrightarrow x_1 < x'_1 \lor (x_1 = x'_1 \land (x_2, \ldots x_n) < (x'_2, \ldots, x'_n))$$

For two matrices $M = (v_1 \mid \cdots \mid v_n)$ and $M' = (v'_1 \mid \cdots \mid v'_n)$ we define

$$M < M' :\Leftrightarrow v_n < v'_n \lor (v_n = v'_n \land (v_1 \mid \cdots \mid v_{n-1}) < (v'_1 \mid \cdots \mid v'_{n-1}))$$

For ordering the schemes we use the common lexicographic order. So we compare two schemes row by row from top to bottom and in each row we compare the matrices from left to right using the order defined above.

Definition 3. *Let $s \in (K^{n \times n})^{r \times 3}$ be a matrix multiplication scheme. We say s is in normal form if $s = \min\{s' \in G * s \mid$ the rank pattern of s' is sorted$\}$, where the minimum is taken with respect to the order defined above.*

Such a normal form clearly exists and it is unique since the group G is finite and the lexicographic order is a total order.

The strategy to compute the normal form is as follows:

Let s be a multiplication scheme and let N be its normal form.

We start by going over all column permutations of s and sort their rows by rank pattern to find a scheme s' with maximal rank pattern. If there are several column permutations that lead to the same maximal rank pattern, we consider each of them separately, since there are at most six.

Then we proceed row by row. For all rows of s' that have maximal rank pattern, we determine the minimal element of their orbit under the action of $\mathrm{GL}(K, n)^3$. From the definition of the normal form, it follows that the smallest row we can produce this way has to be the first row (A, B, C) of N. However, we might be able to reach the first row of N from several different rows and also the choice of U, V and W is in general not unique.

Apart from the first row of N we also compute the stabilizer of the first row, which is the set of all triples $(U, V, W) \in \mathrm{GL}(K, n)^3$ such that $(A, B, C) = (UAV^{-1}, VBW^{-1}, WCU^{-1})$. For each possible row that can be mapped to the first row we compute the *tail*, by which we mean the list of all remaining rows after applying a suitable triple (U, V, W).

We then continue this process iteratively. We go over each tail and determine a row that has maximal rank vector and becomes minimal under the action of the stabilizer. To do this we apply every element of the stabilizer to all possible candidates for the next row. This uniquely determines the next row of the normal form and we get again a list of tails and the stabilizer of the already determined rows.

The full process is listed in Algorithm 1.

Proposition 1. *Algorithm 1 terminates and is correct.*

Proof. The termination of the algorithm is guaranteed, since in line 19 the new tails contain one row less than in the previous step, so eventually the list of tails only contains empty elements.

To prove correctness we first note that the choice of P ensures that it contains a scheme that can be mapped to its normal form without applying further column permutations. From now on we only consider the iteration of the loop in line 3 where s' is this scheme.

Input   : A matrix multiplication scheme s
Output: An equivalent scheme in normal form
1 $P := \{s' \in (S_3 \times S_r) * s \mid s' \text{ has maximal rank pattern in } (S_3 \times S_r) * s\}$
2 $o := s$
3 **for** $s' \in P$ **do**
4      $candidate := ()$
5      $tails = \{s'\}$
6      $stab = \mathrm{GL}(K, n)^3$
7   **while** $tails \neq \{()\}$ **do**
8          $min := (1, \ldots, 1)^T$
9          $newtails := \{\}$
10   **for** $t \in tails$ **do**
11   **for** $r \in t$ *with maximal rank vector* **do**
12                 $g := \mathrm{argmin}_{g \in stab} g * r$
13   **if** $g * r < min$ **then**
14                     $min := g * r$
15                     $newtails := \{\}$
16   **if** $g * r = min$ **then**
17                     $newtails := newtails \cup (g * t \setminus \{g * r\})$
18             $stab := \{g \in stab \mid g * min = min\}$
19             $tails := newtails$
20   append min to $candidate$
21   **if** $candidate < o$ **then**
22         $o := candidate$
23 **return** o

Algorithm 1. Normal Form Computation

It remains to show that after lines 4 to 20 the candidate is in normal form. To this end we prove the following loop invariant for the while loop: *candidate* is an initial segment of the normal form and there is a $g \in stab$ and a $t \in tails$ such that $g * t$ is a permutation of the remaining rows of the normal form.

The lines 4, 5 and 6 ensure that the loop invariant is true at the start of the loop. We now assume that the loop invariant holds at the beginning of an iteration and prove that it is still true after the iteration. Since we know that there are $g \in stab$ and $t \in tails$ such that $g * t$ is a permutation of the remaining part of the normal form and the rank vector is invariant under the group action the lines 10 and 11 will at some point select an r that can be mapped to the next row of the normal form.

Since the normal form is the lexicographically smallest scheme in its equivalence class, the next row must always be the smallest row that has not been added to *candidate* yet. Therefore by choosing g such that $g * r$ is minimal in line 12 we ensure that min is the next row of the normal form.

In line 17 a transformed version $t' = g * t \setminus \{g * r\}$ of the element t with r removed is added to *newtails*. Therefore, *newtails* still contains an element t' that can be mapped to the remaining rows of the normal form.

Finally, we have to show that *stab* still contains a suitable element. Let $g' \in stab$ be such that g' maps t to a permutation of the remaining rows of the normal form. Let g be the element chosen to minimize r in line 11. Since *stab* is a group it must contain $g' \cdot g^{-1}$. Moreover, *newtails* contains $g * t \setminus \{g * r\}$ which is mapped to $g' * t \setminus \{g' * r\}$. Therefore, $g' \cdot g^{-1}$ has the desired property.   □

5   Minimizing the First Row

Algorithm 1 is more efficient than a naive walk through the whole symmetry group G because we can expect the stabilizer to quickly become small during the computation. However, in the first iteration we still go over the full group $\mathrm{GL}(K, n)^3$. In this section we describe how this can be avoided.

The order we have chosen ensures that the first row has a particular form.

Proposition 2. *Let* $G = \mathrm{GL}(K, n)^3$ *and let* $(A, B, C) \in (K^{n \times n})^3$ *be such that* (A, B, C) *is the minimal element of* $G * (A, B, C)$. *Then the following hold:*

1. *A has the form*

$$\begin{pmatrix} 0 & 0 \\ I_r & 0 \end{pmatrix}$$

   where $r = \mathrm{rank} A$.
2. *B is in column echelon form.*
3. *If* $\mathrm{rank} A = n$, *then* $A = I_n$ *and B has the form*

$$\begin{pmatrix} 0 & 0 \\ I_r & 0 \end{pmatrix} \tag{3}$$

   where $r = \mathrm{rank} B$.

Proof. Using Gaussian elimination we can find

$$(A', B', C') = (U, V, W) * (A, B, C)$$

where A' and B' are in the described form. Note that for part 3 we can first determine V and W and then choose $U = VA^{-1}$. To show that (A, B, C) already is in this form we proceed by induction on n. If $n = 1$, then the claims are true. For the induction step assume that the claims are true for $n - 1$.

1. We first consider the special case $\mathrm{rank} A = n$. Denote by $v_1, \ldots, v_n$ the columns of A. Since $A \leq A' = I_n$ there are two cases:
   Case 1: $v_n < e_n$. Then $v_n = 0$ contradicting the assumption that A has full rank.
   Case 2: $v_n = e_n$. Then the last row of A contains only zeros apart from the 1 in the bottom right corner. Otherwise we could use column reduction to make A smaller. Since A is minimal, also the matrix we get when we remove the last

column and row from A has to be minimal. So by the induction hypothesis A has the desired form.

Now suppose rank$A < n$. Since the last column of A' contains only zeros and A is minimal, the last column of A consists only of zeros. We can use row reduction to form a matrix A'' that is equivalent to A, has a zero row and all other rows equal to those of A. So $A'' < A$. We then shift the zero row of A'' to the top. Since this doesn't make A'' bigger, it is still not greater than A. Because of the minimality of A, its first row has then to be zero as well. Now we can remove the last column and first row of A and the resulting matrix must still be minimal. So by the induction hypothesis A is of the desired form.

2. Since we already showed $A = A'$ we can assume $U = V = I_n$. So B' is the column echelon form of B. We write B as $(v_1 \mid \cdots \mid v_n)$ and B' as $(v'_1 \mid \cdots \mid v'_n)$. We again have two cases:

Case 1: $v_n < v'_n$. So $v'_n \neq 0$ and since B' is in column echelon form this implies $v'_n = e_n$. Then $v_n = 0$ which contradicts that B' is the column echelon form of B.

Case 2: $v_n = v'_n$. Since B' is in column echelon form we either have $v_n = e_n$ or $v_n = 0$. We claim that the matrix we get by removing the last column and row from B is minimal. If not, there is a sequence of column operations that makes that matrix smaller. Let $B'' = (v''_1 \mid \cdots \mid v''_n)$ be the matrix we get by applying these operations to B and let i be the index of the right most column that was changed. So v''_i with the last element removed must be smaller than v_i with the last element removed. However, this implies that $v''_i < v_i$ and therefore $B'' < B$, which is a contradiction. So by the induction hypothesis B with the last row and column removed must be in column echelon form.

It remains to show that the last rows of B and B' are equal. There must exist a sequence of column operations that turn B into B'. If $v_n = e_n = v'_n$, then these operations would eliminate all elements in the last row of B, except the one in the bottom right corner. This implies $B' \leq B$ and therefore $B' = B$. If $v_n = 0 = v'_n$, then this sequence cannot change the last row because any column operation not involving the last column would destroy the column echelon form in the upper left part. Therefore $B = B'$.

3. Let rank$A = n$. We have already shown that $A = I_n$. For any choice of V we can choose $U = VA^{-1}$ to ensure $A' = I_n$. So B is minimal under arbitrary row and column permutations. So in this case the claim can be shown the same way as 1. $\qquad\square$

Let $s \in (K^{n \times n})^{r \times 3}$ be a matrix multiplication scheme. Denote by (U, V, W) the element of $\mathrm{GL}(K, n)^3$ used to transform s into normal form and denote by (A_1, B_1, C_1) the first row of the normal form of s. Let (A, B, C) be the row that is mapped to (A_1, B_1, C_1) and assume that the columns of s do not need to be permuted.

Then (A, B, C) must have a maximal rank vector. Therefore, A has the maximal rank of all the matrices in the scheme. So if the scheme contains a matrix of full rank then A has full rank. Moreover, A_1 is the minimal element equivalent to A under the action of $\mathrm{GL}(K, n)^3$.

Input : A triple of $n \times n$ matrices (A, B, C)
Output: A minimal triple equivalent under the action of $\mathrm{GL}(K, n)^3$

1 **if** rank$A = n$ **then**
2     $A_1 := I_n$
3     $C' := CA$
4   **if** rank$B = n$ **then**
5         $B_1 := I_n$
6         $C_1 := \min_{W \in \mathrm{GL}(K,n)} W B^{-1} C' W^{-1}$
7   **else**
8         $B_1 := \min_{V,W \in \mathrm{GL}(K,n)} V B W^{-1}$
9         $S := \{(V, W) \mid V, W \in \mathrm{GL}(K, n) \land V B = B_1 W\}$
10        $C_1 := \min_{(V,W) \in S} W C' V^{-1}$
11 **else**
12     $A_1 := \min_{U,V \in \mathrm{GL}(K,n)} U A V^{-1}$
13     $(U, V) := \mathrm{argmin}_{U,V \in \mathrm{GL}(K,n)} U A V^{-1}$
14     $B' = V B; C' = C U^{-1}$
15     $S := \{(U, V) \mid U, V \in \mathrm{GL}(K, n) \land U A_1 = A_1 V\}$
16     $B_1 := \min_{(U,V) \in S, W \in \mathrm{GL}(K,n)} V B' W^{-1}$
17     $C'' = W C'$, where W is chosen as in the line above
18     $S' := \{(U, V, W) \in \mathrm{GL}(K, n)^3 \mid U A_1 = A_1 V \land V B_1 = B_1 W\}$
19     $C_1 := \min_{(U,V,W) \in S'} W C'' V^{-1}$ **return** (A_1, B_1, C_1)

Algorithm 2. Special treatment of first row

If A has full rank, then $A_1 = I_n$ by Proposition 2. So we consider the scheme $s' = (A^{-1}, I_n, I_n) * s$ instead and update A, B, C and U, V, W accordingly. Then $A = A_1 = I_n$ and therefore $U = V$. So by Proposition 2, B_1 must be of the form 3.

If B also has full rank, then we set $s'' = (B^{-1}, B^{-1}, I_n) * s'$ and adjust A, B, C and U, V, W again. So we have $B = B_1 = I_n$ and $U = V = W$. Now we can determine C_1 and the stabilizer of the first row by iterating over $\mathrm{GL}(K, n)$ and minimizing $W C W^{-1}$.

If B does not have full rank, we determine all invertible matrices V and W such that $V B W^{-1} = B_1$. This can be done by solving the linear system $V B = B_1 W$ and discarding all solutions corresponding to singular matrices. Since $U = V$ we go through all possibilities for V and W and minimize $W C V^{-1}$. This allows us to determine C_1 and the stabilizer of the first row.

If A does not have full rank, we solve the linear system $U A = A_1 V$ and discard all solutions corresponding to singular matrices. The remaining solutions are the possible choices for U and V such that $U A V^{-1} = A_1$. By Proposition 2, B_1 must be in column echelon form. So for all possible choices of U and V we determine W such that $V B W^{-1}$ is in column echelon form. The smallest matrix $V B W^{-1}$ constructed this way must be equal to B_1. Then we go over all such triples (U, V, W) that map B to B_1 and determine those that minimize $W C U^{-1}$. So we find C_1 and the stabilizer of the first row.

The process is summarized in Algorithm 2.

6   Timings and Analysis

All the timing and analysis is done on an extension of Heule et al.'s data set with an implementation of our algorithm for 3×3 matrices over $\mathbb{Z}_2$. This data set contains 64,150 schemes.

For a comparison we have tested the equivalence check of Heule et al. on 10,000 randomly selected pairs from the data set and computed the normal form of 10,000 randomly selected schemes. Checking equivalence of two schemes took on average 0.0092 s. Computing a normal form took on average 1.87 s. The check for syntactic equivalence of the schemes in normal form takes about 0.00002 s, which is negligible. Thus, in our application checking equivalence of a single new scheme against a set of known schemes in normal form is faster than directly checking equivalence as soon as we have at least 204 schemes.

To get an idea how well the algorithm scales for larger values of n we have experimentally determined the size of the stabilizers in Algorithm 2. Instead of iterating over the complete group $\mathrm{GL}(K, n)^3$, Algorithm 2 only iterates over stabilizers from the beginning on. In the case that the scheme contains at least one matrix of full rank, which in the data set are slightly more than half of the schemes, we have to iterate over all elements of $\mathrm{GL}(K, n)^2$ in the worst case. However, this is a very pessimistic upper bound. The size of $\mathrm{GL}(\mathbb{Z}_2, 3)^2$ is 28,224, whereas the average size of the stabilizer we actually iterate over is 460.

In the second case we iterate over S in line 15, which again cannot exceed the size of $\mathrm{GL}(K, n)^2$ and on the data set has on average 135 elements. We also have to iterate over S' in line 18. The size of S' is bounded by $|\mathrm{GL}(K, n)^3|$, which in our case is 4,741,632. However, for a sample of the data set the largest stabilizer that occurred contains 576 elements and on average this stabilizer has 274 elements.

In summary, naively computing a normal form by simply iterating over all elements of $\mathrm{GL}(K, n)^3$ for each row will take time $O(r|\mathrm{GL}(K, n)|^3)$, where r is the length of the scheme. Assuming that after $O(1)$ iterations of Algorithm 1 we are left with a stabilizer of size $O(1)$, the cost of Algorithm 1 is only $O(r + |\mathrm{GL}(K, n)|^3)$. We cannot prove that the stabilizers become so small so quickly, but the assumption is consistent with our experiments. Finally, assuming that the solution space in line 18 of Algorithm 2 has at most $O(|\mathrm{GL}(K, n)|^2)$ elements, Algorithm 2 pushes the total cost of computing a normal form down to $O(r + |\mathrm{GL}(K, n)|^2)$. Again, we cannot prove any such claim about line 18, but the assumption is consistent with our experiments.

References

1. Alman, J., Williams, V.V.: A refined laser method and faster matrix multiplication. In: Proceedings of the 2021 ACM-SIAM Symposium on Discrete Algorithms (SODA), pp. 522–539 (2021). https://doi.org/10.1137/1.9781611976465.32
2. Berger, G.O., Absil, P.A., De Lathauwer, L., Jungers, R.M., Van Barel, M.: Equivalent polyadic decompositions of matrix multiplication tensors. J. Comput. Appl. Math. **406**, 17, Paper no. 113941 (2022). https://doi.org/10.1016/j.cam.2021.113941

3. Bläser, M.: On the complexity of the multiplication of matrices of small formats. J. Complex. **19**(1), 43–60 (2003). https://doi.org/10.1016/S0885-064X(02)00007-9

4. Bürgisser, P., Clausen, M., Shokrollahi, M.A.: Algebraic Complexity Theory, vol. 315. Springer, Heidelberg (2013)

5. Courtois, N.T., Bard, G.V., Hulme, D.: A new general-purpose method to multiply 3×3 matrices using only 23 multiplications (2011). https://doi.org/10.48550/ARXIV.1108.2830

6. de Groote, H.F.: On varieties of optimal algorithms for the computation of bilinear mappings ii. optimal algorithms for 2×2-matrix multiplication. Theor. Comput. Sci. **7**(2), 127–148 (1978). https://doi.org/10.1016/0304-3975(78)90045-2

7. Heule, M.J.H., Kauers, M., Seidl, M.: New ways to multiply 3×3-matrices. J. Symbolic Comput. **104**, 899–916 (2021). https://doi.org/10.1016/j.jsc.2020.10.003

8. Johnson, R.W., McLoughlin, A.M.: Noncommutative bilinear algorithms for 3×3 matrix multiplication. SIAM J. Comput. **15**(2), 595–603 (1986). https://doi.org/10.1137/0215043

9. Laderman, J.D.: A noncommutative algorithm for multiplying 3×3 matrices using 23 multiplications. Bull. Am. Math. Soc. **82**(1), 126–128 (1976). https://doi.org/10.1090/S0002-9904-1976-13988-2

10. Oh, J., Kim, J., Moon, B.R.: On the inequivalence of bilinear algorithms for 3×3 matrix multiplication. Inf. Process. Lett. **113**(17), 640–645 (2013). https://doi.org/10.1016/j.ipl.2013.05.011

11. Rosowski, A.: Fast commutative matrix algorithm (2019). https://doi.org/10.48550/ARXIV.1904.07683

12. Smirnov, A.V.: The bilinear complexity and practical algorithms for matrix multiplication. Comput. Math. Math. Phys. **53**(12), 1781–1795 (2013). https://doi.org/10.1134/S0965542513120129

13. Strassen, V.: Gaussian elimination is not optimal. Numer. Math. **13**, 354–356 (1969). https://doi.org/10.1007/BF02165411

14. Winograd, S.: On multiplication of 2×2 matrices. Linear Algebra Appl. **4**(4), 381–388 (1971). https://doi.org/10.1016/0024-3795(71)90009-7

Bideterministic Weighted Automata

Peter Kostolányi$^{(\boxtimes)}$ (ID)

Department of Computer Science, Comenius University in Bratislava,
842 48 Mlynská dolina, Bratislava, Slovakia
`kostolanyi@fmph.uniba.sk`

Abstract. A deterministic finite automaton is called bideterministic if its transpose is deterministic as well. The study of such automata in a weighted setting is initiated. All trim bideterministic weighted automata over integral domains and positive semirings are proved to be minimal. On the contrary, it is observed that this property does not hold over finite commutative rings in general. Moreover, it is shown that the problem of determining whether a given rational series is realised by a bideterministic automaton is decidable over fields as well as over tropical semirings.

Keywords: Bideterministic weighted automaton · Minimal automaton · Integral domain · Positive semiring · Decidability

1 Introduction

It is well known that – in contrast to the classical case of automata without weights – weighted finite automata might not always be determinisable. Partly due to relevance of deterministic weighted automata for practical applications such as natural language and speech processing [24] and partly due to the purely theoretical importance of the determinisability problem, questions related to deterministic weighed automata – such as the decidability of determinisability, existence of efficient determinisation algorithms, or characterisations of series realised by deterministic weighted automata – have received significant attention. They were studied for weighted automata over specific classes of semi-rings, such as tropical semirings or fields [1,6,18–21,24,25], as well as over strong bimonoids [9], often under certain additional restrictions.

The questions mentioned above are known to be relatively hard. For instance, despite some partial results [18–21], the decidability status of the general determinisability problem for weighted automata is still open over tropical semirings or over the field of rationals [21]. It thus makes sense to take a look at stronger forms of determinism in weighted automata, which may be amenable to a somewhat easier analysis.

The work was supported by the grant VEGA 1/0601/20.

ⓒ The Author(s), under exclusive license to Springer Nature Switzerland AG 2022
D. Poulakis and G. Rahonis (Eds.): CAI 2022, LNCS 13706, pp. 161–174, 2022.
https://doi.org/10.1007/978-3-031-19685-0_12

One possibility is to study deterministic weighted automata with additional requirements on their weights. This includes for instance the research on crisp-deterministic weighted automata by M. Ćirić et al. [9]. Another possibility is to examine the weighted counterpart of some particularly simple subclass of deterministic finite automata without weights – that is, to impose further restrictions not only on weights of deterministic weighted automata, but on the concept of determinism itself. This is a direction that we follow in this article.

More tangibly, this article aims to initiate the study of *bideterministic* finite automata in the weighted setting. A finite automaton is bideterministic if it is deterministic and its transpose – *i.e.*, an automaton obtained by reversing all transitions and exchanging the roles of its initial and terminal states – is deterministic as well. Note that this implies that a bideterministic automaton always contains at most one initial and at most one terminal state. Bideterministic finite automata have been first touched upon from a theoretical perspective by J.-É. Pin [27], as a particular case of reversible finite automata. The fundamental properties of bideterministic finite automata have mostly been explored by H. Tamm and E. Ukkonen [35,36] – in particular, they have shown that a trim bideterministic automaton is always a minimal nondeterministic automaton for the language it recognises, minimality being understood in the strong sense, *i.e.*, with respect to the number of states.[1] An alternative proof of this fact was recently presented by R. S. R. Myers, S. Milius, and H. Urbat [26].

Apart from these studies, bideterministic automata have been – explicitly or implicitly – considered in connection to the star height problem [14,22,23], from the perspective of language inference [3], in the theory of block codes [33], and in connection to presentations of inverse monoids [16,34].

We define *bideterministic weighted automata* over a semiring by analogy to their unweighted counterparts, and study the conditions under which the fundamental property of H. Tamm and E. Ukkonen [35,36] generalises to the weighted setting. Thus, given a semiring S, we ask the following questions: Are all trim bideterministic weighted automata over S minimal? Does every bideterministic automaton over S admit a bideterministic equivalent that is at the same time minimal? We answer both these questions in affirmative when S is an integral domain or a positive – *i.e.*, both zero-sum free and zero-divisor free – semiring. On the other hand, we show that the answer is negative for a large class of commutative semirings including a multitude of *finite* commutative *rings*.

Finally, we consider the problem of deciding whether a weighted automaton over a semiring S admits a bideterministic equivalent, and show that it is decidable when S is a field or a tropical semiring (of nonnegative integers, integers, or rationals). This suggests that the bideterminisability problem for weighted automata might be somewhat easier than the determinisability problem, whose decidability status over fields such as the rationals and over tropical semirings remains open [21].

---

[1] In fact, H. Tamm and E. Ukkonen [35,36] have shown a stronger property: a trim bideterministic automaton is the *only* minimal nondeterministic finite automaton recognising its language.

2   Preliminaries

A *semiring* is a quintuple $(S, +, \cdot, 0, 1)$ such that $(S, +, 0)$ is a commutative monoid, $(S, \cdot, 1)$ is a monoid, multiplication distributes over addition both from left and from right, and $a \cdot 0 = 0 \cdot a = 0$ holds for all $a \in S$; it is said to be *commutative* when $\cdot$ is. A semiring S is *zero-sum free* [13,15] if $a + b = 0$ for some $a, b \in S$ implies $a = b = 0$ and *zero-divisor free* [15], or *entire* [13], if $a \cdot b = 0$ for some $a, b \in S$ implies that $a = 0$ or $b = 0$. A semiring is *positive* [12,17] if it is both zero-sum free and zero-divisor free. A *ring* is a semiring $(R, +, \cdot, 0, 1)$ such that R forms an abelian group with addition. An *integral domain* is a nontrivial zero-divisor free commutative ring. A *field* is an integral domain $(\mathbb{F}, +, \cdot, 0, 1)$ such that $\mathbb{F} \setminus \{0\}$ forms an abelian group with multiplication.

We now briefly recall some basic facts about noncommutative formal power series and weighted automata. More information can be found in [7,10,11,29]. Alphabets are assumed to be finite and nonempty in what follows.

A *formal power series* over a semiring S and alphabet Σ is a mapping $r \colon \Sigma^* \to S$. The value of r upon $w \in \Sigma^*$ is usually denoted by (r, w) and called the *coefficient* of r at w; the coefficient of r at ε, the empty word, is referred to as the *constant coefficient*. The series r itself is written as

$$r = \sum_{w \in \Sigma^*} (r, w)\, w.$$

The set of all formal power series over S and Σ is denoted by $S\langle\langle \Sigma^* \rangle\rangle$.

Given series $r, s \in S\langle\langle \Sigma^* \rangle\rangle$, their *sum* $r + s$ and *product* $r \cdot s$ are defined by $(r + s, w) = (r, w) + (s, w)$ and

$$(r \cdot s, w) = \sum_{\substack{u,v \in \Sigma^* \\ uv = w}} (r, u)(s, v)$$

for all $w \in \Sigma^*$. Every $a \in S$ is identified with a series with constant coefficient a and all other coefficients zero, and every $w \in \Sigma^*$ with a series with coefficient 1 at w and zero coefficients at all $x \in \Sigma^* \setminus \{w\}$. Thus, for instance, $r = 2ab + 3abb$ is a series with $(r, ab) = 2$, $(r, abb) = 3$, and $(r, x) = 0$ for every $x \in \Sigma^* \setminus \{ab, abb\}$. One may observe that $(S\langle\langle \Sigma^* \rangle\rangle, +, \cdot, 0, 1)$ is a semiring again.

For I an index set, a family $(r_i \mid i \in I)$ of series from $S\langle\langle \Sigma^* \rangle\rangle$ is *locally finite* if $I(w) = \{i \in I \mid (r_i, w) \neq 0\}$ is finite for all $w \in \Sigma^*$. The *sum* over the family $(r_i \mid i \in I)$ can then be defined by

$$\sum_{i \in I} r_i = r,$$

where the coefficient (r, w) at each $w \in \Sigma^*$ is given by a *finite* sum

$$(r, w) = \sum_{i \in I(w)} (r_i, w).$$

The *support* of $r \in S\langle\langle\Sigma^*\rangle\rangle$ is the language $\text{supp}(r) = \{w \in \Sigma^* \mid (r, w) \neq 0\}$.
The *left quotient* of $r \in S\langle\langle\Sigma^*\rangle\rangle$ by a word $x \in \Sigma^*$ is a series $x^{-1}r$ such that
$(x^{-1}r, w) = (r, xw)$ for all $w \in \Sigma^*$.

A *weighted (finite) automaton* over a semiring S and alphabet Σ is a quadru-
ple $\mathcal{A} = (Q, \sigma, \iota, \tau)$, where Q is a finite set of states, $\sigma \colon Q \times \Sigma \times Q \to S$ a transition weighting function, $\iota \colon Q \to S$ an initial weighting function, and $\tau \colon Q \to S$
a terminal weighting function. We often assume without loss of generality that
$Q = [n] = \{1, \ldots, n\}$ for some nonnegative integer n; we write $\mathcal{A} = (n, \sigma, \iota, \tau)$
instead of $\mathcal{A} = ([n], \sigma, \iota, \tau)$ in that case.

A *transition* of $\mathcal{A} = (Q, \sigma, \iota, \tau)$ is a triple $(p, c, q) \in Q \times \Sigma \times Q$ such that
$\sigma(p, c, q) \neq 0$. A *run* of $\mathcal{A}$ is a word $\gamma = q_0 c_1 q_1 c_2 q_2 \ldots q_{n-1} c_n q_n \in (Q\Sigma)^* Q$,
for some nonnegative integer n, such that $q_0, \ldots, q_n \in Q$, $c_1, \ldots, c_n \in \Sigma$,
and (q_{k-1}, c_k, q_k) is a transition for $k = 1, \ldots, n$; we also say that γ is a run
from q_0 *to* q_n. Moreover, we write $\lambda(\gamma) = c_1 c_2 \ldots c_n \in \Sigma^*$ for the *label* of γ
and $\sigma(\gamma) = \sigma(q_0, c_1, q_1)\sigma(q_1, c_2, q_2)\ldots\sigma(q_{n-1}, c_n, q_n) \in S$ for the *value* of γ.
The *monomial* $\|\gamma\| \in S\langle\langle\Sigma^*\rangle\rangle$ realised by the run γ is defined by

$$\|\gamma\| = (\iota(q_0)\sigma(\gamma)\tau(q_n))\,\lambda(\gamma).$$

If we denote by $\mathcal{R}(\mathcal{A})$ the set of all runs of the automaton $\mathcal{A}$, then the family
of monomials $(\|\gamma\| \mid \gamma \in \mathcal{R}(\mathcal{A}))$ is obviously locally finite and the *behaviour* of $\mathcal{A}$
can be defined by the infinite sum

$$\|\mathcal{A}\| = \sum_{\gamma \in \mathcal{R}(\mathcal{A})} \|\gamma\|.$$

In particular, $\|\mathcal{A}\| = 0$ if $Q = \emptyset$. A series $r \in S\langle\langle\Sigma^*\rangle\rangle$ is *rational* over S if
$r = \|\mathcal{A}\|$ for some weighted automaton $\mathcal{A}$ over S and Σ.

A state $q \in Q$ of a weighted automaton $\mathcal{A} = (Q, \sigma, \iota, \tau)$ over S and Σ is said
to be *accessible* if there is a run in $\mathcal{A}$ from some $p \in Q$ satisfying $\iota(p) \neq 0$ to q.[2]
Dually, a state $q \in Q$ is *coaccessible* if there is a run in $\mathcal{A}$ from q to some $p \in Q$
such that $\tau(p) \neq 0$. The automaton $\mathcal{A}$ is *trim* if all its states are both accessible
and coaccessible [29].

Given a weighted automaton $\mathcal{A} = (Q, \sigma, \iota, \tau)$ and $q \in Q$, we denote by $\|\mathcal{A}\|_q$
the *future* of q, i.e., the series realised by an automaton $\mathcal{A}_q = (Q, \sigma, \iota_q, \tau)$ where
$\iota_q(q) = 1$ and $\iota_q(p) = 0$ for all $p \in Q \setminus \{q\}$.

Let $S^{m \times n}$ be the set of all $m \times n$ matrices over S. A *linear representation* of
a weighted automaton $\mathcal{A} = (n, \sigma, \iota, \tau)$ over S and Σ is given by $\mathcal{P}_{\mathcal{A}} = (n, \mathbf{i}, \mu, \mathbf{f})$,
where $\mathbf{i} = (\iota(1), \ldots, \iota(n))$, $\mu \colon (\Sigma^*, \cdot) \to (S^{n \times n}, \cdot)$ is a monoid homomorphism
such that for all $c \in \Sigma$ and $i, j \in [n]$, the entry of $\mu(c)$ in the i-th row and j-th
column is given by $\sigma(i, c, j)$, and $\mathbf{f} = (\tau(1), \ldots, \tau(n))^T$. The representation $\mathcal{P}_{\mathcal{A}}$
describes $\mathcal{A}$ unambiguously, and $(\|\mathcal{A}\|, w) = \mathbf{i}\mu(w)\mathbf{f}$ holds for all $w \in \Sigma^*$.

As a consequence of this connection to linear representations, methods of
linear algebra can be employed in the study of weighted automata *over fields*.
This leads to a particularly well-developed theory, including a polynomial-time

---

[2] Note that the value of this run might be zero in case S is not zero-divisor free.

minimisation algorithm, whose basic ideas go back to M.-P. Schützenberger [32] and which has been explicitly described by A. Cardon and M. Crochemore [8]. The reader may consult [7,29,30] for a detailed exposition.

For our purposes, we only note that the gist of this minimisation algorithm lies in an observation that given a weighted automaton $\mathcal{A}$ over a field $\mathbb{F}$ and alphabet Σ with $\mathcal{P}_\mathcal{A} = (n, \mathbf{i}, \mu, \mathbf{f})$, one can find in polynomial time a finite language $L = \{x_1, \ldots, x_m\}$ of words over Σ that is prefix-closed, and the vectors $\mathbf{i}\mu(x_1), \ldots, \mathbf{i}\mu(x_m)$ form a basis of the vector subspace $\mathrm{Left}(\mathcal{A})$ of $\mathbb{F}^{1 \times n}$ generated by the vectors $\mathbf{i}\mu(x)$ with $x \in \Sigma^*$. Such a language L is called a *left basic language* of $\mathcal{A}$. Similarly, one can find in polynomial time a *right basic language* of $\mathcal{A}$ – i.e., a finite language $R = \{y_1, \ldots, y_k\}$ of words over Σ that is suffix-closed, and the vectors $\mu(y_1)\mathbf{f}, \ldots, \mu(y_k)\mathbf{f}$ form a basis of the vector subspace $\mathrm{Right}(\mathcal{A})$ of $\mathbb{F}^{n \times 1}$ generated by the vectors $\mu(y)\mathbf{f}$ with $y \in \Sigma^*$.

The actual minimisation algorithm then consists of two reduction steps. The original weighted automaton $\mathcal{A}$ with representation $\mathcal{P}_\mathcal{A} = (n, \mathbf{i}, \mu, \mathbf{f})$ is first transformed into an equivalent automaton $\mathcal{B}$ with $\mathcal{P}_\mathcal{B} = (k, \mathbf{i}', \mu', \mathbf{f}')$. Here, $k \leq n$ is the size of the right basic language $R = \{y_1, \ldots, y_k\}$ of $\mathcal{A}$ with $y_1 = \varepsilon$,

$$\mathbf{i}' = \mathbf{i}Y, \quad \mu'(c) = Y_\ell^{-1}\mu(c)Y \text{ for all } c \in \Sigma, \quad \text{and} \quad \mathbf{f}' = (1, 0, \ldots, 0)^T, \quad (1)$$

where $Y \in \mathbb{F}^{n \times k}$ is a matrix of full column rank with columns $\mu(y_1)\mathbf{f}, \ldots, \mu(y_k)\mathbf{f}$ and $Y_\ell^{-1} \in \mathbb{F}^{k \times n}$ is its left inverse matrix. The automaton $\mathcal{B}$ is then transformed into a *minimal* equivalent automaton $\mathcal{C}$ with $\mathcal{P}_\mathcal{C} = (m, \mathbf{i}'', \mu'', \mathbf{f}'')$. Here, $m \leq k$ is the size of the left basic language $L = \{x_1, \ldots, x_m\}$ of $\mathcal{B}$ with $x_1 = \varepsilon$,

$$\mathbf{i}'' = (1, 0, \ldots, 0), \quad \mu''(c) = X\mu'(c)X_r^{-1} \text{ for all } c \in \Sigma, \quad \text{and} \quad \mathbf{f}'' = X\mathbf{f}', \quad (2)$$

where $X \in \mathbb{F}^{m \times k}$ is a matrix of full row rank with rows $\mathbf{i}'\mu'(x_1), \ldots, \mathbf{i}'\mu'(x_m)$ and X_r^{-1} is its right inverse matrix. As the vector space $\mathrm{Left}(\mathcal{B})$ – which is the row space of X – is invariant under $\mu'(c)$ for all $c \in \Sigma$, it follows that

$$\mathbf{i}''X = \mathbf{i}', \quad \mu''(c)X = X\mu'(c) \text{ for all } c \in \Sigma, \quad \text{and} \quad \mathbf{f}'' = X\mathbf{f}', \quad (3)$$

showing that the automaton $\mathcal{C}$ is *conjugate* [4,5] to $\mathcal{B}$ by the matrix X. Thus $\mathbf{i}''\mu''(x)X = \mathbf{i}'\mu'(x)$ for all $x \in \Sigma^*$, so that the vector $\mathbf{i}''\mu''(x)$ represents the coordinates of $\mathbf{i}'\mu'(x)$ with respect to the basis $(\mathbf{i}'\mu'(x_1), \ldots, \mathbf{i}'\mu'(x_m))$ of $\mathrm{Left}(\mathcal{B})$. In particular, note that $(\mathbf{i}''\mu''(x_1), \ldots, \mathbf{i}''\mu''(x_m))$ is the standard basis of $\mathbb{F}^m$.

Finally, let us mention that any weighted automaton $\mathcal{A}$ over $\mathbb{F}$ and Σ with $\mathcal{P}_\mathcal{A} = (n, \mathbf{i}, \mu, \mathbf{f})$ gives rise to a linear mapping $\Lambda[\mathcal{A}]\colon \mathrm{Left}(\mathcal{A}) \to \mathbb{F}\langle\langle \Sigma^* \rangle\rangle$, uniquely defined by

$$\Lambda[\mathcal{A}]\colon \mathbf{i}\mu(x) \mapsto \sum_{w \in \Sigma^*} (\mathbf{i}\mu(x)\mu(w)\mathbf{f}) \, w = x^{-1}\|\mathcal{A}\| \quad (4)$$

for all $x \in \Sigma^*$. This mapping is always injective when $\mathcal{A}$ is a minimal automaton realising its behaviour [30].

3   Bideterministic Weighted Automata over a Semiring

In the same way as for finite automata without weights [35,36], we say that a weighted automaton $\mathcal{A}$ is *bideterministic* if both $\mathcal{A}$ and its transpose are deterministic; in particular, $\mathcal{A}$ necessarily contains at most one state with nonzero initial weight and at most one state with nonzero terminal weight. This is made more precise by the following definition.

Definition 1. *Let S be a semiring and Σ an alphabet. A weighted automaton $\mathcal{A} = (Q, \sigma, \iota, \tau)$ over S and Σ is* bideterministic *if all of the following conditions are satisfied:*

(i) There is at most one state $p \in Q$ such that $\iota(p) \neq 0$.
(ii) If $\sigma(p, c, q) \neq 0$ and $\sigma(p, c, q') \neq 0$ for $p, q, q' \in Q$ and $c \in \Sigma$, then $q = q'$.
(iii) There is at most one state $q \in Q$ such that $\tau(q) \neq 0$.
(iv) If $\sigma(p, c, q) \neq 0$ and $\sigma(p', c, q) \neq 0$ for $p, p', q \in Q$ and $c \in \Sigma$, then $p = p'$.

The conditions (i) and (ii) assure that the automaton $\mathcal{A}$ is *deterministic*, while the conditions (iii) and (iv) assure the same property for its transpose.

It has been shown by H. Tamm and E. Ukkonen [35,36] that a trim bideterministic automaton without weights is always a minimal nondeterministic automaton for the language it recognises. As a consequence, every language recognised by some bideterministic automaton also admits a minimal automaton that is bideterministic. Moreover, by uniqueness of minimal deterministic finite automata and existence of efficient minimisation algorithms, it follows that it is decidable whether a language is recognised by a bideterministic automaton.

In what follows, we ask whether these properties generalise to bideterministic weighted automata over some semiring S. That is, given a semiring S, we are interested in the following three questions.[3]

Question 1. Is every trim bideterministic weighted automaton over S necessarily minimal?
Question 2. Does every bideterministic automaton over S admit an equivalent minimal weighted automaton over S that is bideterministic?
Question 3. Is it decidable whether a weighted automaton over S admits a bideterministic equivalent?

An affirmative answer to Question 1 clearly implies an affirmative answer to Question 2 as well. We study the first two questions in Sect. 4 and the last question in Sect. 5.

4   The Minimality Property of Bideterministic Automata

We now study the conditions on a semiring S under which the trim bideterministic weighted automata over S are always minimal, and answer the Question 1,

---

[3] Minimality of an automaton is understood with respect to the number of states in what follows.

as well as the related Question 2, for three representative classes of semirings. In particular, we show that every trim bideterministic weighted automaton over a *field* – or, more generally, over an *integral domain* – is minimal. The same property is observed for bideterministic weighted automata over *positive semirings*, including for instance the *tropical semirings* and *semirings of formal languages*. On the other hand, we prove that both questions have negative answers over a large class of commutative semirings other than integral domains, which also includes numerous *finite commutative rings*.

4.1   Fields and Integral Domains

The minimality property of trim bideterministic weighted automata *over fields* follows by the fact that the Cardon-Crochemore minimisation algorithm for these automata, described in Sect. 2, preserves both bideterminism and the number of useful states of a bideterministic automaton, as we now observe.

Theorem 2. *Let A be a bideterministic weighted automaton over a field $\mathbb{F}$. Then the Cardon-Crochemore minimisation algorithm applied to A outputs a bideterministic weighted automaton C. Moreover, if A trim, then C has the same number of states as A.*

Proof. Let $\mathcal{P} = (n, \mathbf{i}, \mu, \mathbf{f})$ be a linear representation of some bideterministic weighted automaton $\mathcal{D}$. Then there is at most one nonzero entry in each row and column of $\mu(c)$ for each $c \in \Sigma$, and at most one nonzero entry in $\mathbf{i}$ and $\mathbf{f}$.

Moreover, the words $x_1, \ldots, x_m$ of the left basic language of $\mathcal{D}$ correspond bijectively to accessible states of $\mathcal{D}$ and the vector $\mathbf{i}\mu(x_i)$ contains, for $i = 1, \ldots, m$, exactly one nonzero entry at the position determined by the state corresponding to x_i. Similarly, the words $y_1, \ldots, y_k$ of the right basic language of $\mathcal{D}$ correspond to coaccessible states and the vector $\mu(y_i)\mathbf{f}$ contains, for $i = 1, \ldots, k$, exactly one nonzero entry. Thus, using these vectors to form the matrices X and Y as in Sect. 2, we see that one obtains monomial matrices after removing the zero columns from X and the zero rows from Y. As a result, a right inverse X_r^{-1} of X can be obtained by taking the reciprocals of all nonzero entries of X and transposing the resulting matrix, and similarly for a left inverse Y_ℓ^{-1} of Y.

The matrices $X\mu(c)X_r^{-1}$ and $Y_\ell^{-1}\mu(c)Y$ for $c \in \Sigma^*$ clearly contain at most one nonzero entry in each row and column, and the vectors $\mathbf{i}Y$ and $X\mathbf{f}$ contain at most one nonzero entry as well. This means that the reduction step (1) applied to a bideterministic automaton A yields a bideterministic automaton B, and that the reduction step (2) applied to the bideterministic automaton B yields a bideterministic minimal automaton C as an output of the algorithm.

When A is in addition trim, then what has been said implies that the words of the right basic language of A correspond bijectively to states of A, so that the automaton B obtained via (1) has the same number of states as A. This automaton is obviously trim as well, and the words of the left basic language of B correspond bijectively to states of B. Hence, the automaton C obtained via (2) also has the same number of states as A.   □

As every integral domain can be embedded into its field of fractions, the property established above holds for automata over integral domains as well.

Corollary 3. *Every trim bideterministic weighted automaton over an integral domain is minimal.*

4.2 Other Commutative Rings

We now show that the property established above for automata over integral domains *cannot* be generalised to automata over commutative rings, by exhibiting a suitable class of commutative *semirings* S such that bideterministic weighted automata over S do not even always admit a minimal bideterministic equivalent.

Theorem 4. *Let S be a commutative semiring with elements $s, t \in S$ such that $st = 0$ and $s^2 \neq 0 \neq t^2$. Then there is a trim bideterministic weighted automaton $\mathcal{A}$ over S such that none of the minimal automata for $\|\mathcal{A}\|$ is bideterministic.*

Proof. Consider a trim bideterministic weighted automaton $\mathcal{A}$ over S depicted in Fig. 1. Clearly, $\|\mathcal{A}\| = s^2 \cdot aba + t^2 \cdot bb$. The automaton $\mathcal{A}$ is not minimal, as the same series is realised by a smaller automaton $\mathcal{B}$ in Fig. 2:

$$\|\mathcal{B}\| = s^2 \cdot aba + t^2 \cdot bb = \|\mathcal{A}\|.$$

The answer to Question 1 of Sect. 3 is thus negative over S.

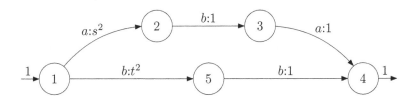

Fig. 1. The trim bideterministic weighted automaton $\mathcal{A}$ over S.

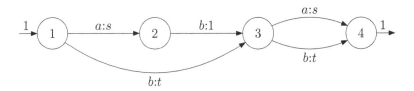

Fig. 2. The four-state weighted automaton $\mathcal{B}$ over S equivalent to $\mathcal{A}$.

We show that $\|\mathcal{A}\|$ is actually not realised by any bideterministic weighted automaton over S with less than five states. This implies that $\mathcal{A}$ is a counterexample to Question 2 of Sect. 3, and eventually completes the proof.

Indeed, consider a bideterministic weighted automaton $\mathcal{C} = (Q, \sigma, \iota, \tau)$ such that $\|\mathcal{C}\| = \|\mathcal{A}\|$. At least one state with nonzero initial weight is needed to realise $\|\mathcal{A}\|$ by $\mathcal{C}$, as $\|\mathcal{A}\| \neq 0$. Let us call this state 1.

As $(\|\mathcal{A}\|, aba) = s^2 \neq 0$, there is a transition on a in $\mathcal{C}$ leading from 1. This cannot be a loop at 1, as otherwise ba would have a nonzero coefficient in $\|\mathcal{C}\|$, contradicting $\|\mathcal{C}\| = \|\mathcal{A}\|$. It thus leads to some new state, say, 2.

There has to be a transition on b leading from 2 and in the same way as above, we observe that it can lead neither to 1, nor to 2, as otherwise a or aa would have a nonzero coefficient in $\|\mathcal{C}\|$. It thus leads to some new state 3.

Exactly the same reasoning gives us existence of another state 4, to which a transition on a leads from 3, and which has a nonzero terminal weight $\tau(4)$.

Existence of one more state has to be established in order to finish the proof. To this end, observe that $(\|\mathcal{A}\|, bb) = t^2 \neq 0$, so that $\mathcal{C}$ has a transition from 1 on b, which cannot be a loop at 1, as otherwise b would have a nonzero coefficient in $\|\mathcal{C}\|$. This transition cannot lead to 2 either, as there already is a transition on b from 2 to 3, so that bb would have coefficient 0 in $\|\mathcal{C}\|$. Likewise, it cannot lead to 3, as there already is a transition on b from 2 to 3 and $\mathcal{C}$ is supposed to be bideterministic. Finally, it also cannot lead to 4, as otherwise there would have to be a loop labelled by b at 4 and b would have a nonzero coefficient in $\|\mathcal{C}\|$. The transition on b from 1 thus indeed leads to some new state 5. □

Note that the class of commutative semirings from Theorem 4 also includes many *finite* commutative *rings*. In particular, the ring $\mathbb{Z}_m$ of integers modulo m falls into this class whenever m has at least two distinct prime factors. The characterisation of commutative rings, over which all trim bideterministic weighted automata are minimal, remains open. It would have been nice to know at least what the situation is over finite rings $\mathbb{Z}_{p^n}$ for p prime and $n \geq 2$.

4.3   Positive Semirings

We now observe that the minimality property does hold for trim bideterministic weighted automata *over positive semirings*. Recall that a semiring is *positive* if it is both zero-sum free and zero-divisor free. This class includes for instance the *tropical semirings*, *semirings of formal languages*, and the *Boolean semiring*.

Theorem 5. *Every trim bideterministic weighted automaton over a positive semiring is minimal.*

Proof. Let $\mathcal{A}$ be a trim bideterministic weighted automaton over a positive semiring S. By positivity of S, the language $\mathrm{supp}(\|\mathcal{A}\|)$ is recognised by a trim bideterministic finite automaton $\mathcal{A}'$ obtained from $\mathcal{A}$ by "forgetting about weights". This is a minimal nondeterministic automaton for $\mathrm{supp}(\|\mathcal{A}\|)$ by the minimality property of trim bideterministic automata without weights [35, 36].

Now, if $\mathcal{A}$ was not minimal, there would be a smaller weighted automaton $\mathcal{B}$ over S such that $\|\mathcal{B}\| = \|\mathcal{A}\|$. By "forgetting about its weights", we would obtain a nondeterministic finite automaton $\mathcal{B}'$ recognising $\mathrm{supp}(\|\mathcal{B}\|) = \mathrm{supp}(\|\mathcal{A}\|)$. However, $\mathcal{B}'$ is smaller than $\mathcal{A}'$, contradicting the minimality of $\mathcal{A}'$. □

5 Decidability of Bideterminisability

Let us now consider the problem of deciding whether a given weighted automaton admits a bideterministic equivalent. While the decidability status of the *determinisability* problem is open both over fields such as the rationals and over tropical semirings [21], we prove that the *bideterminisability* problem is decidable both over effective fields and over tropical semirings (of nonnegative integers, integers, and rationals).

5.1 Fields

We prove decidability of the bideterminisability problem for automata over fields by strengthening Theorem 2 – we show that the Cardon-Crochemore minimisation algorithm outputs a bideterministic automaton not only when applied to a bideterministic automaton, but also when applied to any bideterminisable automaton. To decide bideterminisability, it thus suffices to run this algorithm and find out whether its output is bideterministic.

Lemma 6. *Let $\mathcal{A}$ be a weighted automaton over a field $\mathbb{F}$ such that some of the minimal automata equivalent to $\mathcal{A}$ is deterministic. Then the Cardon-Crochemore algorithm applied to $\mathcal{A}$ outputs a deterministic automaton.*

Proof. Let $\mathcal{C}$ with $\mathcal{P}_{\mathcal{C}} = (m, \mathbf{i}, \mu, \mathbf{f})$ be the output of the Cardon-Crochemore algorithm upon $\mathcal{A}$ and $L = \{x_1, \ldots, x_m\}$ with $x_1 = \varepsilon$ the left basic language used in reduction step (2). Then $\mathbf{i}\mu(x)$ represents, for all $x \in \Sigma^*$, the coordinates of the series $x^{-1}\|\mathcal{A}\|$ with respect to the basis $(x_1^{-1}\|\mathcal{A}\|, \ldots, x_m^{-1}\|\mathcal{A}\|)$ of the vector space $\mathcal{Q}(\|\mathcal{A}\|)$ generated by left quotients of $\|\mathcal{A}\|$ by words.

To see this, recall that $(\mathbf{i}\mu(x_1), \ldots, \mathbf{i}\mu(x_m))$ is the standard basis of $\mathbb{F}^m$ and that the linear mapping $\Lambda[\mathcal{C}]$ given as in (4) is injective by minimality of $\mathcal{C}$. As the image of $\Lambda[\mathcal{C}]$ spans $\mathcal{Q}(\|\mathcal{C}\|) = \mathcal{Q}(\|\mathcal{A}\|)$, we see that

$$\left(x_1^{-1}\|\mathcal{A}\|, \ldots, x_m^{-1}\|\mathcal{A}\|\right) = (\Lambda[\mathcal{C}](\mathbf{i}\mu(x_1)), \ldots, \Lambda[\mathcal{C}](\mathbf{i}\mu(x_m)))$$

is indeed a basis of $\mathcal{Q}(\|\mathcal{A}\|)$. Moreover, given an arbitrary word $x \in \Sigma^*$ with $\mathbf{i}\mu(x) = (a_1, \ldots, a_m) \in \mathbb{F}^m$, we obtain

$$\begin{aligned}
x^{-1}\|\mathcal{A}\| = \Lambda[\mathcal{C}](\mathbf{i}\mu(x)) &= \Lambda[\mathcal{C}](a_1\mathbf{i}\mu(x_1) + \ldots + a_m\mathbf{i}\mu(x_m)) = \\
&= a_1\Lambda[\mathcal{C}](\mathbf{i}\mu(x_1)) + \ldots + a_m\Lambda[\mathcal{C}](\mathbf{i}\mu(x_m)) = \\
&= a_1 x_1^{-1}\|\mathcal{A}\| + \ldots + a_m x_m^{-1}\|\mathcal{A}\|,
\end{aligned}$$

from which the said property follows.

Now, assume for contradiction that $\mathcal{C}$ is not deterministic. By minimality of $\mathcal{C}$, there is some $x \in \Sigma^*$ such that $\mathbf{i}\mu(x)$ contains at least two nonzero entries. However, by our assumptions, there also is an m-state *deterministic* automaton $\mathcal{D}$ such that $\|\mathcal{D}\| = \|\mathcal{A}\|$. Linear independence of $x_1^{-1}\|\mathcal{A}\|, \ldots, x_m^{-1}\|\mathcal{A}\|$ implies that the m states of $\mathcal{D}$ can be labelled as $q_1, \ldots, q_m$ so that $x_i^{-1}\|\mathcal{A}\|$ is a scalar multiple of $\|\mathcal{D}\|_{q_i}$ for $i = 1, \ldots, m$. By determinism of $\mathcal{D}$, every $x^{-1}\|\mathcal{A}\|$ with $x \in \Sigma^*$ is a scalar multiple of some $\|\mathcal{D}\|_{q_i}$ with $i \in [m]$, and hence also of some $x_i^{-1}\|\mathcal{A}\|$. It thus follows that there is some $x \in \Sigma^*$ such that $x^{-1}\|\mathcal{A}\|$ has two different coordinates with respect to $(x_1^{-1}\|\mathcal{A}\|, \ldots, x_m^{-1}\|\mathcal{A}\|)$: a contradiction. $\square$

Theorem 7. *Let $\mathcal{A}$ be a weighted automaton over a field. If $\mathcal{A}$ has a bideterministic equivalent, then the Cardon-Crochemore algorithm applied to $\mathcal{A}$ outputs a bideterministic automaton.*

Proof. Let $\mathcal{A}$ admit a bideterministic equivalent $\mathcal{B}$, and assume that it is trim. Then $\mathcal{B}$ is minimal by Corollary 3, so Lemma 6 implies that the algorithm applied to $\mathcal{A}$ yields a deterministic automaton $\mathcal{D}$. If $\mathcal{D}$ was not bideterministic, then there would be $u, v \in \Sigma^*$ such that $u^{-1}\|\mathcal{D}\|$ is not a scalar multiple of $v^{-1}\|\mathcal{D}\|$ and $\mathrm{supp}(u^{-1}\|\mathcal{D}\|) \cap \mathrm{supp}(v^{-1}\|\mathcal{D}\|) \neq \emptyset$. On the other hand, bideterminism of $\mathcal{B}$ implies[4] $\mathrm{supp}(u^{-1}\|\mathcal{B}\|) \cap \mathrm{supp}(v^{-1}\|\mathcal{B}\|) = \emptyset$ when $u^{-1}\|\mathcal{B}\|$ is not a scalar multiple of $v^{-1}\|\mathcal{B}\|$. This contradicts the assumption that $\|\mathcal{B}\| = \|\mathcal{D}\| = \|\mathcal{A}\|$. $\square$

Corollary 8. *Bideterminisability of weighted automata over effective fields is decidable in polynomial time.*

5.2 Tropical Semirings

We now establish decidability of the bideterminisability problem for weighted automata over the tropical (min-plus) semirings $\mathbb{N}_{\min} = (\mathbb{N} \cup \{\infty\}, \min, +, \infty, 0)$, $\mathbb{Z}_{\min} = (\mathbb{Z} \cup \{\infty\}, \min, +, \infty, 0)$, and $\mathbb{Q}_{\min} = (\mathbb{Q} \cup \{\infty\}, \min, +, \infty, 0)$.

Theorem 9. *Bideterminisability of weighted automata over the semirings $\mathbb{N}_{\min}$, $\mathbb{Z}_{\min}$, and $\mathbb{Q}_{\min}$ is decidable.*

Proof. By positivity of tropical semirings, the minimal deterministic finite automaton $\mathcal{B}$ for $\mathrm{supp}(\|\mathcal{A}\|)$ is bideterministic whenever a tropical automaton $\mathcal{A}$ is bideterminisable. Given $\mathcal{A}$, we may thus remove the weights and minimise the automaton to get $\mathcal{B}$. If $\mathcal{B}$ is not bideterministic, $\mathcal{A}$ is not bideterminisable. If $\mathcal{B}$ is empty, $\mathcal{A}$ is bideterminisable. If $\mathcal{B}$ is bideterministic and nonempty, $\mathcal{A}$ is bideterminisable if and only if it is equivalent to some $\mathcal{B}'$ obtained from $\mathcal{B}$ by assigning weights to its transitions, its initial state, and its terminal state.

We show that existence of such $\mathcal{B}'$ is decidable given $\mathcal{A}$ and $\mathcal{B}$. Denote the unknown weights by $x_1, \ldots, x_N$, and let $\mathbf{x} = (x_1, \ldots, x_N)$. Here, x_1 corresponds to the unknown initial weight, $x_2, \ldots, x_{N-1}$ to the unknown transition weights, and x_N to the unknown terminal weight. Moreover, for each $w \in \mathrm{supp}(\|\mathcal{A}\|)$, let $\Psi(w) = (1, \eta_2, \ldots, \eta_{N-1}, 1)$, where η_i denotes, for $i = 2, \ldots, N-1$, the number of times the unique successful run of $\mathcal{B}$ upon w goes through the transition corresponding to the unknown weight x_i.

---

[4] This is a slight extension of a well-known property of bideterministic automata without weights – see, e.g., L. Polák [28, Section 5].

In order for $\mathcal{B}'$ to exist, the unknown weights have to satisfy the equations $\Psi(w) \cdot \mathbf{x}^T = (\|\mathcal{A}\|, w)$ for all $w \in \text{supp}(\|\mathcal{A}\|)$. If this system has a solution, then its solution set coincides with the one of a *finite* system of equations

$$\Psi(w_i) \cdot \mathbf{x}^T = (\|\mathcal{A}\|, w_i) \qquad \text{for } i = 1, \ldots, M, \tag{5}$$

where $w_1, \ldots, w_M \in \text{supp}(\|\mathcal{A}\|)$ are such that $(\Psi(w_1), \ldots, \Psi(w_M))$ is a basis of the vector space over $\mathbb{Q}$ generated by $\Psi(w)$ for $w \in \text{supp}(\|\mathcal{A}\|)$. This basis can be effectively obtained, e.g., from the representation of $\{\Psi(w) \mid w \in \text{supp}(\|\mathcal{A}\|)\}$ as a semilinear set. Hence, $w_1, \ldots, w_M$ can be found as well.

We may thus solve the system (5) over $\mathbb{N}$, $\mathbb{Z}$, or $\mathbb{Q}$ depending on the semiring considered. While Gaussian elimination is sufficient to solve the system over $\mathbb{Q}$, the solution over $\mathbb{Z}$ and $\mathbb{N}$ requires more sophisticated methods, namely an algorithm for solving systems of linear Diophantine equations in the former case [31], and integer linear programming in the latter case [31].

If there is no solution, $\mathcal{A}$ is not bideterminisable. Otherwise, any solution $\mathbf{x}$ gives us a bideterministic tropical automaton $\mathcal{B}_\mathbf{x}$ obtained from $\mathcal{B}$ by assigning the weights according to $\mathbf{x}$. By what has been said, either all such automata $\mathcal{B}_\mathbf{x}$ are equivalent to $\mathcal{A}$, or none of them is. Equivalence of a deterministic tropical automaton with a nondeterministic one is decidable [2], so we may take any of the automata $\mathcal{B}_\mathbf{x}$ and decide whether $\|\mathcal{B}_\mathbf{x}\| = \|\mathcal{A}\|$. If so, we may set $\mathcal{B}' = \mathcal{B}_\mathbf{x}$ and $\mathcal{A}$ is bideterminisable. Otherwise, $\mathcal{A}$ is not bideterminisable. $\square$

Note that the decision algorithm described makes use of deciding equivalence of a nondeterministic tropical automaton with a deterministic one, which is **PSPACE**-complete [2]. Nevertheless, we leave the complexity of the bideterminisability problem open.

Finally, let us note that it can be shown that the decidability result just established *does not* generalise to all effective positive semirings.

References

1. Allauzen, C., Mohri, M.: Efficient algorithms for testing the twins property. J. Autom. Lang. Comb. **8**(2), 117–144 (2003)
2. Almagor, S., Boker, U., Kupferman, O.: What's decidable about weighted automata? Inf. Comput. **282**, 104651 (2022)
3. Angluin, D.: Inference of reversible languages. J. ACM **29**(3), 741–765 (1982)
4. Béal, M.-P., Lombardy, S., Sakarovitch, J.: On the equivalence of $\mathbb{Z}$-automata. In: Caires, L., Italiano, G.F., Monteiro, L., Palamidessi, C., Yung, M. (eds.) ICALP 2005. LNCS, vol. 3580, pp. 397–409. Springer, Heidelberg (2005). https://doi.org/10.1007/11523468_33
5. Béal, M.-P., Lombardy, S., Sakarovitch, J.: Conjugacy and equivalence of weighted automata and functional transducers. In: Grigoriev, D., Harrison, J., Hirsch, E.A. (eds.) CSR 2006. LNCS, vol. 3967, pp. 58–69. Springer, Heidelberg (2006). https://doi.org/10.1007/11753728_9
6. Bell, J., Smertnig, D.: Noncommutative rational Pólya series. Sel. Math. **27**(3), article 34 (2021)

7. Berstel, J., Reutenauer, C.: Noncommutative Rational Series with Applications. Cambridge University Press, Cambridge (2011)
8. Cardon, A., Crochemore, M.: Détermination de la représentation standard d'une série reconnaissable. Informatique Théorique et Applications **14**(4), 371–379 (1980)
9. Ćirić, M., Droste, M., Ignjatović, J., Vogler, H.: Determinization of weighted finite automata over strong bimonoids. Inf. Sci. **180**, 3497–3520 (2010)
10. Droste, M., Kuich, W., Vogler, H. (eds.): Handbook of Weighted Automata. Springer, Heidelberg (2009). https://doi.org/10.1007/978-3-642-01492-5
11. Droste, M., Kuske, D.: Weighted automata. In: Pin, J.É. (ed.) Handbook of Automata Theory, vol. 1, no. 4, pp. 113–150. European Mathematical Society (2021)
12. Eilenberg, S.: Automata, Languages, and Machines, vol. A. Academic Press, Cambridge (1974)
13. Golan, J.S.: Semirings and their Applications. Kluwer Academic Publishers, Dordrecht (1999)
14. Gruber, H.: Digraph complexity measures and applications in formal language theory. Discrete Math. Theor. Comput. Sci. **14**(2), 189–204 (2012)
15. Hebisch, U., Weinert, H.J.: Semirings. World Scientific (1998)
16. Janin, D.: Free inverse monoids up to rewriting. Technical report, LaBRI - Laboratoire Bordelais de Recherche en Informatique (2015). Available at https://hal.archives-ouvertes.fr/hal-01182934
17. Kirsten, D.: An algebraic characterization of semirings for which the support of every recognizable series is recognizable. Theoret. Comput. Sci. **534**, 45–52 (2014)
18. Kirsten, D., Lombardy, S.: Deciding unambiguity and sequentiality of polynomially ambiguous min-plus automata. In: Symposium on Theoretical Aspects of Computer Science, STACS 2009, pp. 589–600 (2009)
19. Kirsten, D., Mäurer, I.: On the determinization of weighted automata. J. Autom. Lang. Comb. **10**(2–3), 287–312 (2005)
20. Kostolányi, P.: Determinisability of unary weighted automata over the rational numbers. Theoret. Comput. Sci. **898**, 110–131 (2022)
21. Lombardy, S., Sakarovitch, J.: Sequential? Theoret. Comput. Sci. **356**, 224–244 (2006)
22. McNaughton, R.: The loop complexity of pure-group events. Inf. Control **11**(1–2), 167–176 (1967)
23. McNaughton, R.: The loop complexity of regular events. Inf. Sci. **1**(3), 305–328 (1969)
24. Mohri, M.: Finite-state transducers in language and speech processing. Comput. Linguist. **23**(2), 269–311 (1997)
25. Mohri, M.: Weighted automata algorithms. In: Droste, M., Kuich, W., Vogler, H. (eds.) Handbook of Weighted Automata. Monographs in Theoretical Computer Science. An EATCS Series, pp. 213–254. Springer, Berlin (2009). https://doi.org/10.1007/978-3-642-01492-5_6
26. Myers, R.S.R., Milius, S., Urbat, H.: Nondeterministic syntactic complexity. In: Foundations of Software Science and Computation Structures, FOSSACS 2021, pp. 448–468 (2021)
27. Pin, J.É.: On reversible automata. In: Latin American Symposium on Theoretical Informatics, LATIN 1992, pp. 401–416 (1992)
28. Polák, L.: Minimalizations of NFA using the universal automaton. Int. J. Found. Comput. Sci. **16**(5), 999–1010 (2005)
29. Sakarovitch, J.: Elements of Automata Theory. Cambridge University Press, Cambridge (2009)

30. Sakarovitch, J.: Rational and recognisable power series. In: Droste, M., Kuich, W., Vogler, H. (eds.) Handbook of Weighted Automata. Monographs in Theoretical Computer Science. An EATCS Series, pp. 105–174. Springer, Berlin (2009). https://doi.org/10.1007/978-3-642-01492-5_4
31. Schrijver, A.: Theory of Linear and Integer Programming. Wiley, Hoboken (1986)
32. Schützenberger, M.P.: On the definition of a family of automata. Inf. Control **4**(2–3), 245–270 (1961)
33. Shankar, P., Dasgupta, A., Deshmukh, K., Rajan, B.S.: On viewing block codes as finite automata. Theoret. Comput. Sci. **290**(3), 1775–1797 (2003)
34. Stephen, J.B.: Presentations of inverse monoids. J. Pure Appl. Algebra **63**(1), 81–112 (1990)
35. Tamm, H., Ukkonen, E.: Bideterministic automata and minimal representations of regular languages. In: Implementation and Application of Automata, CIAA 2003, pp. 61–71 (2003)
36. Tamm, H., Ukkonen, E.: Bideterministic automata and minimal representations of regular languages. Theoret. Comput. Sci. **328**(1–2), 135–149 (2004)

How to Decide Functionality of Compositions of Top-Down Tree Transducers

Sebastian Maneth[1], Helmut Seidl[2], and Martin Vu[1(✉)]

[1] Universität Bremen, Bremen, Germany
{maneth,martin.vu}@uni-bremen.de
[2] TU München, Munich, Germany
seidl@in.tum.de

Abstract. We prove that functionality of compositions of top-down tree transducers is decidable by reducing the problem to the functionality of one top-down tree transducer with look-ahead.

1 Introduction

Tree transducers are fundamental devices that were invented in the 1970's in the context of compilers and mathematical linguistics. Since then they have been applied in a huge variety of contexts such as, e.g., programming languages [14], security [10], or XML databases [9].

The perhaps most basic type of tree transducer is the top-down tree transducer [15,16] (for short *transducer*). One important decision problem for transducers concerns *functionality*: given a (nondeterministic) transducer, does it realize a function? This problem was shown to be decidable by Ésik [8] (even in the presence of look-ahead); note that this result also implies the decidability of equivalence of deterministic transducers [8], see also [7,11].

A natural and fundamental question is to ask whether functionality can also be decided for *compositions of transducers*. It is well known that compositions of transducers form a proper hierarchy, more precisely: compositions of $n + 1$ transducers are strictly more expressive than compositions of n transducers [6]. Even though transducers are well studied, the question of deciding functionality for compositions of transducers has remained open. In this paper we fill this gap and show that the question can be answered affirmatively.

Deciding functionality for compositions of transducers has several applications. For instance, if an arbitrary composition of (top-down and bottom-up) tree transducers is functional, then an equivalent deterministic transducer with look-ahead can be constructed [5]. Together with our result this implies that it is decidable for such a composition whether or not it is definable by a deterministic transducer with look-ahead; note that the construction of such a single deterministic transducer improves efficiency, because it removes the need of computing intermediate results of the composition. Also other recent definability results can

© The Author(s), under exclusive license to Springer Nature Switzerland AG 2022
D. Poulakis and G. Rahonis (Eds.): CAI 2022, LNCS 13706, pp. 175–191, 2022.
https://doi.org/10.1007/978-3-031-19685-0_13

now be generalized to compositions: for instance, given such a composition we can now decide whether or not an equivalent linear transducer or an equivalent homomorphism exists [12] (and if so, construct it).

Let us now discuss the idea of our proof in detail. Initially, we consider a composition τ of two transducers T_1 and T_2. Given τ, we construct a 'candidate' transducer with look-ahead M with the property that M is functional if and only if τ is functional. Our construction of M is an extension of the product construction in [2, p. 195]. The latter constructs a transducer N (*without* look-ahead) that is obtained by translating the right-hand sides of the rules of T_1 by the transducer T_2. It is well-known that in general, the transducer N is *not* equivalent to τ [2] and thus N may not be functional even though τ is. This is due to the fact that the transducer T_2 may

- copy or
- delete input subtrees.

Copying of an input tree means that the tree is translated several times and in general by different states. Deletion means that in a translation rule a particular input subtrees is not translated at all.

Imagine that T_2 copies and translates an input subtree in two different states q_1 and q_2, so that the domains D_1 and D_2 of these states differ and moreover, T_1 nondeterministically produces outputs in the union of D_1 and D_2. Now the problem that arises in the product construction of N is that N needs to guess the output of T_1, however, the two states corresponding to q_1 and q_2 *cannot* guarantee that the same guess is used. However, the same guess may be used. This means that N (seen as a binary relation) is a superset of τ. To address this problem we show that it suffices to change T_1 so that it only outputs trees in the intersection of D_1 and D_2. Roughly speaking this can be achieved by changing T_1 so that it runs several tree automata in parallel, in order to carry out the necessary domain checks.

Imagine now a transducer T_1 that translates two input subtrees in states q_1 and q_2, respectively, but has no rules for state q_2. This means that the translation of T_1 (and of τ) is empty. However, the transducer T_2 deletes the position of q_2. This causes the translation of N to be non-empty. To address this problem we equip N with look-ahead. The look-ahead checks if the input tree is in the domains of all states of T_1 translating the current input subtree.

Finally, we are able to generalize the result to arbitrary compositions of transducers $T_1, \ldots, T_n$. For this, we apply the extended composition described above to the transducers T_{n-1} and T_n, giving us the transducer with look-ahead M. The look-ahead of M can be removed and incorporated into the transducer T_{n-2} using a composition result of [2]. The resulting composition of $n-1$ transducers is functional if and only if the original composition is.

An extended version of this paper including all details of our proofs can be found in [13].

2 Top-Down Tree Transducers

For $k \in \mathbb{N}$, we denote by $[k]$ the set $\{1, \ldots, k\}$. Let $\Sigma = \{e_1^{k_1}, \ldots, e_n^{k_n}\}$ be a *ranked alphabet*, where $e_j^{k_j}$ means that the symbol e_j has *rank* k_j. By Σ_k we denote the set of all symbols of Σ which have rank k. The set T_Σ of *trees over* Σ consists of all strings of the form $a(t_1, \ldots, t_k)$, where $a \in \Sigma_k$, $k \geq 0$, and $t_1, \ldots, t_k \in T_\Sigma$. Instead of $a()$ we simply write a. We fix the set X of *variables* as $X = \{x_1, x_2, x_3, \ldots\}$.

Let B be an arbitrary set. We define $T_\Sigma[B] = T_{\Sigma'}$ where Σ' is obtained from Σ by $\Sigma'_0 = \Sigma_0 \cup B$ while for all $k > 0$, $\Sigma'_k = \Sigma_k$. In the following, let A, B be arbitrary sets. We let $A(B) = \{a(b) \mid a \in A, b \in B\}$.

Definition 1. *A* top-down tree transducer *T (or* transducer *for short) is a tuple of the form $T = (Q, \Sigma, \Delta, R, q_0)$ where Q is a finite set of* states, *Σ and Δ are the* input *and* output *ranked alphabets, respectively, disjoint with Q, R is a finite set of* rules, *and $q_0 \in Q$ is the* initial state. *The rules contained in R are of the form $q(a(x_1, \ldots, x_k)) \to t$, where $q \in Q$, $a \in \Sigma_k$, $k \geq 0$ and t is a tree in $T_\Delta[Q(X)]$.*

If $q(a(x_1, \ldots, x_k)) \to t \in R$ then we call t a right-hand side of q and a. The rules of R are used as rewrite rules in the natural way, as illustrated by the following example.

Example 1. Consider the transducer $T = (\{q_0, q\}, \Sigma, \Delta, R, q_0)$ where $\Sigma_0 = \{e\}$, $\Sigma_1 = \{a\}$, $\Delta_0 = \{e\}$, $\Delta_1 = \{a\}$ and $\Delta_2 = \{f\}$ and R consists the following rules (numbered 1 to 4):

$$1: q_0(a(x_1)) \to f(q(x_1), q_0(x_1)) \quad 2: q_0(e) \to e$$
$$3: q(a(x_1)) \to a(q(x_1)) \qquad\qquad 4: q(e) \to e.$$

On input $a(a(e))$, the transducer T produces the output tree $f(a(e), f(e, e))$ as follows

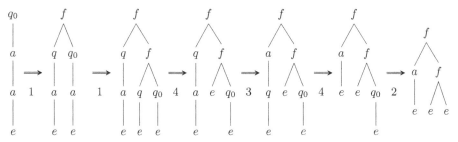

Informally, when processing a tree $s \in T_\Sigma$, the transducer T produces a tree t in which all proper subtrees of s occur as disjoint subtrees of t, 'ordered' by size. As the reader may realize, given an input tree s of size n, the transducer T produces an output tree that is of size $(n^2 + n)/2$. Hence, this translation has quadratic size increase, i.e., the size of the output tree is a most quadratic in size of the input tree. Note that transducers can have polynomial or exponential size increase [1]. □

Let $s \in T_\Sigma$. Then $T(s)$ contains all trees in T_Δ obtainable from $q_0(s)$ by applying rules of T.

Clearly, T defines a binary relation over T_Σ and T_Δ. In the following, we denote by $\mathcal{R}(T)$ the binary relation that the transducer T defines. We say that the transducer T is *functional* if the relation $\mathcal{R}(T)$ is a function. Let q be a state of T. We denote by $\text{dom}(q)$ the *domain* of q, i.e., the set of all trees $s \in T_\Sigma$ for which some tree $t \in T_\Delta$ is obtainable from $q(s)$ by applying rules of T. We define the *domain* of T by $\text{dom}(T) = \text{dom}(q_0)$. For instance in Example 1, $\text{dom}(T) = T_\Sigma$. However, if we remove the rule 1 for instance then the domain of T shrinks to the set $\{e\}$.

A transducer $T = (Q, \Sigma, \Delta, R, q)$ is a *top-down tree automaton* (for short *automaton*) if $\Sigma = \Delta$ and all rules of T are of the form $q(a(x_1, \ldots, x_k)) \rightarrow a(q_1(x_1), \ldots q_k(x_k))$ where $a \in \Sigma_k$, $k \geq 0$.

Let T_1 and T_2 be transducers. As $\mathcal{R}(T_1)$ and $\mathcal{R}(T_2)$ are relations, they can be composed. Hence,

$$\mathcal{R}(T_1) \circ \mathcal{R}(T_2) = \{(s, u) \mid \text{for some } t, (s, t) \in \mathcal{R}(T_1) \text{ and } (t, u) \in \mathcal{R}(T_2)\}.$$

If the output alphabet of T_1 and the input alphabet of T_2 coincide then the transducers T_1 and T_2 can be composed as well. The composition $T_1 \hat{\circ} T_2$ of the transducers T_1 and T_2 is a machine that on input s generates an output tree as follows. On input s, the tree s is first translated by T_1. Afterwards, the tree produced by T_1 is translated by T_2 which yields the output tree. Clearly, $T_1 \hat{\circ} T_2$ computes the relation $\mathcal{R}(T_1) \circ \mathcal{R}(T_2)$. We say that the composition $T_1 \hat{\circ} T_2$ is *functional* if the relation $\mathcal{R}(T_1) \circ \mathcal{R}(T_2)$ is a function.

3   Functionality of Two-Fold Compositions

In this section we show that for a composition τ of two transducers, a transducer M with look-ahead can be constructed such that M is functional if and only if τ is functional. Before formally introducing the construction for M and proving its correctness, we explain how to solve the challenges described in Sect. 1, i.e., we show how to handle copying and deleting rules. In the following, we call the product construction in [2, p. 195] simply the *p-construction*.

To see how precisely we handle copying rules, consider the transducers T_1 and T_2. Let the transducer T_1 consist of the rules

$$q_1(a(x_1)) \rightarrow b(q_1(x_1)) \quad q_1(e) \rightarrow e_i \mid i = 1, 2, 3$$

while transducer T_2 consists of the rules

$$q_2(b(x_1)) \rightarrow f(q_2'(x_1), q_2''(x_1)) \quad q_2'(e_j) \rightarrow e \mid j = 1, 2$$
$$q_2''(e_3) \quad \rightarrow e' \qquad\qquad\qquad q_2''(e_j) \rightarrow e \mid j = 1, 2.$$

The composition $\tau = T_1 \hat{\circ} T_2$ defines a relation that only contains a single pair, that is τ only translates the tree $a(e)$ into $f(e, e)$. Therefore, τ is functional. For T_1 and T_2, the p-construction yields the transducer N with the rules

$$(q_1, q_2)(a(x_1)) \rightarrow f((q_1, q_2')(x_1), (q_1, q_2'')(x_1)) \quad (q_1, q_2')(e) \rightarrow e$$
$$(q_1, q_2'')(e) \quad\quad \rightarrow e' \quad\quad\quad\quad\quad\quad\quad\quad\quad\quad (q_1, q_2'')(e) \rightarrow e.$$

On input $a(e)$, the transducer N can produce either $f(e, e)$ or $f(e, e')$. Therefore, N and τ are clearly not equivalent. Furthermore, the transducer N is obviously not functional even though the composition τ is.

In order to obtain a better understanding of why this phenomenon occurs, we analyze the behavior of N and τ on input $a(e)$ in the following.

In the translation of τ, the states q_2' and q_2'' process the same tree produced by q_1 on input e due to the copying rule $q_2(b(x_1)) \rightarrow f(q_2'(x_1), q_2''(x_1))$. Furthermore, q_2' and q_2'' process a tree in $\text{dom}(q_2') \cap \text{dom}(q_2'')$. More precisely, q_2' and q_2'' both process either e_1 or e_2.

In the translation of N on the other hand, due to the rule $(q_1, q_2)(a(x_1)) \rightarrow f((q_1, q_2')(x_1), (q_1, q_2'')(x_1))$, the states (q_1, q_2') and (q_1, q_2'') process e by 'guessing independently' from each other what q_1 might have produced on input e. In particular, the problem is that (q_1, q_2'') can apply the rule $(q_1, q_2'')(e) \rightarrow e'$ which eventually leads to the production of $f(e, e')$. Applying this rule means that (q_1, q_2'') guesses that e_3 is produced by q_1. While this guess is valid, i.e., e_3 is producible by q_1 on input e, quite clearly $e_3 \notin \text{dom}(q_2')$.

In general, guesses performed by states of N cannot be 'synchronized', i.e., we cannot guarantee that states guess the same tree. Our solution to fix this issue is to restrict (q_1, q_2') and (q_1, q_2'') such that either state is only allowed to guess trees in $\text{dom}(q_2') \cap \text{dom}(q_2'')$. To understand why this approach works in general consider the following example.

Example 2. Let T_1 and T_2 be arbitrary transducers. Let $\tau = T_1 \,\hat{\circ}\, T_2$ be functional. Let T_1 on input s produce either $b(t_1)$ or $b(t_2)$. Let T_2 contain the rule

$$q_2(b(x_1)) \rightarrow f(q_2^1(x_1), q_2^2(x_1))$$

where q_2 is the initial state of T_2. The application of this rule effectively means that the states q_2^1 and q_2^2 process the same subtree produced by T_1. Let $t_1, t_2 \in \text{dom}(q_2^1) \cap \text{dom}(q_2^2)$. Informally speaking, it does not matter whether the state q_2^1 processes t_1 or t_2; for either input q_2^1 produces the same output tree r and nothing else, otherwise, the functionality of τ is contradicted. The same holds for q_2^2. □

Informally, Example 2 suggests that if (q_1, q_2') and (q_1, q_2'') only guess trees in $\text{dom}(q_2') \cap \text{dom}(q_2'')$, then it does not matter which tree exactly those states guess if the composition is functional. The final result in either case is the same. Quite clearly this is the case in our example. (In effect, q_2'' is forbidden to guess e_3.) Thus, restricting (q_1, q_2') and (q_1, q_2'') basically achieves the same result as synchronizing their guesses if the composition is functional.

Now the question is how exactly do we restrict the states of N? Consider the states (q_1, q_2') and (q_1, q_2'') of N in our example. The trick is to restrict q_1 such that q_1 can only produce trees in $\text{dom}(q_2') \cap \text{dom}(q_2'')$. Thus any guess is guaranteed to be in $\text{dom}(q_2') \cap \text{dom}(q_2'')$. In order to restrict which output trees T_1 can produce, we compose T_1 with the *domain automaton* of T_2.

For an arbitrary transducer $T = (Q, \Sigma, \Delta, R, q)$, the *domain automaton* A of T is constructed analogous to the automaton in [4, Theorem 3.1]. The set of states of A is the power set of Q where $\{q\}$ is the initial state of A. The idea is that if in a translation of T on input s, the states $q_1 \ldots, q_n$ process the node v of s then $\{q_1 \ldots, q_n\}$ processes the node v of s in a computation of A. The rules of A are thus defined as follows.

Let $S = \{q_1 \ldots, q_n\}$, $n > 0$, and $a \in \Sigma_k$. In the following, we denote by $\mathrm{rhs}_T(q_j, a)$, where $j \in [n]$, the set of all right-hand sides of q_j and a. For all non-empty subsets $\Gamma_1 \subseteq \mathrm{rhs}_T(q_1, a), \ldots, \Gamma_n \subseteq \mathrm{rhs}_T(q_n, a)$, we define a rule

$$S(a(x_1, \ldots, x_k)) \to a(S_1(x_1), \ldots, S_k(x_k))$$

where for $i \in [k]$, S_i is defined as the set $\bigcup_{j=1}^n \Gamma_j \langle x_i \rangle$. We denote by $\Gamma_j \langle x_i \rangle$ the set of all states q' such that $q'(x_i)$ occurs in some tree γ in Γ_j; e.g., for

$$\Gamma_j = \{a(q(x_1), q'(x_2)), \ a(a(q_1(x_1), q_2(x_2)), q_3(x_1))\},$$

we have $\Gamma_j \langle x_1 \rangle = \{q, q_1, q_3\}$ and $\Gamma_j \langle x_2 \rangle = \{q', q_2\}$. We define that the state $\emptyset$ of A realizes the identity. Hence, the rules for the state $\emptyset$ are defined in the obvious way.

We now explain why subsets Γ_j of right-hand sides are used for the construction of rules of A. Recall that the idea is that if in a translation of T on input s, the states $q_1 \ldots, q_n$ process the node v of s then $\{q_1 \ldots, q_n\}$ processes the node v of s in a computation of A. Due to copying rules, multiple instances of a state q_1 may access v. Two instance of q_1 may process v in different manners. This necessitates the use of subsets Γ_j of right-hand sides. For a better understanding, consider the following example.

Example 3. Let $T = (\{q_0, q\}, \Sigma, \Delta, R, q_0)$ where $\Sigma_0 = \Delta_0 = \{e\}$, $\Sigma_1 = \Delta_1 = \{a\}$ and $\Sigma_2 = \Delta_2 = \{f\}$. The set R contains the following rules:

$$\begin{array}{ll}
q_0(a(x_1)) \quad \to f(q_0(x_1), q_0(x_1)) & q(a(x_1)) \quad \to e' \\
q_0(f(x_1, x_2)) \to q_0(x_1) & q(f(x_1, x_2)) \to e' \\
q_0(f(x_1, x_2)) \to f(q(x_1), q(x_2)) & q(e) \quad\quad\; \to e' \\
q_0(e) \quad\quad\quad\; \to e. &
\end{array}$$

Consider the input tree $s = a(f(e, e))$. Clearly, on input s, the tree $f(e, f(e', e'))$ is producible by T. In this translation, two instances of the state q_0 process the subtree $f(e, e)$ of s, however the instances of q_0 do not process $f(e, e)$ in the same way. The first instance of q_0 produces e on input $f(e, e)$ while the second instance produces $f(e', e')$. These translations mean that the states q_0 and q process the leftmost e of s.

Consider the domain automaton A of T. By definition, A contains the rule $\{q_0\}(a(x_1)) \to a(\{q_0\}(x_1))$ which is obtained from the right-hand side of the rule $q_0(a(x_1)) \to f(q_0(x_1), q_0(x_1))$ of T. To simulate that the states q_0 and q process the leftmost e of s in the translation from s to $f(e, f(e', e'))$, we clearly require the rule $\{q_0\}(f(x_1, x_2)) \to f(\{q_0, q\}(x_1), \{q\}(x_2))$ obtained from the right-hand sides of the rules $q_0(f(x_1, x_2)) \to q_0(x_1)$ and $q_0(f(x_1, x_2)) \to f(q(x_1), q(x_2))$ of T.

For completeness, we list the remaining rules of A. The automaton A also contains the rules

$$
\begin{aligned}
\{q_0\}\ (f(x_1, x_2)) &\to f(\{q\}(x_1), \{q\}(x_2)) & \{q\}\ (a(x_1)) &\to a(\emptyset(x_1)) \\
\{q_0\}\ (f(x_1, x_2)) &\to f(\{q_0\}(x_1), \emptyset(x_2)) & \{q\}\ (f(x_1, x_2)) &\to f(\emptyset(x_1), \emptyset(x_2)) \\
\{q_0\}\ (e) &\to e & \{q\}\ (e) &\to e. \\
\emptyset\ (f(x_1, x_2)) &\to f(\emptyset(x_1), \emptyset(x_2)) & \emptyset\ (a(x_1)) &\to a(\emptyset(x_1)) \\
\emptyset\ (e) &\to e.
\end{aligned}
$$

For the rules of the state $\{q_0, q\}$ consider the following. The right-hand sides of rules of $\{q_0, q\}$ are identical to the right-hand sides of rules of $\{q_0\}$, i.e., the rules for $\{q_0, q\}$ are obtained by substituting $\{q_0\}$ on the left-hand-side of rules of A by $\{q_0, q\}$. □

The automaton A has the following property.

Lemma 1. Let $S \neq \emptyset$ be a state of A. Then $s \in dom(S)$ if and only if $s \in \bigcap_{q \in S} dom(q)$.

Obviously, Lemma 1 implies that A recognizes the domain of T.

Using the domain automaton A of T_2, we transform T_1 into the transducer $\hat{T}_1$. Formally, the transducer $\hat{T}_1$ is obtained from T_1 and A using the p-construction. In our example, the transducer $\hat{T}_1$ obtained from T_1 and T_2 includes the following rules

$$
\begin{aligned}
(q_1, \{q_2\})\ (a(x_1)) &\to b((q_1, \{q_2', q_2''\})(x_1)) \\
(q_1, \{q_2', q_2''\})\ (e) &\to e_j
\end{aligned}
$$

where $j = 1, 2$. The state $(q_1, \{q_2\})$ is the initial state of $\hat{T}_1$. Informally, the idea is that in a translation of $\hat{\tau} = \hat{T}_1 \hat{\circ} T_2$, a tree produced by a state (q, S) of $\hat{T}_1$ is only processed by states in S. The following result complements this idea.

Lemma 2. If the state (q, S) of $\hat{T}_1$ produces the tree t and $S \neq \emptyset$ then $t \in \bigcap_{q_2 \in S} dom(q_2)$.

We remark that if a state of the form $(q, \emptyset)$ occurs then it means that in a translation of $\hat{\tau}$, no state of T_2 will process a tree produced by $(q, \emptyset)$. Note that as A is nondeleting and linear, $\hat{T}_1$ defines the same relation as $T_1 \hat{\circ} A$ [2, Th. 1]. Informally, the transducer $\hat{T}_1$ is a restriction of the transducer T_1 such that $range(\hat{T}_1) = range(T_1) \cap dom(T_2)$. Therefore, the following holds.

Lemma 3. $\mathcal{R}(T_1) \circ \mathcal{R}(T_2) = \mathcal{R}(\hat{T}_1) \circ \mathcal{R}(T_2)$.

Due to Lemma 3, we focus on $\hat{T}_1$ instead of T_1 in the following.

Consider the transducer $\hat{N}$ obtained from $\hat{T}_1$ and T_2 using the p-construction. By construction, the states of $\hat{N}$ are of the form $((q, S), q')$ where (q, S) is a state of $\hat{T}_1$ and q' is a state of T_2. In the following, we write (q, S, q') instead for better readability. Informally, the state (q, S, q') implies that in a translation of $\hat{\tau}$ the state q' is supposed to process a tree produced by (q, S). Because trees produced by (q, S) are only supposed to be processed by states in S, we only consider states (q, S, q') where $q' \in S$. For $\hat{T}_1$ and T_2, we obtain the transducer $\hat{N}$ with the following rules

$$(q_1, \{q_2\}, q_2) \ (a(x_1)) \rightarrow f((q_1, S, q_2')(x_1), (q_1, S, q_2'')(x_1))$$
$$(q_1, S, q_2') \ (e) \qquad \rightarrow e$$
$$(q_1, S, q_2'') \ (e) \qquad \rightarrow e$$

where $S = \{q_2', q_2''\}$ and $i = 1, 2$. The initial state of $\hat{N}$ is $(q_1, \{q_2\}, q_2)$. Obviously, $\hat{N}$ computes the relation $\mathcal{R}(T_1) \circ \mathcal{R}(T_2)$.

In the following, we briefly explain our idea. In a translation of $\hat{N}$ on input $a(e)$, the subtree e is processed by (q_1, S, q_2') and (q_1, S, q_2''). Note that in a translation of $\hat{\tau}$ the states q_2' and q_2'' would process the same tree produced by (q_1, S) on input e. Consider the state (q_1, S, q_2'). If (q_1, S, q_2'), when reading e, makes a *valid* guess, i.e., (q_1, S, q_2') guesses a tree t that is producible by (q_1, S) on input e, then $t \in \text{dom}(q_2')$ by construction of $\hat{T}_1$. Due to previous considerations (cf. Example 2), it is thus sufficient to ensure that all guesses of states of $\hat{N}$ are valid. While obviously in the case of $\hat{N}$, all guesses are indeed valid, guesses of transducers obtained from the p-construction are in general not always valid; in particular if *deleting rules* are involved.

To be more specific, consider the following transducers T_1' and T_2'. Let T_1' contain the rules

$$q_1(a(x_1, x_2)) \rightarrow b(q_1'(x_1), q_1''(x_2), q_1'''(x_2)) \qquad q_1'(e) \rightarrow e$$

where $\text{dom}(q_1'')$ consists of all trees whose left-most leaf is labeled by e while $\text{dom}(q_1''')$ consists of all trees whose left-most leaf is labeled by c. Let T_2' contain the rules

$$q_2(b(x_1, x_2, x_3)) \rightarrow q_2(x_1) \qquad q_2(e) \rightarrow e_j \mid j = 1, 2.$$

As the translation of T_1' is empty, obviously the translation of $\tau' = T_1' \hat{\circ} T_2'$ is empty as well. Thus, τ' is functional. However, the p-construction yields the transducer N' with the rules

$$(q_1, q_2)(a(x_1, x_2)) \rightarrow (q_1', q_2)(x_1) \qquad (q_1', q_2)(e) \rightarrow e_j \mid j = 1, 2$$

Even though $\tau' = T_1' \hat{\circ} T_2'$ is functional, the transducer N' is not. More precisely, on input $a(e, s)$, where s is an arbitrary tree, N' can produce either e_1 or e_2 while τ' would produce nothing. The reason is that in the translation of N', the tree $a(e, s)$ is processed by the state (q_1, q_2) by applying the deleting rule $\eta = (q_1, q_2)(a(x_1, x_2)) \rightarrow (q_1', q_2)(x_1)$. Applying η means that (q_1, q_2) guesses that on input $a(e, s)$, the state q_1 produces a tree of the form $b(t_1, t_2, t_3)$ by applying the rule $q_1(a(x_1, x_2)) \rightarrow b(q_1'(x_1), q_1''(x_2), q_1'''(x_2))$ of T_1. However, this guess is not valid, i.e., q_1 does not produce such a tree on input $a(e, s)$, as by definition $s \notin \text{dom}(q_1'')$ or $s \notin \text{dom}(q_1''')$. The issue is that N' itself cannot verify the validity of this guess because, due to the deleting rule η, N' does not read s.

As the reader might have guessed our idea is that the validity of each guess is verified using look-ahead. First, we need to define look-ahead.

A *transducer with look-ahead* (or *la-transducer*) M' is a transducer that is equipped with an automaton called the *la-automaton*. Formally, M' is a tuple $M' = (Q, \Sigma, \Delta, R, q, B)$ where Q, Σ, Δ and q are defined as for transducers and

B is the la-automaton. The rules of R are of the form $q(a(x_1 : l_1, \ldots, x_k : l_k)) \to t$ where for $i \in [k]$, l_i is a state of B. Consider the input s. The la-transducer M' processes s in two phases: First each input node of s is annotated by the states of B at its children, i.e., an input node v labeled by $a \in \Sigma_k$ is relabeled by $\langle a, l_1, \ldots, l_k \rangle$ if B arrives in the state l_i when processing the i-th subtree of v. Relabeling the nodes s provides M' with additional information about the subtrees of s, e.g., if the node v is relabeled by $\langle a, l_1, \ldots, l_k \rangle$ then the i-th subtree of v is a tree in $\mathrm{dom}(l_i)$. The relabeled tree is then processed by M'. To this end a rule $q(a(x_1 : l_1, \ldots, x_k : l_k)) \to t$ is interpreted as $q(\langle a, l_1, \ldots, l_k \rangle (x_1, \ldots, x_k)) \to t$.

In our example, the idea is to equip N' with an la-automaton to verify the validity of guesses. In particular, the la-automaton is the domain automaton A' of T_1'. Recall that a state of A' is a set consisting of states of T_1'. To process relabeled trees the rules of N' are as follows

$$(q_1, q_2)(a(x_1 : \{q_1'\}, x_2 : \{q_1'', q_1'''\})) \to (q_1', q_2)(x_1) \quad (q_1', q_2)(e) \to e_j \mid j = 1, 2$$

Consider the tree $a(e, s)$, where s is an arbitrary tree. The idea is that if the root of $a(e, s)$ is relabeled by $\langle a, \{q_1'\}, \{q_1'', q_1'''\} \rangle$, then due to Lemma 1, $e \in \mathrm{dom}(q_1')$ and $s \in \mathrm{dom}(q_1'') \cap \mathrm{dom}(q_1''')$ and thus on input $a(e, s)$ a tree of the form $b(t_1, t_2, t_3)$ is producible by q_1 using the rule $q_1(a(x_1, x_2)) \to b(q_1'(x_1), q_1''(x_2), q_1'''(x_2))$. Quite clearly, the root of $a(e, s)$ is *not* relabeled. Thus, the translation of N' equipped with the la-automaton A' is empty as the translation of τ' is.

3.1   Construction of the LA-Transducer M

Recall that for a composition τ of two transducers T_1 and T_2, we aim to construct an la-transducer M such that M is functional if and only if τ is functional.

In the following we show that combining the ideas presented above yields the la-transducer M. For T_1 and T_2, we obtain M by first completing the following steps.

1. Construct the domain automaton A of T_2
2. Construct the transducer $\hat{T}_1$ from T_1 and A using the p-construction
3. Construct the transducer N from $\hat{T}_1$ and T_2 using the p-construction

We then obtain M by extending N into a transducer with look-ahead. Note that the states of N are written as (q, S, q') instead of $((q, S), q')$ for better readability, where (q, S) is a state of $\hat{T}_1$ and q' is a state of T_2. Recall that (q, S, q') means that q' is supposed to process a tree generated by (q, S). Furthermore, recall that S is a set of states of T_2 and that the idea is that trees produced by (q, S) are only supposed to be processed by states in S. Thus, we only consider states (q, S, q') of N where $q' \in S$.

The transducer M with look-ahead is constructed as follows. The set of states of M and the initial state of M are the states of N and the initial state of N, respectively. The la-automaton of M is the domain automaton $\hat{A}$ of $\hat{T}_1$.

We now define the rules of M. First, recall that a state of $\hat{A}$ is a set consisting of states of $\hat{T}_1$. Furthermore, recall that for a set of right-hand sides Γ and a variable x, we denote by $\Gamma\langle x\rangle$ the set of all states q such that $q(x)$ occurs in some $\gamma \in \Gamma$. For a right-hand side γ, the set $\gamma\langle x\rangle$ is defined analogously. For all rules

$$\eta = (q, S, q')(a(x_1, \ldots, x_k)) \to \gamma$$

of N we proceed as follows: If η is obtained from the rule $(q, S)(a(x_1, \ldots, x_k)) \to \xi$ of $\hat{T}_1$ and subsequently translating ξ by the state q' of T_2 then we define the rule

$$(q, S, q')(a(x_1 : l_1, \ldots, x_k : l_k)) \to \gamma$$

for M where for $i \in [k]$, l_i is a state of $\hat{A}$ such that $\xi\langle x_i\rangle \subseteq l_i$. Recall that relabeling a node v, that was previously labeled by a, by $\langle a, l_1, \ldots, l_k\rangle$ means that the i-th subtree of v is a tree in $\mathrm{dom}(l_i)$. By Lemma 1, $s \in \mathrm{dom}(l_i)$ if and only if $s \in \bigcap_{\hat{q}\in l_i} \mathrm{dom}(\hat{q})$. Thus, if the node v of a tree s is relabeled by $\langle a, l_1, \ldots, l_k\rangle$ then it means that (q, S) can process subtree of s rooted at v using the rule $(q, S)(a(x_1, \ldots, x_k)) \to \xi$.

In the following, we present a detailed example for the construction of M for two transducers T_1 and T_2.

Example 4. Let the transducer T_1 contain the rules

$$
\begin{aligned}
&q_0(f(x_1, x_2)) \to f(q_1(x_1), q_2(x_2)) && q_0(f(x_1, x_2)) \to q_3(x_2) \\
&q_1(f(x_1, x_2)) \to f(q_1(x_1), q_1(x_2)) && q_2(f(x_1, x_2)) \to f(q_2(x_1), q_1(x_2)) \\
&q_1(f(x_1, x_2)) \to f'(q_1(x_1), q_1(x_2)) && q_2(f(x_1, x_2)) \to f'(q_2(x_1), q_1(x_2)) \\
&q_1(e) \qquad\qquad \to e && q_2(e) \qquad\qquad \to e \\
&q_1(d) \qquad\qquad \to d && q_3(d) \qquad\qquad \to d
\end{aligned}
$$

and let the initial state of T_1 be q_0. Informally, when reading the symbol f, the states q_1 and q_2 nondeterministically decide whether or not to relabel f by f'. However, the domain of q_2 only consists of trees whose leftmost leaf is labeled by e. The state q_3 only produces the tree d on input d. Thus, the domain of T_1 only consists of trees of the form $f(s_1, s_2)$ where s_1 and s_2 are trees and either the leftmost leaf of s_2 is e or $s_2 = d$.

The initial state of the transducer T_2 is $\hat{q}_0$ and T_2 contains the rules

$$
\begin{aligned}
&\hat{q}_0(f(x_1, x_2)) \to f(\hat{q}_1(x_1), \hat{q}_2(x_1)) && \hat{q}_0(d) \qquad\qquad \to d \\
&\hat{q}_1(f(x_1, x_2)) \to f(\hat{q}_1(x_1), \hat{q}_1(x_2)) && \hat{q}_2(f(x_1, x_2)) \to f(\hat{q}_2(x_1), \hat{q}_2(x_2)) \\
&\hat{q}_1(f'(x_1, x_2)) \to f'(\hat{q}_1(x_1), \hat{q}_2(x_2)) && \hat{q}_2(e) \qquad\qquad \to e \\
&\hat{q}_1(e) \qquad\qquad \to e && \hat{q}_2(d) \qquad\qquad \to d \\
&\hat{q}_1(d) \qquad\qquad \to d.
\end{aligned}
$$

Informally, on input s, the state $\hat{q}_2$ produces s if the symbol f' does not occur in s; otherwise $\hat{q}_2$ produces no output. The state $\hat{q}_1$ realizes the identity. In conjunction with the rules of $\hat{q}_0$, it follows that the domain of T_2 only consists of the tree d and trees $f(s_1, s_2)$ with no occurrences of f' in s_1.

Consider the composition $\tau = T_1 \,\hat{\circ}\, T_2$. On input s, the composition τ yields $f(s_1, s_1)$ if s is of the form $f(s_1, s_2)$ and the leftmost leaf of s_2 is labeled by e. If the input tree is of the form $f(s_1, d)$, the output tree d is produced. Clearly, τ is functional. We remark that both phenomena described in Sect. 3 occur in the composition τ. More precisely, simply applying the p-construction to T_1 and T_2 yields a nondeterministic transducer due to 'independent guessing'. Furthermore, not checking the validity of guesses causes nondeterminism on input $f(s_1, d)$.

In the following, we show how to construct the la-automaton M from the transducers T_1 and T_2.

Construction of the Domain Automaton A. We begin by constructing the domain automaton A of T_2. Denote by Q_2 the set of states of T_2. The set of states of A is the power set of Q_2 and the initial state of A is $\{\hat{q}_0\}$. The rules of A are

$$
\begin{aligned}
\{\hat{q}_0\} \ (f(x_1, x_2)) &\to f(S(x_1), \emptyset(x_2)) \\
\{\hat{q}_0\} \ (d) &\to d \\
S \ (f(x_1, x_2)) &\to f(S(x_1), S(x_2)) \\
S \ (e) &\to e \\
S \ (d) &\to d
\end{aligned}
$$

where $S = \{\hat{q}_1, \hat{q}_2\}$. The state $\emptyset$ realizes the identity. The rules for the state $\emptyset$ are straight forward and hence omitted here. By construction of the domain automaton, A also contains for instance the state $\{\hat{q}_0, \hat{q}_1\}$. However as $\{\hat{q}_0, \hat{q}_1\}$ is *unreachable*, its rules are omitted.

Informally, a state q of a transducer T is called *reachable*, if it is the initial state of T or if it occurs on the right-hand side of a reachable state.

For the following transducers, we only consider states that are reachable and their rules. Unreachable states and their rules are omitted.

Construction of the Transducer $\hat{T}_1$. For T_1 and A, the p-construction yields the following transducer $\hat{T}_1$. The transducer $\hat{T}_1$ contains the rules

$$
\begin{aligned}
(q_0, \{\hat{q}_0\}) \ (f(x_1, x_2)) &\to f((q_1, S)(x_1), q_2(x_2)) \\
(q_0, \{\hat{q}_0\}) \ (f(x_1, x_2)) &\to (q_3, \{\hat{q}_0\})(x_2) \\
q_1 \ (f(x_1, x_2)) &\to f(q_1(x_1), q_1(x_2)) \\
q_1 \ (f(x_1, x_2)) &\to f'(q_1(x_1), q_1(x_2)) \\
q_1 \ (e) &\to e \\
q_1 \ (d) &\to d \\
(q_1, S) \ (f(x_1, x_2)) &\to f((q_1, S)(x_1), (q_1, S)(x_2)) \\
(q_1, S) \ (e) &\to e \\
(q_1, S) \ (d) &\to d \\
q_2 \ (f(x_1, x_2)) &\to f(q_2(x_1), q_1(x_2)) \\
q_2 \ (f(x_1, x_2)) &\to f'(q_2(x_1), q_1(x_2)) \\
q_2 \ (e) &\to e \\
(q_3, \{\hat{q}_0\}) \ (d) &\to d
\end{aligned}
$$

and the initial state of $\hat{T}_1$ is $(q_0, \{\hat{q}_0\})$. For better readability, we just write q_1 and q_2 instead of $(q_1, \emptyset)$ and $(q_2, \emptyset)$, respectively.

Construction of the Transducer N. For $\hat{T}_1$ and T_2, we construct the transducer N containing the rules

$$
\begin{aligned}
(q_0, \{\hat{q}_0\}, \hat{q}_0) \ (f(x_1, x_2)) &\to f((q_1, S, \hat{q}_1)(x_1), (q_1, S, \hat{q}_2)(x_1)) \\
(q_0, \{\hat{q}_0\}, \hat{q}_0) \ (f(x_1, x_2)) &\to (q_3, \{\hat{q}_0\}, \hat{q}_0)(x_2) \\
(q_1, S, \hat{q}_1) \ (f(x_1, x_2)) &\to f((q_1, S, \hat{q}_1)(x_1), (q_1, S, \hat{q}_1)(x_2)) \\
(q_1, S, \hat{q}_1) \ (e) &\to e \\
(q_1, S, \hat{q}_1) \ (d) &\to d \\
(q_1, S, \hat{q}_2) \ (f(x_1, x_2)) &\to f((q_1, S, \hat{q}_2)(x_1), (q_1, S, \hat{q}_2)(x_2)) \\
(q_1, S, \hat{q}_2) \ (e) &\to e \\
(q_1, S, \hat{q}_2) \ (d) &\to d \\
(q_3, \{\hat{q}_0\}, \{\hat{q}_0\}) \ (d) &\to d
\end{aligned}
$$

The initial state of N is $(q_0, \{\hat{q}_0\}, \hat{q}_0)$. We remark that though no nondeterminism is caused by 'independent guessing', N is still nondeterministic on input $f(s_1, d)$ as the validity of guesses cannot be checked. To perform validity checks for guesses, we extend N with look-ahead.

Construction of the Look-Ahead Automaton Â. Recall that the look-ahead automaton of M is the domain automaton $\hat{A}$ of $\hat{T}_1$. The set of states of $\hat{A}$ is the power set of the set of states of $\hat{T}_1$. The initial state of $\hat{A}$ is $\{(q_0, \{\hat{q}_0\})\}$ and $\hat{A}$ contains the following rules.

$$
\begin{aligned}
\{(q_0, \{\hat{q}_0\})\} \ (f(x_1, x_2)) &\to f(\{(q_1, S)\}(x_1), \{q_2\})\}(x_2)) \\
\{(q_0, \{\hat{q}_0\})\} \ (f(x_1, x_2)) &\to f(\emptyset(x_1), \{(q_3, \{\hat{q}_0\})\}(x_2)) \\
\{q_1\} \ (f(x_1, x_2)) &\to f(\{q_1\}(x_1), \{q_1\}(x_2)) \\
\{q_1\} \ (e) &\to e \\
\{q_1\} \ (d) &\to d \\
\{(q_1, S)\} \ (f(x_1, x_2)) &\to f(\{(q_1, S)(x_1)\}, \{(q_1, S)\}(x_2)) \\
\{(q_1, S)\} \ (e) &\to e \\
\{(q_1, S)\} \ (d) &\to d \\
\{q_2\} \ (f(x_1, x_2)) &\to f(\{q_2\}(x_1), \{q_1\}(x_2)) \\
\{q_2\} \ (e) &\to e \\
(q_3, \{\hat{q}_0\}, \{\hat{q}_0\}) \ (d) &\to d
\end{aligned}
$$

For better readability, we again just write q_1 and q_2 instead of $(q_1, \emptyset)$ and $(q_2, \emptyset)$, respectively. We remark that, by construction of the domain automaton, $\hat{A}$ also contains the rule

$$
\{(q_0, \{\hat{q}_0\})\}(f(x_1, x_2)) \to f(\{(q_1, S)\}(x_1), \{q_2, (q_3, \{\hat{q}_0\})\}(x_2)),
$$

however, since no rules are defined for the state $\{q_2, (q_3, \{\hat{q}_0\})\}$, this rule can be omitted.

Construction of the LA-Transducer M. Finally, we construct the la-transducer M. The initial state of M is $(q_0, \{\hat{q}_0\}, \hat{q}_0)$ and the rules of M are

$$
\begin{aligned}
(q_0, \{\hat{q}_0\}, \hat{q}_0) \ (f(x_1\!:\!\{(q_1, S)\}, x_2\!:\!\{q_2\})) & \rightarrow f((q_1, S, \hat{q}_1)(x_1), (q_1, S, \hat{q}_2)(x_1)) \\
(q_0, \{\hat{q}_0\}, \hat{q}_0) \ (f(x_1\!:\!\emptyset, x_2\!:\!\{q_3, \{\hat{q}_0\}\})) & \rightarrow (q_3, \{\hat{q}_0\}, \hat{q}_0)(x_2) \\
(q_1, S, \hat{q}_1) \ (f(x_1\!:\!\{(q_1, S)\}, x_2\!:\!\{(q_1, S)\})) & \rightarrow f((q_1, S, \hat{q}_1)(x_1), (q_1, S, \hat{q}_1)(x_2)) \\
(q_1, S, \hat{q}_1) \ (e) & \rightarrow e \\
(q_1, S, \hat{q}_1) \ (d) & \rightarrow d \\
(q_1, S, \hat{q}_2) \ (f(x_1\!:\!\{(q_1, S)\}, x_2\!:\!\{(q_1, S)\})) & \rightarrow f((q_1, S, \hat{q}_2)(x_1), (q_1, S, \hat{q}_2)(x_2)) \\
(q_1, S, \hat{q}_2) \ (e) & \rightarrow e \\
(q_1, S, \hat{q}_2) \ (d) & \rightarrow d \\
(q_3, \{\hat{q}_0\}, \{\hat{q}_0\}) \ (d) & \rightarrow d
\end{aligned}
$$

By construction, the transducer N contains the rule

$$
\eta = (q_0, \{\hat{q}_0\}, \hat{q}_0)(f(x_1, x_2)) \rightarrow f((q_1, S, \hat{q}_1)(x_1), (q_1, S, \hat{q}_2)(x_1)).
$$

This rule is obtained from the rule $(q_0, \{\hat{q}_0\})(f(x_1, x_2)) \rightarrow f((q_1, S)(x_1), q_2(x_2))$ of $\hat{T}_1$.

Consider the input tree $f(s_1, s_2)$ where s_1 and s_2 are arbitrary ground trees. Clearly, translating $f(s_1, s_2)$ with N begins with the rule η. Recall that the transducer N is equipped with look-ahead in order to guarantee that guesses performed by states of N are valid. In particular, to guarantee that the guess corresponding to η is valid, we need to test whether or not $s_1 \in \mathrm{dom}(q_1, S)$ and $s_2 \in \mathrm{dom}(q_2)$. Therefore, M contains the rule

$$
(q_0, \{\hat{q}_0\}, \hat{q}_0)(f(x_1\!:\!\{(q_1, S)\}, x_2\!:\!\{q_2\})) \rightarrow f((q_1, S, \hat{q}_1)(x_1), (q_1, S, \hat{q}_2)(x_1)).
$$

Recall that if f is relabeled by $\langle f, \{q_1, S\}, \{q_2\}\rangle$ via the la-automaton $\hat{A}$, this means precisely that $s_1 \in \mathrm{dom}(q_1, S)$ and $s_2 \in \mathrm{dom}(q_2)$. We remark that by definition, M also contains rules of the form

$$
(q_0, \{\hat{q}_0\}, \hat{q}_0)(f(x_1\!:\!l_1, x_2\!:\!l_2)) \rightarrow f((q_1, S, \hat{q}_1)(x_1), (q_1, S, \hat{q}_2)(x_1)),
$$

where l_1 and l_2 are states of $\hat{A}$ such that $\{q_1, S\} \subseteq l_1$ and $\{q_2\} \subseteq l_2$ and l_1 or l_2 is a proper superset. However, as none such states l_1 and l_2 are reachable by $\hat{A}$, we have omitted rules of this form. Other rules are omitted for the same reason. $\square$

3.2 Correctness of the LA-Transducer M

In the following we prove the correctness of our construction. More precisely, we prove that M is functional if and only if $T_1 \hat{\circ} T_2$ is. By Lemma 3, it is sufficient to show that M is functional if and only if $\hat{T}_1 \hat{\circ} T_2$ is.

First, we prove that the following claim: If M is functional then $\hat{T}_1 \hat{\circ} T_2$ is functional. More precisely, we show that $\mathcal{R}(\hat{T}_1) \circ \mathcal{R}(T_2) \subseteq \mathcal{R}(M)$. Obviously, this implies our claim. First of all, consider the transducers N and N' obtained from the p-construction in our examples in Sect. 3. Notice that the relations defined by N and N' are supersets of $\mathcal{R}(T_1) \circ \mathcal{R}(T_2)$ and $\mathcal{R}(T_1') \circ \mathcal{R}(T_2')$, respectively.

In the following, we show that this observation can be generalized. Consider arbitrary transducers T and T'. We claim that the transducer $\check{N}$ obtained from the p-construction for T and T' always defines a superset of the composition $\mathcal{R}(T) \circ \mathcal{R}(T')$. To see that our claim holds, consider a translation of $T \hat{\circ} T'$ in which the state q' of T' processes a tree t produced by the state q of T on input s. If the corresponding state (q, q') of $\check{N}$ processes s then (q, q') can guess that q has produced t and proceed accordingly. Thus $\check{N}$ can effectively simulate the composition $T \hat{\circ} T'$.

As M is in essence obtained from the p-construction extended with look-ahead, M 'inherits' this property. Note that the addition of look-ahead does not affect this property. Therefore our claim follows.

Lemma 4. $\mathcal{R}(\hat{T}_1) \circ \mathcal{R}(T_2) \subseteq \mathcal{R}(M)$.

In fact an even stronger result holds.

Lemma 5. *Let (q_1, S) be a state of $\hat{T}_1$ and q_2 be a state of T_2. If on input s, (q_1, S) can produce the tree t and on input t, q_2 can produce the tree r then (q_1, S, q_2) can produce r on input s.*

Consider a translation of $\hat{T}_1 \hat{\circ} T_2$ in which T_2 processes the tree t produced by T_1 on input s. We call a translation of M *synchronized* if the translation simulates a translation of $\hat{T}_1 \hat{\circ} T_2$, i.e., if a state (q, S, q') of M processes the subtree s' of s and the corresponding state of q' of T_2 processes the subtree t' of t and t' is produced by (q, S) on input s', then (q, S, q') guesses t'.

We now show that if $\hat{T}_1 \hat{\circ} T_2$ is functional, then so is M. Before we prove our claim consider the following auxiliary results.

Lemma 6. *Consider an arbitrary input tree s. Let $\hat{s}$ be a subtree of s. Assume that in an arbitrary translation of M on input s, the state (q_1, S, q_2) processes $\hat{s}$. Then, a synchronized translation of M on input s exists in which the state (q_1, S, q_2) processes the subtree $\hat{s}$.*

It is easy to see that the following result holds for arbitrary transducers.

Proposition 1. *Let $\tau = T_1 \hat{\circ} T_2$ where T_1 and T_2 are arbitrary transducers. Let s be a tree such that $\tau(s) = \{r\}$ is a singleton. Let t_1 and t_2 be distinct trees produced by T_1 on input s. If t_1 and t_2 are in the domain of T_2 then $T_2(t_1) = T_2(t_2) = \{r\}$.*

Using Lemma 6 and Proposition 1, we now show that the following holds. Note that in the following t/v, where t is some tree and v is a node, denotes the subtree of t rooted at the node v.

Lemma 7. *Consider an arbitrary input tree s. Let $\hat{s}$ be a subtree of s. Let the state (q_1, S, q_2) process $\hat{s}$ in a translation M on input s. If $\hat{T}_1 \hat{\circ} T_2$ is functional then (q_1, S, q_2) can only produce a single output tree on input $\hat{s}$.*

Proof. Assume to the contrary that (q_1, S, q_2) can produce distinct trees r_1 and r_2 on input $\hat{s}$. For r_1, it can be shown that a tree t_1 exists such that

1. on input $\hat{s}$, the state (q_1, S) of $\hat{T}_1$ produces t_1 and
2. on input t_1, the state q_2 of T_2 produces r_1.

It can be shown that a tree t_2 with the same properties exists for r_2. Informally, this means that r_1 and r_2 are producible by (q_1, S, q_2) by simulating the 'composition of (q_1, S) and q_2'.

Due to Lemma 6, a synchronized translation of M on input s exists in which the state (q_1, S, q_2) processes the subtree $\hat{s}$ of s. Let g be the node at which (q_1, S, q_2) processes $\hat{s}$. Let $\hat{q}_1, \ldots, \hat{q}_n$ be all states of M of the form (q_1, S, q_2'), where q_2' is some state of T_2, that occur in the synchronized translation of M and that process $\hat{s}$. Note that by definition $q_2' \in S$. Due to Lemmas 2 and 5, we can assume that in the synchronized translation, the states $\hat{q}_1, \ldots, \hat{q}_n$ all guess that the tree t_1 has been produced by the state (q_1, S) of $\hat{T}_1$ on input $\hat{s}$. Hence, we can assume that at the node g, the output subtree r_1 is produced. Therefore, a synchronized translation of M on input s exists, that yields an output tree $\hat{r}_1$ such that $\hat{r}_1/g = r_1$, where $\hat{r}_1/g$ denotes the subtree of $\hat{r}_1$ rooted at the node g. Analogously, it follows that a synchronized translation of M on input s exists, that yields an output tree $\hat{r}_2$ such that $\hat{r}_2/g = r_2$.

As both translation are synchronized, i.e., 'simulations' of translations of $\hat{T}_1 \hat{\circ} T_2$ on input s, it follows that the trees $\hat{r}_1$ and $\hat{r}_2$ are producible by $\hat{T}_1 \hat{\circ} T_2$ on input s. Due to Proposition 1, $\hat{r}_1 = \hat{r}_2$ and therefore $r_1 = \hat{r}_1/g = \hat{r}_2/g = r_2$. □

Lemma 4 implies that if M is functional then $\hat{T}_1 \hat{\circ} T_2$ is functional as well. Lemma 7 implies that if $\hat{T}_1 \hat{\circ} T_2$ is functional then so is M. Therefore, we deduce that due Lemmas 4 and 7 the following holds.

Corollary 1. $\hat{T}_1 \hat{\circ} T_2$ *is functional if and only if M is functional.*

In fact, Corollary 1 together with Lemma 4 imply that $\hat{T}_1 \hat{\circ} T_2$ and M are equivalent if $\hat{T}_1 \hat{\circ} T_2$ is functional, since it can be shown that $\mathrm{dom}(\hat{T}_1 \hat{\circ} T_2) = \mathrm{dom}(M)$.

Since functionality for transducers with look-ahead is decidable [8], Corollary 1 implies that it is decidable whether or not $\hat{T}_1 \hat{\circ} T_2$ is functional. Together with Lemma 3, we obtain:

Theorem 1. *Let T_1 and T_2 be top-down tree transducers. It is decidable whether or not $T_1 \hat{\circ} T_2$ is functional.*

3.3   Functionality of Arbitrary Compositions

In this section, we show that the question whether or not an arbitrary composition is functional can be reduced to the question of whether or not a two-fold composition is functional.

Lemma 8. *Let τ be a composition of transducers. Then two transducers T_1, T_2 can be constructed such that $T_1 \hat{\circ} T_2$ is functional if and only if τ is functional.*

Proof. Consider a composition of n transducers $T_1', \ldots, T_n'$. W.l.o.g. assume that $n > 2$. For $n \leq 2$, our claim follows trivially. Let τ be the composition of $T_1', \ldots, T_n'$. We show that transducer $\hat{T}_1, \ldots, \hat{T}_{n-1}$ exist such that $\hat{T}_1 \hat{\circ} \cdots \hat{\circ} \hat{T}_{n-1}$ is functional if and only if τ is.

Consider an arbitrary input tree s. Let t be a tree produced by the composition $T_1' \,\hat{\circ}\, \cdots \,\hat{\circ}\, T_{n-2}'$ on input s. Analogously as in Proposition 1, the composition $T_{n-1}' \,\hat{\circ}\, T_n'$, on input t, can only produce a single output tree if τ is functional. For the transducers T_{n-1}' and T_n', we construct the la-transducer M according to our construction in Sect. 3.1. It can be shown that, the la-transducer M our construction yields has the following properties regardless of whether or not $T_{n-1} \,\hat{\circ}\, T_n$ is functional

(a) $\mathrm{dom}(M) = \mathrm{dom}(T_{n-1} \,\hat{\circ}\, T_n)$ and
(b) on input t, M only produces a single output tree if and only if $T_{n-1} \,\hat{\circ}\, T_n$ does

Therefore, $\tau(s)$ is a singleton if and only if $T_1' \,\hat{\circ}\, \cdots \,\hat{\circ}\, T_{n-2}' \,\hat{\circ}\, M(s)$ is a singleton. Engelfriet has shown that every transducer with look-ahead can be decomposed to a composition of a deterministic bottom-up relabeling and a transducer (Theorem 2.6 of [4]). It is well known that (nondeterministic) relabelings are independent of whether they are defined by bottom-up transducers or by top-down transducers (Lemma 3.2 of [3]). Thus, any transducer with look-ahead can be decomposed into a composition of a nondeterministic top-down relabeling and a transducer. Let R and T be the relabeling and the transducer such that M and $R \,\hat{\circ}\, T$ define the same relation. Then obviously, $\tau(s)$ is a singleton if and only if $T_1' \,\hat{\circ}\, \cdots \,\hat{\circ}\, T_{n-2}' \,\hat{\circ}\, R \,\hat{\circ}\, T(s)$ is a singleton.

Consider arbitrary transducers $\bar{T}_1$ and $\bar{T}_2$. Baker has shown that if $\bar{T}_2$ is non-deleting and linear then a transducer T can be constructed such that T and $\bar{T}_1 \,\hat{\circ}\, \bar{T}_2$ are equivalent (Theorem 1 of [2]). By definition, any relabeling is non-deleting and linear. Thus, we can construct a transducer $\tilde{T}$ such that $\tilde{T}$ and $T_{n-2}' \,\hat{\circ}\, R$ are equivalent. Therefore, it follows that $\tau(s)$ is a singleton if and only if $T_1' \,\hat{\circ}\, \cdots \,\hat{\circ}\, T_{n-3}' \,\hat{\circ}\, \tilde{T} \,\hat{\circ}\, T(s)$ is a singleton. This yields our claim.   □

Lemma 8 and Theorem 1 yield that functionality of compositions of transducers is decidable.

Engelfriet has shown that any la-transducer can be decomposed into a composition of a nondeterministic top-down relabeling and a transducer [3,4]. Recall that while la-transducers generalize transducers, bottom-up transducers and la-transducers are incomparable [4]. Baker, however, has shown that the composition of n bottom-up-transducers can be realized by the composition of $n+1$ top-down transducers [2]. For any functional composition of transducers an equivalent deterministic la-transducer can be constructed [5]. Therefore we obtain our following main result.

Theorem 2. *Functionality for arbitrary compositions of top-down and bottom-up tree transducers is decidable. In the affirmative case, an equivalent deterministic top-down tree transducer with look-ahead can be constructed.*

4   Conclusion

We have presented a construction of an la-transducer for a composition of transducers which is functional if and only if the composition of the transducers is functional—in which case it is equivalent to the composition. This construction is remarkable since transducers are not closed under composition in general, neither does functionality of the composition imply that each transducer occurring therein, is functional. By Engelfriet's construction in [5], our construction provides the key step to an efficient implementation (i.e., a deterministic transducer, possibly with look-ahead) for a composition of transducers – whenever possible (i.e., when their translation is functional). As an open question, it remains to see how large the resulting functional transducer necessarily must be, and whether the construction can be simplified if for instance only compositions of linear transducers are considered.

References

1. Aho, A.V., Ullman, J.D.: Translations on a context-free grammar. Inf. Control **19**(5), 439–475 (1971)
2. Baker, B.S.: Composition of top-down and bottom-up tree transductions. Inf. Control **41**(2), 186–213 (1979)
3. Engelfriet, J.: Bottom-up and top-down tree transformations - a comparison. Math. Syst. Theory **9**(3), 198–231 (1975)
4. Engelfriet, J.: Top-down tree transducers with regular look-ahead. Math. Syst. Theory **10**, 289–303 (1977)
5. Engelfriet, J.: On tree transducers for partial functions. Inf. Process. Lett. **7**(4), 170–172 (1978)
6. Engelfriet, J.: Three hierarchies of transducers. Math. Syst. Theory **15**(2), 95–125 (1982)
7. Engelfriet, J., Maneth, S., Seidl, H.: Deciding equivalence of top-down XML transformations in polynomial time. J. Comput. Syst. Sci. **75**(5), 271–286 (2009)
8. Ésik, Z.: Decidability results concerning tree transducers I. Acta Cybern. **5**(1), 1–20 (1980)
9. Hakuta, S., Maneth, S., Nakano, K., Iwasaki, H.: XQuery streaming by forest transducers. In: ICDE 2014, Chicago, USA, 31 March–4 April 2014, pp. 952–963 (2014)
10. Küsters, R., Wilke, T.: Transducer-based analysis of cryptographic protocols. Inf. Comput. **205**(12), 1741–1776 (2007)
11. Maneth, S.: A survey on decidable equivalence problems for tree transducers. Int. J. Found. Comput. Sci. **26**(8), 1069–1100 (2015)
12. Maneth, S., Seidl, H., Vu, M.: Definability results for top-down tree transducers. In: Moreira, N., Reis, R. (eds.) DLT 2021. LNCS, vol. 12811, pp. 291–303. Springer, Cham (2021). https://doi.org/10.1007/978-3-030-81508-0_24
13. Maneth, S., Seidl, H., Vu, M.: How to decide functionality of compositions of top-down tree transducers. arXiv:2209.01044 (2022)
14. Matsuda, K., Inaba, K., Nakano, K.: Polynomial-time inverse computation for accumulative functions with multiple data traversals. High. Order Symb. Comput. **25**(1), 3–38 (2012)
15. Rounds, W.C.: Mappings and grammars on trees. Math. Syst. Theory **4**(3), 257–287 (1970)
16. Thatcher, J.W.: Generalized sequential machine maps. J. Comput. Syst. Sci. **4**(4), 339–367 (1970)

Computation of Solutions to Certain Nonlinear Systems of Fuzzy Relation Inequations

Ivana Micić⬥, Zorana Jančić⬥, and Stefan Stanimirović$^{(\boxtimes)}$⬥

Faculty of Science and Mathematics, University of Niš, Niš, Serbia
{ivana.micic,zorana.jancic,stefan.stanimirovic}@pmf.edu.rs

Abstract. Although fuzzy relation equations and inequations have a broad field of application, it is common that they have no solutions or have only the trivial solution. Therefore, it is desirable to study new types of fuzzy relation inequations similar to the well-studied ones and with nontrivial solutions. This paper studies fuzzy relation inequations that include the degree of subsethood and the degree of equality of fuzzy sets. We provide formulae for determining the greatest solutions to systems of such fuzzy relation inequations. We provide alternative ways to compute these solutions when we cannot run the methods based on these formulae.

Keywords: Fuzzy relation equation · Fuzzy relation · Degree of equality

1 Introduction

In many scientific and technological areas, such as image processing, fuzzy control and data processing, drawing conclusions based on vague and imprecise data is required. As a primary mechanism to formalize the connection between such fuzzy data sets, fuzzy relation equations (FREs) and fuzzy relation inequations (FRIs) have been widely studied.

Linear systems of fuzzy relation equations and inequations, i.e., systems in which the unknown fuzzy relation appears only on one side of the sign $=$ or $\leqslant$, were introduced and studied in [13,14]. Linear systems were firstly considered over the Gödel structure, and later, the same systems were studied over broader sets of truth values, such as complete residuated lattices [1–3,7,18]. Afterward, nonlinear systems of fuzzy relation inequations have been examined. Among others, the so-called homogeneous and heterogeneous weakly linear systems have been introduced and studied by Ignjatović et al. [8–10] over a complete residuated

This research was supported by the Science Fund of the Republic of Serbia, GRANT No 7750185, Quantitative Automata Models: Fundamental Problems and Applications - QUAM, and by Ministry of Education, Science and Technological Development, Republic of Serbia, Contract No. 451-03-68/2022-14/200124.

ⓒ The Author(s), under exclusive license to Springer Nature Switzerland AG 2022
D. Poulakis and G. Rahonis (Eds.): CAI 2022, LNCS 13706, pp. 192–202, 2022.
https://doi.org/10.1007/978-3-031-19685-0_14

lattice. These are the systems where an unknown fuzzy relation appears on both sides of the sign $=$ or $\leqslant$. Also, the authors have provided procedures for determining the greatest solutions for such systems. Although these systems have applications in many areas, including concurrency theory and social network analysis, it is common that only the trivial solution exists for such systems. Accordingly, it is preferable to modify the criteria given in [9].

In [11,15], Stanimirović et al. have introduced approximate bisimulations for fuzzy automata over a complete Heyting algebra. Following this idea, Micić et al. have introduced in [12] approximate regular and approximate structural fuzzy relations for fuzzy social networks defined also over a complete Heyting algebra. These notions are generalizations of bisimulations and regular and structural fuzzy relations, and are defined as solutions to certain generalizations of weakly linear systems of FRIs. However, the observed underlying structure of truth values is a complete Heyting algebras, a special type of complete residuated lattice with an idempotent multiplication. In this paper, we study systems of FRIs introduced in [11,12,15], but we study them over an arbitrary complete residuated lattice.

Our results are the following: We show that the set of all solutions to these systems of FRIs form a complete lattice, and therefore, there exists the greatest solution for every such system. We study further properties of such systems. We give a procedure for computing this greatest solution. Unfortunately, this procedure suffers the same shortcoming as the ones developed in [8–10]. That is, it may not finish in a finite number of steps for every complete residuated lattice. Thus, we propose two alternatives in such cases. First, we show that fuzzy relations generated by this procedure converge to the solution of these systems in the case when a complete residuated lattice is a BL-algebra defined over the [0, 1] interval. And second, we show that we can compute the greatest crisp solution to such systems. In this case, a procedure always terminates in a finite number of steps. However, such solutions are smaller than or equal to the solutions obtained by the procedure for computing the greatest solutions.

2   Preliminaries

Since we study FRIs over complete residuated lattices, we emphasize the basic characteristics of this structure. An algebra $\mathcal{L} = (L, \wedge, \vee, \otimes, \rightarrow, 0, 1)$ where:

1) $(L, \wedge, \vee, 0, 1)$ is a lattice bounded by 0 and 1;
2) $(L, \otimes, 1)$ is a commutative monoid in which 1 as neutral element for $\otimes$;
3) operations $\otimes$ and $\rightarrow$ satisfy condition:

$$x \otimes y \leqslant z \quad \text{iff} \quad x \leqslant y \rightarrow z, \quad \text{for each } x, y, z \in L. \tag{1}$$

is called a *residuated lattice*. In addition, (1) is often called the *adjunction property*, while we say that $\otimes$ and $\rightarrow$ form the *adjoint pair*. Residuated lattice in which $(L, \wedge, \vee)$ is a complete lattice, is called a *complete residuated lattice*.

For presenting conjunction and implication, operations $\otimes$ (called *multiplication*) and $\rightarrow$ (called *residuum*) are used, and for the general and existential quantifier the infimum ($\bigwedge$) and supremum ($\bigvee$) are used, respectively. The equivalence of truth values is presented by operation *biresiduum* (or *biimplication*, denoted by $\leftrightarrow$ and defined by:

$$x \leftrightarrow y = (x \rightarrow y) \wedge (y \rightarrow x), \quad x, y \in L.$$

The well-known properties of operators $\otimes$ and $\rightarrow$ are: operator $\otimes$ is non-decreasing with respect to $\leqslant$ in both arguments, while operator $\rightarrow$ is non-decreasing with respect to $\leqslant$ in the second argument, whereas it is non-increasing in the first argument. The most studied and applied structures of truth values are the *product (Goguen) structure*, the *Gödel structure* and the *Łukasiewicz structure*. For their definitions, as well for other properties of complete residuated lattices, we refer to [4,5].

If in a residuated lattice for every $x, y \in L$, the following holds:

1. $(x \rightarrow y) \vee (y \rightarrow x) = 1$ (*prelinearity*);
2. $x \otimes (x \rightarrow y) = x \wedge y$ (*divisibility*);

then it is called a *BL-algebra*. The algebra $([0,1], \min, \max, \otimes, \rightarrow, 0, 1)$, where operation $\rightarrow$ is given by following formula:

$$x \rightarrow y = \bigvee \{z \in L \mid x \otimes z \leqslant y\},$$

is a BL-algebra if and only if operation $\otimes$ is a continuous t-norm (cf. [5, Theorem 1.40]). Specially, the Goguen (product) structure is a BL-algebra, as well as Gödel and Łukasiewicz.

In the sequel, we assume that L is a support set of some complete residuated lattice $\mathcal{L}$. For a nonempty set A, a mapping from A into L is called a *fuzzy subset* of a set A over $\mathcal{L}$, or just a *fuzzy subset* of A. The inclusion (ordering) and the equality of fuzzy sets are defined coordinate-wise (cf. [4,5]). The set of all fuzzy subsets of A is denoted with L^A. For two fuzzy sets $\alpha_1, \alpha_2 \in L^A$ the meet and the join of α_1 and α_2 are defined as fuzzy subsets of A in the following way:

$$(\alpha_1 \wedge \alpha_2)(a) = \alpha_1(a) \wedge \alpha_2(a), \quad \text{and} \quad (\alpha_1 \vee \alpha_2)(a) = \alpha_1(a) \vee \alpha_2(a),$$

for every $a \in A$. A fuzzy subset of $\alpha \in L^A$, such that rang of α is $\{0,1\} \subseteq L$, is called a *crisp subset* of the set A. Further, with 2^A we denote the set $\{\alpha \mid \alpha \subseteq A\}$. For a fuzzy subset $\alpha \in L^A$, the crisp part of α is the crisp set $\alpha^c \in 2^A$ defined by $\alpha^c(a) = 1$, if $\alpha(a) = 1$, and $\alpha^c(a) = 0$, if $\alpha(a) \neq 1$, for every $a \in A$. In other words, we have that $\alpha^c = \{a \in A \mid \alpha(a) = 1\}$.

A *fuzzy relation* on a set A is any fuzzy subset of $A \times A$, or in other words, it is any function from $A \times A$ to L. The set of all fuzzy relations on A is denoted with $L^{A \times A}$. A crisp relation is a fuzzy relation that takes values only in the set $\{0,1\}$. The set of all crisp relations on a set A is denoted by $2^{A \times A}$. The universal relation on a set A, denoted by u_A, is defined as $u_A(a_1, a_2) = 1$ for all $a_1, a_2 \in A$.

The composition of fuzzy relations $\varphi, \phi \in L^{A \times A}$ is a fuzzy relation $\varphi \circ \phi \in L^{A \times A}$ defined by:

$$(\varphi \circ \phi)(a_1, a_2) = \bigvee_{a_3 \in A} \varphi(a_1, a_3) \otimes \phi(a_3, a_2), \quad \text{for every } a_1, a_2 \in A. \quad (2)$$

Note that the composition of fuzzy relations is an associative operation on a set $L^{A \times A}$. For $x \in L$ and $\varphi \in L^{A \times A}$, we define fuzzy relations $x \otimes \varphi \in L^{A \times A}$ and $x \rightarrow \varphi \in L^{A \times A}$ as

$$(x \otimes \varphi)(a_1, a_2) = x \otimes \varphi(a_1, a_2), \quad \text{and} \quad (x \rightarrow \varphi)(a_1, a_2) = x \rightarrow \varphi(a_1, a_2), \quad (3)$$

for every $a_1, a_2 \in A$. We assume that $\circ$ has a higher precedence than $\otimes$ and $\rightarrow$ defined by (3). For $\varphi, \phi_1, \phi_2 \in L^{A \times A}$, family $\varphi_i \in L^{A \times A} (i \in I)$ and $x \in L$, the following holds:

$$\phi_1 \leqslant \phi_2 \quad \text{implies} \quad \varphi \circ \phi_1 \leqslant \varphi \circ \phi_2 \text{ and } \phi_1 \circ \varphi \leqslant \phi_2 \circ \varphi, \quad (4)$$

$$\phi \circ \left(\bigvee_{i \in I} \varphi_i \right) = \bigvee_{i \in I} (\phi \circ \varphi_i), \quad \left(\bigvee_{i \in I} \varphi_i \right) \circ \phi = \bigvee_{i \in I} (\varphi_i \circ \phi). \quad (5)$$

For $\alpha_1, \alpha_2 \in L^A$, the *degree of subsethood* of α_1 in α_2, denoted with $S(\alpha_1, \alpha_2) \in L$, is defined by:

$$S(\alpha_1, \alpha_2) = \bigwedge_{a \in A} \alpha_1(a) \rightarrow \alpha_2(a), \quad (6)$$

while the *degree of equality of* α_1 *and* α_2, denoted with $E(\alpha_1, \alpha_2) \in L$, is defined as:

$$E(\alpha_1, \alpha_2) = \bigwedge_{a \in A} \alpha_1(a) \leftrightarrow \alpha_2(a). \quad (7)$$

Intuitively, $S(\alpha_1, \alpha_2)$ can be understood as a truth degree of the statement that if some element of A belongs to α_1, then this element belongs to α_2. Also, $E(\alpha_1, \alpha_2)$ can be understood as a truth degree of the statement that an element of A belongs to α_1 if and only if it belongs to α_2. (see [5] for more details). It can easily be proved that $E(\alpha_1, \alpha_2) = S(\alpha_1, \alpha_2) \wedge S(\alpha_2, \alpha_1)$.

It should be noted that the degree of subsethood is a kind of residuation operation that assigns a scalar to a pair of fuzzy sets. That operation is the residual of the operation of multiplying the fuzzy set by a scalar. Thus, the fuzzy relation inequations considered in this paper can be connected with fuzzy relation equations defined using residuals known in the literature.

Let $\varphi, \psi \in L^{A \times A}$ be fuzzy relations. Then we define the *right residual* $\varphi \backslash \psi \in L^{A \times A}$ of ψ by φ and the *left residual* $\psi / \varphi \in L^{A \times A}$ of ψ by φ, respectively, by the following formulae:

$$(\varphi \backslash \psi)(a_1, a_2) = S(\varphi a_1, \psi a_2), \quad \text{for every } a_1, a_2 \in A$$
$$(\psi / \varphi)(a_1, a_2) = S(a_2 \varphi, a_1 \psi), \quad \text{for every } a_1, a_2 \in A.$$

It can be shown that the right and left residuals satisfy the following two adjunction properties:

$$\varphi \circ \chi \leqslant \psi \quad \text{iff} \quad \chi \leqslant \varphi \backslash \psi, \tag{8}$$

$$\chi \circ \varphi \leqslant \psi \quad \text{iff} \quad \chi \leqslant \psi / \varphi, \tag{9}$$

where $\chi \in L^{A \times A}$ an arbitrary fuzzy relation on A. For other properties of residuals of fuzzy relations we address to [8–10]. In addition, we define crisp relations $\varphi \diagdown \psi \in 2^{A \times A}$ and $\psi \diagup \varphi \in 2^{A \times A}$, called the *Boolean right residual of* ψ *by* φ and the *Boolean left residual of* ψ *by* φ), respectively, in the following way:

$$\varphi \diagdown \psi = (\varphi \backslash \psi)^c = \{(a_1, a_2) \in A \times A | \varphi a_1 \leqslant \psi a_2\}, \quad \text{for every } a_1, a_2 \in A,$$

$$\psi \diagup \varphi = (\psi / \varphi)^c = \{(a_1, a_2) \in A \times A | a_2 \varphi \leqslant a_1 \psi\}, \quad \text{for every } a_1, a_2 \in A.$$

Again, if $X \in 2^{A \times A}$, then the following two adjunction properties hold:

$$\varphi \circ X \leqslant \psi \quad \text{iff} \quad X \leqslant \varphi \diagdown \psi, \tag{10}$$

$$X \circ \varphi \leqslant \psi \quad \text{iff} \quad X \leqslant \psi \diagup \varphi, \tag{11}$$

For properties on Boolean right and Boolean left residuals in the context of matrices over additively idempotent semirings, we refer to [6, 16].

3   Certain Types of Fuzzy Relation Inequations and Their Solutions

For a given nonempty set A, a family $\{\varrho_i\}_{i \in I}$ of fuzzy relations on A, and a scalar $\lambda \in L$ from a complete residuated lattice $\mathcal{L}$, consider the following systems of fuzzy relation inequations:

$$S(\varphi \circ \varrho_i, \varrho_i \circ \varphi) \geqslant \lambda \quad \text{for every } i \in I, \tag{12}$$

$$S(\varrho_i \circ \varphi, \varphi \circ \varrho_i) \geqslant \lambda \quad \text{for every } i \in I, \tag{13}$$

$$E(\varphi \circ \varrho_i, \varrho_i \circ \varphi) \geqslant \lambda \quad \text{for every } i \in I, \tag{14}$$

where $\varphi \in L^{A \times A}$ is an unknown fuzzy relation on A. Note that (14) is equivalent to the conjunction of (12) and (13). According to the definition (6) and the adjunction property, we conclude that a fuzzy relation φ is a solution to (12) if and only if:

$$\lambda \otimes \varphi \circ \varrho_i \leqslant \varrho_i \circ \varphi \quad \text{for all } i \in I, \tag{15}$$

and similarly, φ is a solution to (13) if and only if:

$$\lambda \otimes \varrho_i \circ \varphi \leqslant \varphi \circ \varrho_i \quad \text{for all } i \in I. \tag{16}$$

Theorem 1. *The sets of all solutions to* (12), (13) *and* (14) *that are subsets of* $\varphi_0 \in L^{A \times A}$ *is a complete lattice. Accordingly, there exists the greatest solution* φ *to* (12), (13) *and* (14) *such that* $\varphi \leqslant \varphi_0$.

Proof. We prove that this statement holds only for (14), since the case (12) and (13) can be shown in a similar way. Note that the empty relation is a solution to the system (14), for every $\lambda \in L$, and hence the set of all solutions to (14) has at least one element. With $\{\varphi_j\}_{j \in J}$ we label the family of all solutions. Let $\psi = \bigvee_{j \in J} \varphi_j$. Using (5) we show that ψ is also a solution to the system (14), and hence $\{\varphi_j\}_{j \in J}$ is a complete lattice such that $\bigvee_{j \in J} \varphi_j$ is its greatest element.

Lemma 1. *Let $\{\varrho_i\}_{i \in I}$ a family of fuzzy relations on a set A, and let $\varphi_1, \varphi_2 \in L^{A \times A}$ such that φ_1 is the greatest solution to (14) when $\lambda = \lambda_1$, and φ_2 is the greatest solutions to (14) when $\lambda = \lambda_2$, for some $\lambda_1, \lambda_2 \in L$ such that $\lambda_1 \leqslant \lambda_2$. Then $\varphi_2 \leqslant \varphi_1$.*

Proof. Denote with R_{λ_1} and R_{λ_2} the set of all fuzzy relations that are solutions to (14) when $\lambda = \lambda_1$ and $\lambda = \lambda_2$, respectively. Evidently every solution from R_{λ_2} belongs also to R_{λ_1}. Therefore, the greatest solution from R_{λ_2} is contained in R_{λ_1}. This implies $\varphi_2 \leqslant \varphi_1$.

Lemma 2. *Let $\{\varrho_i\}_{i \in I}$ a family of fuzzy relations on a set A. Let $\lambda_1, \lambda_2 \in L$ be two values, such that $\lambda_1 < \lambda_2$, and the greatest solution to (14) when $\lambda = \lambda_1$ is the same as the greatest solution to (14) when $\lambda = \lambda_2$. Then the greatest solution to (14) for every $\lambda \in L$ such that $\lambda_1 \leqslant \lambda \leqslant \lambda_2$, is the same fuzzy relation.*

Proof. Denote with φ_1 (resp. φ_2) the greatest solution to (14) when $\lambda = \lambda_1$ (resp. $\lambda = \lambda_2$). According to the previous lemma, $\lambda_1 \leqslant \lambda \leqslant \lambda_2$ implies $\varphi_1 \leqslant \varphi_2 \leqslant \varphi_1$, and hence $\varphi_1 = \varphi_2$.

Theorem 1 states that, for all systems of the form (12), (13) and (14), there exists the greatest solution, but it does not provide a way to compute it. Here, we provide a method for determining these greatest solutions. Precisely, we propose a function for finding the greatest solution to (14) contained in a given fuzzy relation. The greatest solutions to (12) and (13) can be obtained in the similar way. A variant of the following result is proven in [17], so we omit its proof.

For a given family $\{\varrho_i\}_{i \in I}$ and an element $\lambda \in L$, define two functions $\mathcal{Q}_1(\lambda), \mathcal{Q}_2(\lambda) : L^{A \times A} \to L^{A \times A}$, for every $\varphi \in L^{A \times A}$, as:

$$\mathcal{Q}_1(\lambda)(\varphi) = \bigwedge_{i \in I} (\lambda \to \varrho_i \circ \varphi)/\varrho_i, \tag{17}$$

$$\mathcal{Q}_2(\lambda)(\varphi) = \bigwedge_{i \in I} \varrho_i \backslash (\lambda \to \varphi \circ \varrho_i). \tag{18}$$

Theorem 2. *Let $\varphi_0 \in L^{A \times A}$ be a fuzzy relation on A, and let $\lambda \in L$. Consider a procedure for computing the array $\{\varphi_n\}_{n \in \mathbb{N}_0}$ of fuzzy relations on A by:*

$$\varphi_{n+1} = \varphi_n \wedge \mathcal{Q}_1(\lambda)(\varphi_n) \wedge \mathcal{Q}_2(\lambda)(\varphi_n), \tag{19}$$

for every $n \in \mathbb{N}_0$. Then the following holds:

a) *φ_k is the greatest solution to (14) contained in φ_0 if and only if $\varphi_k = \varphi_{k+1}$.*
b) *If $\mathcal{L}(\{\varrho\}_{i \in I}, \lambda)$ is a finite subalgebra of $\mathcal{L}$, procedure (19) produces the greatest solution to (14), contained in φ_0, in a finite number of steps.*

According to the previous result we obtain a function for computing the greatest solution to (14) contained in a fuzzy relation φ_0, for a given $\lambda \in L$ and $\varphi_0 \in L^{A \times A}$, formalized by Function 1.

Function 1. ComputeTheGreatestSolution($\{\varphi_n\}_{n \in \mathbb{N}_0}, \varphi_0, \lambda$)

    Input: *a family* $\{\varrho_i\}_{i \in I}$*, a fuzzy relation* $\varphi_0 \in L^{A \times A}$ *and an element* $\lambda \in L$.

    Output: *a fuzzy relation that is the greatest solution to system* (14) *contained in* φ_0.

1. $\phi = \varphi_0$;
2. *do*:
3. $\varphi = \phi$;
4. $\phi = \phi \wedge \mathcal{Q}_1(\lambda)(\phi) \wedge \mathcal{Q}_2(\lambda)(\phi)$;
5. *while*$(\phi \neq \varphi)$;
6. *return* ϕ;

If the previous function is unable to compute the output in a finite number of steps, we discuss alternative ways to obtain this fuzzy relations. Precisely, the exact solution to (14), contained in φ_0 is obtained when the above function finishes after a finite number of steps. If that is not the case, then when $\mathcal{L}$ is a BL-algebra, then the array of fuzzy relations computed using Function 1 is convergent, and the limit value of this array is the sought solution. According to this fact, in the case when Function 1 does not finish after limited number of steps, we can use Theorem 3 to compute the greatest solution to system (14) contained in a given fuzzy relation.

Theorem 3. *Let* $\varphi_0 \in L^{A \times A}$ *be a fuzzy relation on* A*, and let* $\lambda \in L$*, where* $\mathcal{L}$ *is a BL-algebra over the interval* $[0, 1]$*. Then the sequence* $\{\varphi_n\}_{n \in \mathbb{N}}$ *of fuzzy relations on* A*, defined by* (19)*, is convergent. If we denote* $\lim_{n \to \infty} \varphi_n = \widetilde{\varphi}$*, then* $\widetilde{\varphi}$ *is the greatest solution to* (14) *contained in* φ_0.

Proof. By the construction of the sequence $\{\varphi_n\}_{n \in \mathbb{N}}$, we can easily conclude that it is monotonic decreasing. In addition, since every element of this sequence is greater than the zero relation and less than the universal relation, we conclude that it is also bounded. The array $\{\varphi_n\}_{n \in \mathbb{N}}$ is convergent because it is monotonic and bounded. Denote with $\widetilde{\varphi}$ a fuzzy relation that is the limit value of this array.

Since the array $\{\varphi_n\}_{n \in \mathbb{N}}$ is non-increasing, we have that $\widetilde{\varphi} \leqslant \varphi_n$ holds for every $n \in \mathbb{N}$, and thus, $\widetilde{\varphi}$ is contained in φ_0. For proving that $\widetilde{\varphi}$ is a solution to (14), we need to prove that $\widetilde{\varphi}$ satisfies (15) and (16). From the definition of the sequence $\{\varphi_n\}_{n \in \mathbb{N}}$, we conclude that for every $n \in \mathbb{N}$ it holds:

$$\varphi_{n+1} \leqslant (\lambda \to \varrho_i \circ \varphi_n)/\varrho_i \quad \text{and} \quad \varphi_{n+1} \leqslant \varrho_i \backslash (\lambda \to \varphi_n \circ \varrho_i), \quad \text{for every } i \in I.$$

According to the definitions of right and left residuals, the previous inequations are equivalent to:

$$\varphi_{n+1} \circ \varrho_i \leqslant \lambda \to \varrho_i \circ \varphi_n \quad \text{and} \quad \varrho_i \circ \varphi_{n+1} \leqslant \lambda \to \varphi_n \circ \varrho_i, \quad \text{for every } i \in I,$$

which are, by the adjunction property, equivalent to:

$$\lambda \otimes \varphi_{n+1} \circ \varrho_i \leqslant \varrho_i \circ \varphi_n \quad \text{and} \quad \lambda \otimes \varrho_i \circ \varphi_{n+1} \leqslant \varphi_n \circ \varrho_i, \quad \text{for every } i \in I.$$

According to the fact that the t-norm $\otimes$ is continuous in BL-algebras, from the previous inequalities we obtain that for every $i \in I$ the following inequality holds:

$$\lambda \otimes \tilde{\varphi} \circ \varrho_i = \lambda \otimes \left(\lim_{n \to \infty} \varphi_n \right) \circ \varrho_i = \lim_{n \to \infty} (\lambda \otimes \varphi_n \circ \varrho_i)$$

$$\leqslant \lim_{n \to \infty} (\varrho_i \circ \varphi_{n-1}) = \varrho_i \circ \left(\lim_{n \to \infty} \varphi_{n-1} \right) = \varrho_i \circ \tilde{\varphi}.$$

Thus, $\tilde{\varphi}$ is solution to (12). Analogously, we prove that $\tilde{\varphi}$ is solution to (13), which means that $\tilde{\varphi}$ is solution to (14).

We now show that an arbitrary solution to (14), contained in φ_0, is less than every fuzzy relation φ_n ($n \in \mathbb{N}$), obtained by procedure (19). Let ψ be a solution to (14), contained in φ_0. Evidently, $\psi \leqslant \varphi_0$. Suppose that $\psi \leqslant \varphi_n$, for some $n \in \mathbb{N}^0$. Then for every $i \in I$ we have $\lambda \otimes \psi \circ \varrho_i \leqslant \varrho_i \circ \psi \leqslant \varrho_i \circ \varphi_n$, which implicate $\psi \leqslant (\lambda \to \varrho_i \circ \varphi_n)/\varrho_i$, and similarly $\psi \leqslant \lambda \to \varphi_n \circ \varrho_i$ for every $i \in I$, we have:

$$\psi \leqslant \varphi_n \wedge \bigwedge_{i \in I}(\lambda \to \varrho_i \circ \varphi_n)/\varrho_i \wedge \bigwedge_{i \in I} \varrho_i \setminus (\lambda \to \varphi_n \circ \varrho_i) = \varphi_{n+1}.$$

According to the mathematical induction it follows $\psi \leqslant \varphi_n$, for every $n \in \mathbb{N}$. Hence, $\psi \leqslant \lim_{n \to \infty} \varphi_n = \tilde{\varphi}$.

The following example shows the case of a system (14) when Function 1 doesn't terminate in a finite number of steps by putting $\lambda = 1$. On the other hand, by putting $\lambda = 0.7$, Function 1 is able to compute the greatest solution to (14).

Example 1. Consider the system (14) over the product structure, where $\{\varrho_i\}_{i \in I} = \{\varrho\}$, and relation $\varrho \in L^{A \times A}$ is defined with:

$$\varrho = \begin{bmatrix} 0.9 & 0 & 0 & 0 & 0.5 & 0 \\ 0 & 0.8 & 0 & 0.3 & 0 & 0.2 \\ 0 & 0 & 0.8 & 0.4 & 0 & 0.4 \\ 0 & 0 & 0.8 & 0.2 & 0.2 & 0 \\ 0 & 1 & 0 & 1 & 0.2 & 0 \\ 0 & 0 & 0.9 & 0 & 0 & 0.1 \end{bmatrix}. \tag{20}$$

Let $\lambda = 0.7$. Then the greatest solution to (14), contained in u_A, outputted by Function 1, is given by:

$$\varphi = \begin{bmatrix} 1 & 1 & 1 & 1 & 50/63 & 40/63 \\ 1 & 1 & 1 & 1 & 5/7 & 4/7 \\ 1 & 1 & 1 & 1 & 50/63 & 40/63 \\ 1 & 1 & 1 & 1 & 5/7 & 4/7 \\ 1 & 1 & 1 & 1 & 1 & 1 \\ 1 & 1 & 1 & 1 & 1 & 1 \end{bmatrix}. \tag{21}$$

On the other hand, if we set $\lambda = 1$, we get that the sequence $\{\varphi_n\}_{n \in \mathbb{N}}$, generated by formula (19), is infinite. Therefore, Function 1 cannot be used to compute the greatest solution to (14). Note that the greatest solution to (14) is equal to

$$\varphi = \lim_{n \to \infty} \varphi_n = \begin{bmatrix} 1 & 50/81 & 400/729 & 50/81 & 5/9 & 1600/6561 \\ 0 & 1 & 64/81 & 32/81 & 8/81 & 32/81 \\ 0 & 8/9 & 1 & 2/5 & 1/10 & 2/5 \\ 0 & 8/9 & 4/5 & 1 & 1/5 & 32/81 \\ 0 & 1 & 80/81 & 1 & 1 & 320/729 \\ 0 & 1 & 9/10 & 9/20 & 1/5 & 1 \end{bmatrix}.$$

As we have stated, Function 1 for computing the greatest solution to (14) does not necessary terminate in the finite number of steps. In that case, it is possible to adjust the procedure to compute the greatest crisp solution to (14) contained in a given crisp relation.

For a given family $\{\varrho_i\}_{i \in I}$ and an element $\lambda \in L$, define two functions $\mathcal{Q}_1^c(\lambda), \mathcal{Q}_2^c(\lambda) : L^{A \times A} \to L^{A \times A}$, for every $\varphi \in L^{A \times A}$, as:

$$\mathcal{Q}_1^c(\lambda)(\varphi) = \bigwedge_{i \in I} (\lambda \to \varrho_i \circ \varphi) \nearrow \varrho_i, \tag{22}$$

$$\mathcal{Q}_2^c(\lambda)(\varphi) = \bigwedge_{i \in I} \varrho_i \diagdown (\lambda \to \varphi \circ \varrho_i). \tag{23}$$

Theorem 4. *Let $\varphi_0 \in 2^A$ be a crisp relation on A and $\lambda \in L$. Define an array $\{\varphi_n\}_{n \in \mathbb{N}}$ of crisp relations on A by:*

$$\varphi_{n+1} = \varphi_n \wedge \mathcal{Q}_1^c(\lambda)(\varphi_n) \wedge \mathcal{Q}_2^c(\lambda)(\varphi_n), \quad n \in N. \tag{24}$$

Then:

1. *Relation φ_k is the greatest crisp solution to (14), such that $\varphi_k \subseteq \varphi_0$ if and only if $\varphi_k = \varphi_{k+1}$.*
2. *Procedure (24) produces the greatest crisp solution to (14) such that $\varphi_k \subseteq \varphi_0$, after finitely many iterations.*

Proof. The proof follows similar as the proof for Theorem 2, so it is omitted.

According to Theorem 4, we give an algorithm for computing the greatest crisp solution to (14), formalized by Function 2.

Function 2. ComputeTheGreatestCrispSolution($\{\varphi_n\}_{n \in \mathbb{N}_0}, \varphi_0, \lambda$)
  Input: *a family $\{\varrho_i\}_{i \in I}$, a crisp relation $\varphi_0 \in 2^A$ and an element $\lambda \in L$.*
  Output: *a crisp relation that is the greatest contained in φ_0 to system* (14).

1. *$\phi = \varphi_0$;*
2. *do:*
3. *$\varphi = \phi$;*
4. *$\phi = \phi \wedge \mathcal{Q}_1^c(\lambda)(\phi) \wedge \mathcal{Q}_2^c(\lambda)(\phi)$;*
5. *while($\phi \neq \varphi$);*
6. *return ϕ.*

The following example demonstrates the case when it is impossible to compute the greatest solution to (14), but it is possible to determine the greatest solution to (14) by means of Theorems 3 and 4.

Example 2. Consider the system (14) over the product structure, where $\{\varrho_i\}_{i \in I} = \{\varrho_1, \varrho_2\}$, where fuzzy relations ϱ_1 and ϱ_2 are given by:

$$\varrho_1 = \begin{bmatrix} 1 & 0.95 & 0.9 & 1 & 0.9 & 0.9 & 0.8 & 0.8 \\ 0 & 0.4 & 0 & 0 & 0 & 0 & 0 & 0 \\ 0 & 0.4 & 1 & 0.4 & 0 & 0 & 0.95 & 0.4 \\ 0 & 0.95 & 0 & 0 & 0.4 & 0 & 0.9 & 0.95 \\ 0.8 & 0.95 & 0.9 & 0.95 & 0.95 & 0.8 & 1 & 0.4 \\ 0.4 & 1 & 0.6 & 0 & 0.6 & 0.6 & 0.9 & 0.9 \\ 0 & 0.95 & 0.4 & 0.9 & 1 & 0.9 & 0.8 & 0.95 \\ 0 & 0.4 & 1 & 0.8 & 0.9 & 1 & 0.95 & 0 \end{bmatrix}, \tag{25}$$

$$\varrho_2 = \begin{bmatrix} 0.8 & 0.95 & 1 & 0.9 & 0.95 & 0.9 & 0 & 0.9 \\ 0.95 & 0.9 & 0 & 0.6 & 0 & 1 & 0.9 & 0 \\ 0.4 & 1 & 0.9 & 0.95 & 0.8 & 0 & 0.4 & 0.9 \\ 0.8 & 0.6 & 1 & 0.9 & 0.9 & 0.6 & 0.6 & 0.95 \\ 0.8 & 0.6 & 0 & 0.4 & 0.4 & 0 & 0.8 & 0.95 \\ 0 & 0.6 & 0.8 & 0.4 & 1 & 0 & 0.9 & 1 \\ 0.6 & 0 & 0.95 & 0.8 & 0.4 & 0 & 0.4 & 0.8 \\ 1 & 0 & 0.95 & 0.8 & 0.8 & 0 & 0 & 0.8 \end{bmatrix}. \tag{26}$$

Take $\lambda = 0.9$. Then Function 1 does not stop in a finite number of steps, thus, it is impossible to determine the greatest solution to (14) by means of this Function. But, by using Theorem 3, we conclude that the greatest solution to (14) is given by:

$$\varphi = \lim_{n \to \infty} \varphi_n = \begin{bmatrix} 1 & 1 & 0 & 40/57 & 0 & 1 & 1 & 0 \\ 0 & 1 & 0 & 0 & 0 & 0 & 0 & 0 \\ 1 & 1 & 1 & 1 & 1 & 1 & 1 & 1 \\ 1 & 1 & 0 & 1 & 0 & 1 & 1 & 0 \\ 1 & 1 & 1 & 1 & 1 & 1 & 1 & 1 \\ 1 & 1 & 0 & 2/3 & 0 & 1 & 1 & 0 \\ 1 & 1 & 0 & 20/27 & 0 & 1 & 1 & 0 \\ 1 & 1 & 1 & 1 & 1 & 1 & 1 & 1 \end{bmatrix}.$$

On the other hand, if we employ Function 2, it ends after three iterations, and the output of this function is a crisp relation given by:

$$R = \begin{bmatrix} 1 & 1 & 0 & 0 & 0 & 1 & 1 & 0 \\ 0 & 1 & 0 & 0 & 0 & 0 & 0 & 0 \\ 1 & 1 & 1 & 1 & 1 & 1 & 1 & 1 \\ 1 & 1 & 0 & 1 & 0 & 1 & 1 & 0 \\ 1 & 1 & 1 & 1 & 1 & 1 & 1 & 1 \\ 1 & 1 & 0 & 0 & 0 & 1 & 1 & 0 \\ 1 & 1 & 0 & 0 & 0 & 1 & 1 & 0 \\ 1 & 1 & 1 & 1 & 1 & 1 & 1 & 1 \end{bmatrix}.$$

References

1. Bartl, E.: Minimal solutions of generalized fuzzy relational equations: probabilistic algorithm based on greedy approach. Fuzzy Sets Syst. **260**, 25–42 (2015)
2. Bartl, E., Belohlavek, R.: Sup-t-norm and inf-residuum are a single type of relational equations. Int. J. Gen Syst **40**(6), 599–609 (2011)
3. Bartl, E., Klir, G.J.: Fuzzy relational equations in general framework. Int. J. Gen Syst **43**(1), 1–18 (2014)
4. Bělohlávek, R.: Fuzzy Relational Systems: Foundations and Principles. Kluwer, New York (2002)
5. Bělohlávek, R., Vychodil, V.: Fuzzy Equational Logic. Studies in Fuzziness and Soft Computing. Springer, Berlin-Heidelberg (2005). https://doi.org/10.1007/b105121
6. Damljanović, N., Ćirić, M., Ignjatović, J.: Bisimulations for weighted automata over an additively idempotent semiring. Theoret. Comput. Sci. **534**, 86–100 (2014)
7. Díaz-Moreno, J.C., Medina, J., Turunen, E.: Minimal solutions of general fuzzy relation equations on linear carriers. An algebraic characterization. Fuzzy Sets Syst. **311**(C), 112–123 (2017)
8. Ignjatović, J., Ćirić, M.: Weakly linear systems of fuzzy relation inequalities and their applications: a brief survey. Filomat **26**(2), 207–241 (2012)
9. Ignjatović, J., Ćirić, M., Bogdanović, S.: On the greatest solutions to weakly linear systems of fuzzy relation inequalities and equations. Fuzzy Sets Syst. **161**(24), 3081–3113 (2010)
10. Ignjatović, J., Ćirić, M., Damljanović, N., Jančić, I.: Weakly linear systems of fuzzy relation inequalities: the heterogeneous case. Fuzzy Sets Syst. **199**, 64–91 (2012)
11. Micić, I., Nguyen, L.A., Stanimirović, S.: Characterization and computation of approximate bisimulations for fuzzy automata. Fuzzy Sets Syst. **442**, 331–350 (2022)
12. Micić, I., Stanimirović, S., Jančić, Z.: Approximate positional analysis of fuzzy social networks. Fuzzy Sets Syst. (2022). https://doi.org/10.1016/j.fss.2022.05.008
13. Sanchez, E.: Equations de relations floues. Ph.D. thesis, Faculté de Médecine de Marseille (1974)
14. Sanchez, E.: Resolution of composite fuzzy relation equations. Inf. Control **30**(1), 38–48 (1976)
15. Stanimirović, S., Micić, I., Ćirić, M.: Approximate bisimulations for fuzzy automata over complete heyting algebras. IEEE Trans. Fuzzy Syst. **30**, 437–447 (2022)
16. Stanimirović, S., Stamenković, A., Ćirić, M.: Improved algorithms for computing the greatest right and left invariant boolean matrices and their application. Filomat **33**(9), 2809–2831 (2019)
17. Stanimirović, S., Micić, I.: On the solvability of weakly linear systems of fuzzy relation equations. Inf. Sci. **607**, 670–687 (2022)
18. Turunen, E.: Necessary and sufficient conditions for the existence of solution of generalized fuzzy relation equations $A \Leftrightarrow X = B$. Inf. Sci. **536**, 351–357 (2020)

When Variable-Length Codes Meet the Field of Error Detection

Jean Néraud[(✉)] [iD]

Université de Rouen, Laboratoire d'Informatique, de Traitement de l'Information et des Systèmes, Avenue de l'Université, 76800 Saint-Étienne-du-Rouvray, France
neraud.jean@gmail.com
http://neraud.jean.free.fr

Abstract. Given a finite alphabet A and a binary relation $\tau \subseteq A^* \times A^*$, a set X is *τ-independent* if $\tau(X) \cap X = \emptyset$. Given a quasi-metric d over A^* (in the meaning of [27]) and $k \geq 1$, we associate the relation $\tau_{d,k}$ defined by $(x, y) \in \tau_{d,k}$ if, and only if, $d(x, y) \leq k$ [3]. In the spirit of [10, 20], the error detection-correction capability of variable-length codes can be expressed in term of conditions over $\tau_{d,k}$. With respect to the prefix metric, the factor one, and every quasi-metric associated to (anti-)automorphisms of the free monoid, we examine whether those conditions are decidable for a given regular code.

Keywords: Anti-reflexive · Automaton · Automorphism · Anti-automorphism · Bernoulli measure · Binary relation · Channel · Code · Codeword · Complete · Distance · Error correction · Error detection · Embedding · Factor · Free monoid · Homomorphism · Independent · Input word · Kraft inequality · Maximal · Measure · Metric · Monoid · Output word · Prefix · Quasi-metric · Regular · Subsequences · Suffix · Synchronization constraint · Transducer · Variable-length code · Word

1 Introduction

In Computer Science, the transmission of finite sequences of symbols (the so-called *words*) via some channel constitutes one of the most challenging research fields. With the notation of the free monoid theory, some classical models may be informally described as indicated in the following:

Two finite *alphabets*, say A and B, are required, every information being modeled by a unique word, say u, in B^* (the *free monoid* generated by B). Usually, in order to facilitate the transmission, beforehand u is transformed in $w \in A^*$, the so-called *input word*: this is done by applying some fixed one-to-one *coding* mapping $\phi : B^* \longrightarrow A^*$. In numerous cases, ϕ is a *monoid homomorphism*, whence $X = \phi(B)$ is a *variable-length code* (for short, a *code*): equivalently every equation among the words of X is necessarily trivial. Such a translation is particularly illustrated by the well-known examples of the Morse and Huffman codes. Next, w is transmitted via a fixed *channel* into $w' \in A^*$, the so-called *output word*: should

© The Author(s), under exclusive license to Springer Nature Switzerland AG 2022
D. Poulakis and G. Rahonis (Eds.): CAI 2022, LNCS 13706, pp. 203–222, 2022.
https://doi.org/10.1007/978-3-031-19685-0_15

w' be altered by some *noise* and then the resulting word $\phi^{-1}(w') \in B^*$ could be different from the initial word u. In the most general model of transmission, the channel is represented by some *probabilistic transducer*. However, in the framework of error detection, most of the models only require that highly likely errors need to be taken into account: in the present paper, we assume the transmission channel modeled by some *binary word relation*, namely $\tau \subseteq A^* \times A^*$. In order to retrieve u, the homomorphism ϕ, and thus the code X, must satisfy specific constraints, which of course depend of the channel τ: in view of some formalization, we denote by $\hat{\tau}$ the *reflexive closure* of τ, and by $\underline{\tau}$ its *anti-reflexive restriction* that is, $\tau \setminus \{(w, w)|w \in A^*\}$.

About the channel itself, the so-called *synchronization constraint* appears mandatory: it states that, for each input word factorized $w = x_1 \cdots x_n$, where $x_1, \cdots, x_n$ are *codewords* in X, every output word has to be factorized $w' = x'_1 \cdots x'_n$, with $(x_1, x'_1), \cdots, (x_n, x'_n) \in \hat{\tau}$. In order to ensure such a constraint, as for the Morse code, some pause symbol could be inserted after each codeword x_i.

With regard to the code X, in order to minimize the number of errors, in most cases some close neighbourhood constraint over $\hat{\tau}(X)$ is applied. In the most frequent use, such a constraint consists of some minimal distance condition: the smaller the distance between the input codeword $x \in X$ and any of its corresponding output words $x' \in \hat{\tau}(X)$, the more optimal is error detection. In view of that, we fix over A^* a *quasi-metric* d, in the meaning of [27] (the difference with a *metric* is that d needs not to satisfy the symmetry axiom). As outlined in [3], given an error tolerance level $k \geq 0$, a corresponding binary word relation, that we denote by $\tau_{d,k}$, can be associated in such a way that $(w, w') \in \tau_{d,k}$ is equivalent to $d(w, w') \leq k$. Below, in the spirit of [10,20], we draw some specification regarding error detection-correction capability. Recall that a subset X of A^* is *independent* with respect to $\tau \subseteq A^* \times A^*$ (for short, τ-*independent*) whenever $\tau(X) \cap X = \emptyset$: this notion, which appears dual with the one of *closed code* [20], relies to the famous *dependence systems* [4,10]. Given a family of codes, say $\mathcal{F}$, a code $X \in \mathcal{F}$ is *maximal in* $\mathcal{F}$ whenever $X \subseteq Y$, with $Y \in \mathcal{F}$, implies $Y = X$. We introduce the four following conditions:

(c1) *Error detection:* X is $\tau_{d,k}$-*independent*.
(c2) *Error correction:* $x, y \in X$ and $\tau_{d,k}(x) \cap \tau_{d,k}(y) \neq \emptyset$ implies $x = y$.
(c3) X is maximal in the family of $\tau_{d,k}$-*independent* codes.
(c4) $\widehat{\tau_{d,k}}(X)$ is a code.

A few comments on Conds. (c1)–(c4):

– By definition, Cond. (c1) is satisfied if, and only if, the distance between different elements of X is greater than k that is, the code X can detect at most k errors in the transmission of every codeword $x \in X$.
– Cond. (c2) states a classical definition: equivalently, for every codeword x we have $\tau_{d,k}^{-1}(\tau_{d,k}(x)) \cap X = \{x\}$ whenever $\tau_{d,k}(x)$ is non-empty [20].
– With Cond. (c3), in the family of $\tau_{d,k}$-independent codes, X cannot be improved. From this point of view, fruitful investigations have been done in several classes determined by code properties [9,11,14,15]. On the other hand, according to the famous Kraft inequality, given a *positive Bernoulli*

measure over A^*, say μ, for every code X we have $\mu(X) \leq 1$. According to a famous result due to Schützenberger, given a *regular* code X, the condition $\mu(X) = 1$ itself corresponds to X being maximal in the whole family of codes, or equivalently X being *complete* that is, every word in A^* is a *factor* of some word in X^*. From this last point of view, no part of X^* appears spoiled.

– At last, Cond. (c4) expresses that the factorization of every output message over the set $\widehat{\tau_{d,k}}(X)$ is done in a unique way. Since d is a quasimetric, the corresponding relation $\tau_{d,k}$ is reflexive, therefore Cond. (c4) is equivalent to $\tau_{d,k}(X)$ being a code.

As shown in [10, 20], there are regular codes satisfying (c1) that cannot satisfy (c2). Actually, in most of the cases it could be very difficult, even impossible, to satisfy all together Conds. (c1)–(c4): necessarily some compromise has to be adopted. In view of this, given a regular code X, a natural question consists in examining whether each of those conditions is satisfied in the framework of some special quasi-metric. From this point of view, in [20], we considered the so-called *edit relations*, some peculiar compositions of one-character *deletion*, *insertion*, and *substitution*: such relations involve the famous Levenshtein and Hamming metrics [8, 13, 17], which are prioritary related to subsequences in words. In the present paper, we focuse on quasimetrics rather involving factors:

– The *prefix* metric is defined by $d_P(w, w') = |w| + |w'| - 2|w \wedge w'|$, where $w \wedge w'$ stands for the longest common *prefix* of w and w': we set $\mathcal{P}_k = \tau_{d_P,k}$.
– The *factor* metric itself is defined by $d_F(w, w') = |w| + |w'| - 2|f|$, where f is a maximum length common factor of w, w': we set $\mathcal{F}_k = \tau_{d_F,k}$.
– A third type of topology can be introduced in connection with *monoid automorphisms* or *anti-automorphisms* (for short, we write *(anti-)automorphisms*): such a topology particularly concerns the domain of DNA sequence comparison. By anti-automorphism of the free monoid, we mean a one-to-one mapping onto A^*, say θ, such that the equation $\theta(uv) = \theta(v)\theta(u)$ holds for every $u, v \in A^*$ (for involvements of such mappings in the dual notion of closed code, see [21]). With every (anti-)automorphism θ we associate the quasi-metric d_θ, defined as follows:
  (1) $d_\theta(w, w') = 0$ is equivalent to $w = w'$;
  (2) we have $d_\theta(w, w') = 1$ whenever $w' = \theta(w)$ holds, with $w \neq w'$;
  (3) in all other cases we set $d_\theta(w, w') = 2$.

By definition we have $\tau_{d_\theta,1} = \widehat{\theta}$ and $\tau_{\theta,1} = \underline{\theta}$. In addition, a code X is $\underline{\theta}$-independent if, and only if, for every pair of different words $x, y \in X$, we have $d_\theta(x, y) = 2$ that is, X is capable to detect at most one error.

We will establish the following result:

Theorem. *With the preceding notation, given a regular code $X \subseteq A^*$, it can be decided whether X satisfies each of the following conditions:*

(i) Conds. (c1)–(c4) wrt. $\mathcal{P}_k$ or $\widehat{\theta}$, for any (anti-)automorphism θ of A^*.
(ii) Conds. (c3), (c4) wrt. $\mathcal{F}_k$.
(iii) If X is finite, Conds. (c1)–(c2) wrt. $\mathcal{F}_k$.

Some comments regarding the proof:

- For proving that X satisfies Cond. (c1) wrt. $\mathcal{P}_k$, the main argument consists in establishing that $\underline{\mathcal{P}_k}$ is a *regular* relation [24, Ch. IV] that is, $\underline{\mathcal{P}_k}$ can be simulated by a finite transducer.
- Once more wrt. $\mathcal{P}_k$, we actually prove that Cond. (c2) is equivalent to $(X \times X) \cap \underline{\mathcal{P}_{2k}} = \emptyset$. Since $X \times X$ is a *recognizable* relation [24, Ch. IV], it can be decided whether that equation holds.
- Regarding Cond. (c3), the critical step is reached by proving that, wrt. each of the quasi-metrics raised in the paper, for a regular code X, being maximal in the family of $\tau_{d,k}$-independent codes is equivalent to being complete that is, $\mu(X) = 1$. This is done by proving that every non-complete $\tau_{d,k}$-independent code, say X, can be embedded into some complete $\tau_{d,k}$-independent one: in other words, X cannot be maximal in such a family of codes. In order to establish such a property, in the spirit of [2,16,18,19,21,28], we provide specific regularity-preserving embedding formulas, whose schemes are based upon the method from [5]. Notice that, in [9,12,14,15,26], wrt. peculiar families of sets, algorithmic methods for embedding a set into some maximal one were also provided.
- With regard to Cond. (c4), for each of the preceding relations, the set $\widehat{\tau_{d,k}}(X) = \tau_{d,k}(X)$ is regular therefore, in any case, by applying the famous Sardinas and Patterson algorithm [25], one can decide whether that condition is satisfied.

We now shorty describe the contents of the paper:

- Sect. 2 is devoted to the preliminaries: we recall fundamental notions over codes, regular (resp., *recognizable*) relations, and automata.
- The aim of Sect. 3 is to study $\mathcal{P}_k$. We prove that, in any case, the corresponding relation $\underline{\mathcal{P}_k}$ is itself regular. Furthermore, given a regular code X, one can decide whether X satisfies any of Conds. (c1)–(c4). Some remarks are also formulated regarding the so-called *suffix* metric.
- Sect. 4 is concerned with the factor metric. We prove that, given a finite code, one can decide whether it satisfies any of Conds. (c1)–(c4). For a non-finite regular codes, we prove that one can decide whether it satisfies Conds. (c3), (c4), however, the question of the decidability of Conds. (c1), (c2) remains open.
- Sect. 5 is devoted to (anti-)automorphisms. We obtain results similar to those involving the relation $\mathcal{P}_k$.
- In Sect. 6, the paper concludes with some possible directions for further research.

2 Preliminaries

Several definitions and notation has already been stated in the introduction. In the whole paper, we fix a finite alphabet A, with $|A| \geq 2$, and we denote by ε the word of length 0. Given two words $v, w \in A^*$, v is a *prefix* (resp., *suffix*, *factor*) of

w if words u, u' exist such that $w = vu$ (resp., $w = u'v$, $w = u'vu$). We denote by $P(w)$ (resp., $S(w)$, $F(w)$) the set of the words that are prefix (resp., suffix, factors) of w. More generally, given a set $X \subseteq A^*$, we set $P(X) = \bigcup_{w \in X} P(w)$; the sets $S(X)$ and $F(X)$ are defined in a similar way. A word $w \in A^*$, is *overlapping-free* if $wv \in A^*w$, with $|v| \leq |w| - 1$, implies $v = \varepsilon$.

Variable-Length Codes

We assume that the reader has a fundamental understanding with the main concepts of the theory of variable-length codes: we suggest, if necessary, that he (she) refers to [1]. Given a subset X of A, and $w \in X^*$, let $x_1, \cdots, x_n \in X$ such that w is the result of the concatenation of the words $x_1, x_2, \cdots, x_n$, in this order. In view of specifying the factorization of w over X, we use the notation $w = (x_1)(x_2) \cdots (x_n)$, or equivalently: $w = x_1 \cdot x_2 \cdots x_n$. For instance, over the set $X = \{a, ab, ba\}$, the word $bab \in X^*$ can be factorized as $(ba)(b)$ or $(b)(ab)$ (equivalently denoted by $ba \cdot b$ or $b \cdot ab$).

A set X is a *variable-length code* (a *code* for short) if for any pair of finite sequences of words in X, say $(x_i)_{1 \leq i \leq n}$, $(y_j)_{1 \leq j \leq p}$, the equation $x_1 \cdots x_n = y_1 \cdots y_p$ implies $n = p$, and $x_i = y_i$ for each integer $i \in [1, n]$ (equivalently, the submonoid X^* is *free*). In other words, every element of X^* has a unique factorization over X. A set $X \neq \{\varepsilon\}$ is a *prefix* (resp., *suffix*) code if $x \in P(y)$ (resp., $x \in S(y)$) implies $x = y$, for every pair of words $x, y \in X$; X is a *bifix* code if it is both a prefix code and a suffix one. A set $X \subseteq A^*$ is *uniform* if all its elements have a common length. In the case where we have $X \neq \{\varepsilon\}$, the uniform code X is bifix.

Given a regular set X, the Sardinas and Patterson algorithm allows to decide whether or not X is a code. Since it will be used several times through the paper, it is convenient to shortly recall it. Actually, some ultimately periodic sequence of sets, namely $(U_n)_{n \geq 0}$, is computed, as indicated in the following:

$$U_0 = X^{-1}X \setminus \{\varepsilon\} \quad \text{and}: \quad (\forall n \geq 0) \quad U_{n+1} = U_n^{-1}X \cup X^{-1}U_n. \tag{1}$$

The algorithm necessarily stops: this corresponds to either $\varepsilon \in U_n$ or $U_n = U_p$, for some pair of different integers $p < n$: X is a code if, and only if, the second condition holds.

A positive *Bernoulli distribution* consists in some total mapping μ from the alphabet A into $\mathbb{R}_+ = \{x \in \mathbb{R} : x \geq 0\}$ (the set of the non-negative real numbers) such that $\sum_{a \in A} \mu(a) = 1$; that mapping is extended into a unique monoid homomorphism from A^* into $(\mathbb{R}_+, \times)$, which is itself extended into a unique positive measure $\mu : 2^{A^*} \longrightarrow \mathbb{R}_+$. In order to do this, for each word $w \in A^*$, we set $\mu(\{w\}) = \mu(w)$; in addition given two disjoint subsets X, Y of A^*, we set $\mu(X \cup Y) = \mu(X) + \mu(Y)$. In the whole paper, we take for μ the so-called *uniform* Bernoulli measure: it is determined by $\mu(a) = 1/|A|$, for each $a \in A$.

The following results are classical: the first one is due to Schützenberger and the second provides some answer to a question actually stated in [23].

Theorem 1 [1, Theorem 2.5.16]. *Given a regular code $X \subseteq A^*$, the following properties are equivalent:*

(i) X is complete;
(ii) X is a maximal code;
(iii) we have $\mu(X) = 1$.

Theorem 2 [5]. *Let $X \subseteq A^*$ be non-complete code, $z \notin \mathrm{F}(X^*)$ overlapping-free, $U = A^* \setminus (X^* \cup A^* z A^*)$, and $Y = (zU)^* z$. Then $Z = X \cup Y$ is a complete code.*

Clearly, if X is a regular set then the same holds for the resulting set Z.

Regular Relations, Recognizable Relations
We also assume the reader to be familiar with the theory of regular relations and automata: if necessary, we suggest that he (she) refers to [6] or [24, Ch. IV].

Given two monoids, say M, N, a binary relation from M into N consists in any subset $\tau \subseteq M \times N$. For $(w, w') \in \tau$, we also set $w' \in \tau(w)$, and we set $\tau(X) = \{\tau(x) : x \in X\}$. The composition in this order of τ by τ' is defined by $\tau \cdot \tau'(x) = \tau'(\tau(x))$ (the notation τ^k refers to that operation); τ^{-1}, the *inverse* of τ, is defined by $(w, w') \in \tau^{-1}$ whenever $(w', w) \in \tau$. We denote by $\bar{\tau}$ the *complement* of τ, i.e. $(M \times N) \setminus \tau$.

A family of subsets of M, say $\mathcal{F} \subseteq 2^M$, is *regularly closed* if, and only if, the sets $X \cup Y$, XY, and X^* belong to $\mathcal{F}$, whenever we have $X, Y \in \mathcal{F}$. Given a family $\mathcal{F} \subseteq 2^M$, its *regular closure* is the smallest (wrt. the inclusion) subset of 2^M that contains $\mathcal{F}$ and which is regularly closed. A binary relation $\tau \subseteq M \times N$ is *regular* (or equivalently, *rational*) if, and only if, it belongs to the regular closure of the finite subsets of $M \times N$. Equivalently there is some finite $M \times N$-*automaton* (or equivalently, *transducer*), say $\mathcal{R}$, with *behavior* $|\mathcal{R}| = \tau$ [7,24]. The family of regular relations is closed under inverse and composition.

The so-called *recognizable* relations constitute a noticeable subfamily in regular relations: a subset $R \subseteq M \times N$ is *recognizable* if, and only if, we have $R = R \cdot \phi \cdot \phi^{-1}$, for some monoid homomorphism $\phi : M \times N \longrightarrow P$, with P finite. Equivalently, finite families of recognizable subsets of M and N, namely $(T_i)_{\in I}$ and $(U_i)_{i \in I}$, exist such that $R = \bigcup_{i \in I} T_i \times U_i$ [24, Corollary II.2.20].

In the paper, we focus on $M = N = A^*$. With this condition, recognizable relations are closed under composition, complement and intersection, their intersection with a regular relation being itself regular [24, Sect. IV.1.4]. According to [24, Theorem IV.1.3], given a regular relation $\tau \subseteq A^* \times A^*$, and a regular (equivalently, recognizable) set $X \subseteq A^*$, the sets $\tau(X)$ and $\tau^{-1}(X)$ are regular. If X, Y are recognizable subsets of A^*, the same holds for $X \times Y$. At last the relation $id_{A^*} = \{(w, w) | w \in A^*\}$ and its complement $\overline{id_{A^*}}$, are regular but non-recognizable.

3   Error Detection and the Prefix Metric

Given $w, w' \in A^*$, a unique pair of words u, u' exist such that $w = (w \wedge w')u$ and $w' = (w \wedge w')u'$: by definition, we have $d_{\mathrm{P}}(w, w') = |u| + |u'|$ (see Fig. 1).

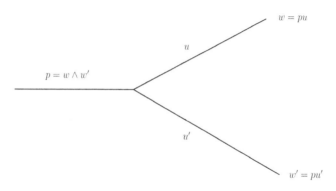

Fig. 1. We have $(w, w') \in \mathcal{P}_k$ iff. $|u| + |u'| \leq k$; similarly, $(w, w') \in \underline{\mathcal{P}_k}$ is equivalent to $1 \leq |u| + |u'| \leq k$.

In addition $\mathcal{P}_k$ is reflexive and symmetric that is, the equality $\mathcal{P}_k^{-1} = \mathcal{P}_k$ holds, and we have $\mathcal{P}_k \subseteq \mathcal{P}_{k+1}$. Below, we provide some example:

Example 1. Let $A = \{a, b\}$. (1) The finite prefix code $X = \{a, ba, b^2\}$ satisfies Cond. (c1) wrt. $\mathcal{P}_1$ (in other words X is 1-error-detecting). Indeed, it follows from $d_P(a, ba) = d_P(a, bb) = 3$ and $d_P(ba, bb) = 2$ that we have $(x, y) \notin \mathcal{P}_1$, for each pair of different words $x, y \in X$, that is $\underline{\mathcal{P}_1}(X) \cap X = \emptyset$. Cond. (c2) is not satisfied by X. Indeed, we have:
$$\mathcal{P}_1^{-1}\mathcal{P}_1(ba) \cap X = \{ba, b, ba^2, bab, \varepsilon, b^2, ba^3, ba^2b, baba, bab^2\} \cap X = \{ba, b^2\}.$$
Classically, X is a maximal code, therefore, since it is $\underline{\mathcal{P}_1}$-independent, X is maximal in the family of $\underline{\mathcal{P}_1}$-independent codes that is, X satisfies Cond. (c3). Consequently, since we have $X \subsetneq \mathcal{P}_1(X)$, the code X cannot satisfy Cond. (c4) (we verify that we have $\varepsilon \in \mathcal{P}_1(X)$).

(2) For $n \geq k + 1$, the complete uniform code $Y = A^n$ satisfies Conds. (c1), (c3). Since Y is a maximal code, it cannot satisfies Cond. (c4) (we have $Y \subsetneq \mathcal{P}_1(Y)$). Cond. (c2) is no more satisfied by Y: indeed, given two different characters a, b, we have $\mathcal{P}_k(a^n) \neq \emptyset$ and $a^{n-1}b \in \mathcal{P}_k^{-1}(\mathcal{P}_k(a^n)) \cap Y$ thus $\mathcal{P}_k^{-1}(\mathcal{P}_k(a^n)) \cap Y \neq \{a^n\}$.

(3) The regular bifix code $Z = \{ab^n a : n \geq 0\} \cup \{ba^n b : n \geq 0\}$ satisfies Cond. (c1) wrt. $\mathcal{P}_1$. Indeed, we have:
$$\underline{\mathcal{P}_1}(Z) = \bigcup_{n \geq 0}\{ab^n, ab^n a^2, ab^n ab, ba^n, ba^n ba, ba^n b^2\}, \text{ thus } \underline{\mathcal{P}_1}(Z) \cap Z = \emptyset.$$
For $n \neq 0$ we have $\mathcal{P}_1^{-1}\mathcal{P}_1(ab^n a) = \{ab^n, ab^{n-1}, ab^n a, ab^{n+1}, ab^n a^2, ab^n a^3, ab^n ab,$ $ab^n a^2 b, ab^n aba, ab^n ab^2\}$, moreover we have $\mathcal{P}_1^{-1}\mathcal{P}_1(a^2) = \mathcal{P}_1(\{a, a^3, a^2 b\})$, thus $\mathcal{P}_1^{-1}\mathcal{P}_1(a^2) = \{a, \varepsilon, a^2, ab, a^3, a^4, a^3 b, a^2 b, a^2 ba, a^2 b^2\}$: in any case we obtain $Z \cap \mathcal{P}_1^{-1}\mathcal{P}_1(ab^n a) = \{ab^n a\}$. Similarly, we have $Z \cap \mathcal{P}_1^{-1}\mathcal{P}_1(ba^n b) = \{ba^n b\}$, hence Z satisfies Cond. (c2). At last, we have $\mu(X) = 2 \cdot 1/4 \sum_{n \geq 0}(1/2)^n = 1$ therefore, according to Theorem 1, Z is a maximal code, whence it is maximal in the family of $\mathcal{P}_1$-independent codes (Cond. (c3)). Since we have $Z \subsetneq \mathcal{P}_1(Z)$, Z cannot satisfies Cond. (c4) (we verify that have $a, a^2 \in \mathcal{P}_1(Z) \subseteq \widehat{\mathcal{P}_1}(Z)$).

In the sequel, we will prove that, given a regular code X, one can decide whether any of Conds. (c1)–(c4) holds. Beforehand we establish the following property which, regarding Cond. (c1), plays a prominent part:

Proposition 1. *For every $k \geq 1$, both the relations $\mathcal{P}_k$ and $\underline{\mathcal{P}}_k$ are regular.*

Proof. In what follows we indicate the construction of a finite automaton with behavior $\underline{\mathcal{P}}_k$ (see Fig. 2). This construction is based on the different underlying configurations in Fig. 1. Firstly, we denote by E the finite set of all the pairs of non-empty words (u, u'), with different initial characters, and such that $|u| + |u'| \leq k$. In addition, F (resp., G) stands for the set of all the pairs (u, ε) (resp., (ε, u)), with $1 \leq |u| \leq k$. Secondly, we construct a finite tree-like $A^* \times A^*$-automaton with behavior $E \cup F \cup G$. Let 0 be the initial state, the other states being terminal. We complete the construction by adding the transitions $(0, (a, a), 0)$, for all $a \in A$. Let $\mathcal{R}_{P,k}$ be the resulting automaton.

By construction we have $|\mathcal{R}_{P,k}| = id_{A^*}(E \cup F \cup G)$. More precisely, $|\mathcal{R}_{P,k}|$ is the set of all the pairs (w, w') such that there are $v, u, u' \in A^*$ satisfying each of the three following conditions:

(1) $w = pu$, $w' = pu'$;
(2) if both the words u, u' are non-empty, their initial characters are different;
(3) $1 \leq |u| + |u'| \leq k$.

In other words, $|\mathcal{R}_{P,k}|$ is the sets of all the pairs (w, w') such that $1 \leq d_P(w, w') = |u| + |u'| \leq k$, therefore we have $\underline{\mathcal{P}}_k = |\mathcal{R}_{P,k}|$. Consequently $\underline{\mathcal{P}}_k$ and $\mathcal{P}_k = \underline{\mathcal{P}}_k \cup id_{A^*}$ are regular relations. $\qquad \square$

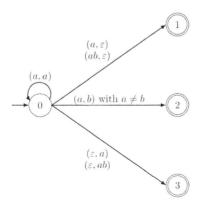

Fig. 2. The case where we have $k = 2$: in the automaton $\mathcal{R}_{P,k}$, the arrows are mutilabelled (a, b stand for all pairs of characters in A) and terminal states are represented with double circles.

Remark 1. In [22] the author introduces a peculiar $(A^* \times A^*) \times \mathbb{N}$-automaton: in view of this, for every $(w, w') \in A^* \times A^*$, the distance $d_P(w, w')$ is the least $d \in \mathbb{N}$ for which $((w, w'), d)$ is the label of some successful path. Furthermore, some alternative proof of the regularity of $\mathcal{P}_k$ can be obtained. However, we note that such a construction cannot involve the relation $\underline{\mathcal{P}}_k$ itself that is, it does not affect Cond. (c1).

The following property is also used in the proof of Proposition 2:

Lemma 1. *Given a positive integer k we have $\mathcal{P}_{2k} = \mathcal{P}_k^2$.*

Proof. In order to prove that $\mathcal{P}_k^2 \subseteq \mathcal{P}_{2k}$, we consider a pair of words $(w, w') \in \mathcal{P}_k^2$. By definition, some word $w'' \in A^*$ exists such that we have $(w, w''), (w'', w') \in \mathcal{P}_k$ that is, $d_P(w, w'') \le k$, $d_P(w'', w') \le k$. This implies $d_P(w, w') \le d_P(w, w'') + d_P(w'', w') \le 2k$ that is, $(w, w') \in \mathcal{P}_{2k}$.

Conversely, let $(w, w') \in \mathcal{P}_{2k}$, and let $p = w \wedge w'$. Regarding the integers $|w| - |p|$, $|w'| - |p|$, exactly one of the two following conditions occurs:

(a) Firstly, at least one of the integers $|w| - |p|$, $|w'| - |p|$ belongs to $[k + 1, 2k]$. Since $\mathcal{P}_{2k}$ is a symmetric relation, without loss of generality, we assume $k + 1 \le |w| - |p| \le 2k$. With this condition a non-empty word v exists such that $w = pvA^k$ that is, $pv \in \mathcal{P}_k(w)$. On the other hand, we have $p = pv \wedge w'$, thus $d_P(pv, w') = (|w| - k) + |w'| - 2|p| = d_P(w, w') - k$. It follows from $d_P(w, w') \le 2k$ that $d_P(pv, w') \le k$, thus $w' \in \mathcal{P}_k(pv)$: this implies $w' \in \mathcal{P}_k(\mathcal{P}_k(w))$, thus $(w, w') \in \mathcal{P}^2$.
(b) Secondly, in the case where we have $|w| - |p| \le k$ and $|w'| - |p| \le k$, by definition we have $p \in \mathcal{P}_k(w)$, $w' \in \mathcal{P}_k(w)$, thus $(w, w') \in \mathcal{P}_k^2$. $\square$

We are now ready to establish the following result:

Proposition 2. *Given a regular code $X \subseteq A^*$, wrt. $\mathcal{P}_k$, it can be decided whether X satisfies any of Conds. (c1), (c2), and (c4).*

Proof. Let X be a regular code. We consider one by one our Conds. (c1), (c2), and (c4):

- *Cond.* (c1) According to Proposition 1, $\mathcal{P}_k(X)$ is a regular set, hence $\mathcal{P}_k(X) \cap X$ itself is regular, therefore one can decide whether Cond. (c1) holds.
- *Cond.* (c2) Since $\mathcal{P}_k$ is a symmetric binary relation, and according to Lemma 1, we have: $\mathcal{P}_k^{-1}(\mathcal{P}_k(X)) \cap X = \mathcal{P}_k^2(X) \cap X = \mathcal{P}_{2k}(X) \cap X$. In addition $x \in X$ implies $\mathcal{P}_k(x) \ne \emptyset$, therefore Cond. (c2) is equivalent to $(X \times X) \cap \mathcal{P}_{2k} \subseteq id_{A^*}$. This last condition is equivalent to $(X \times X) \cap (\mathcal{P}_{2k} \cap \overline{id_{A^*}}) = \emptyset$, thus $(X \times X) \cap \mathcal{P}_{2k} = \emptyset$. According to Proposition 1 and since $X \times X$ is a recognizable relation, the set $(X \times X) \cap \mathcal{P}_{2k}$ is regular, therefore one can decide whether X satisfies Cond. (c2).
- *Cond.* (c4) According to Proposition 1, the set $\mathcal{P}_k(X)$ itself is regular. Consequently, by applying Sardinas and Patterson algorithm, it can be decided whether X satisfies Cond. (c4). $\square$

It remains to study the bahaviour of X wrt. Cond. (c3). In order to do this, we proceed by establishing the two following results:

Proposition 3. *Let $X \subseteq A^*$ be a non-complete regular $\mathcal{P}_k$-independent code. Then a complete regular $\mathcal{P}_k$-independent code containing X exists.*

Proof. Beforehand, in view of Theorem 2, we indicate the construction of a convenient word $z \in A^* \setminus F(X^*)$. Since X is a non-complete set, by definition some word z_0 exists in $A^* \setminus F(X^*)$: without loss of generality, we assume $|z_0| \geq k$ (otherwise, we substitute to z_0 any word in $z_0 A^{k-|z_0|}$). Let a be the initial character of z_0, b be a character different of a, and $z = z_0 a b^{|z_0|}$. Classically, z is overlapping-free (see e.g. [1, Proposition 1.3.6]): set $U = A^* \setminus (X^* \cup A^* z A^*)$, $Y = z(Uz)^*$, and $Z = X \cup Y$.

According to Theorem 2, the set Z is a (regular) complete code. For proving that Z is $\underline{\mathcal{P}_k}$-independent, we argue by contradiction. By assuming that $\underline{\mathcal{P}_k}(Z) \cap Z \neq \emptyset$, according to the construction of Z, exactly one of the two following cases occurs:

(a) Firstly, $x \in X$, $y \in Y$ exist such that $(x, y) \in \underline{\mathcal{P}_k}$. With this condition we have $d_P(x, y) = d_P(y, x) = \left|(x \wedge y)^{-1} x\right| + \left|(x \wedge y)^{-1} y\right| \leq k$. According to the construction of Y, the word $z = z_0 a b^{|z_0|}$ is a suffix of y. It follows from $\left|(x \wedge y)^{-1} y\right| \leq k \leq |z_0|$ that $z_0 a b^{|z_0|-1} \in F(x \wedge y)$, thus $z_0 \in F(x)$: a contradiction with $z_0 \notin F(X^*)$.

(b) Secondly, $y, y' \in Y$ exist such that $(y, y') \in \underline{\mathcal{P}_k}$. Let $p = y \wedge y'$, $u = p^{-1} y$, and $u' = p^{-1} y'$.

(b1) At first, assume $p \in \{y, y'\}$ that is, without loss of generality, $y' = p$. With such a condition, y' is a prefix of y. Since we have $y, y' \in z(Uz)^*$, and since z is an overlapping-free word, necessarily a word $v \in (Uz)^*$ exists such that $y = y' v$. It follows from $|v| = d_P(y, y') \leq k \leq |z| - 1$ that $v = \varepsilon$, thus $y = y'$: a contradiction with $\mathcal{P}_k$ being antireflexive.

(b2) Consequently we have $p \notin \{y, y'\}$ that is, $1 \leq |u| \leq k$ and $1 \leq |u'| \leq k$ (see Fig. 3). By construction, we have $b^{|z_0|} \in S(z) \subseteq S(y) \cap S(y')$: it follows from $1 \leq |u| \leq k \leq |z_0|$ and $1 \leq |u'| \leq k \leq |z_0|$ that $u, u' \in S(b^{|z_0|}) \setminus \{\varepsilon\}$, thus $u, u' \in bb^*$: a contradiction with $p = y \wedge y'$.

In any case we obtain a contradiction, therefore Z is $\underline{\mathcal{P}_k}$-independent. $\square$

Proposition 4. *Let $X \subseteq A^*$ be a regular code. Then X is maximal in the family of $\underline{\mathcal{P}_k}$-independent codes if, and only if, we have $\mu(X) = 1$.*

Proof. According to Theorem 1, $\mu(X) = 1$ implies X being a maximal code, thus X being maximal as a $\underline{\mathcal{P}_k}$-independent code. For the converse, we argue by contrapositive. Once more according to Theorem 1, $\mu(X) \neq 1$ implies X non-complete. According to Proposition 3, some $\underline{\mathcal{P}_k}$-independent code strictly containing X exists, hence X is not maximal as a $\underline{\mathcal{P}_k}$-independent code. $\square$

If X is a regular set, $\mu(X)$ can be computed by making use of some rational series. As a consequence, we obtain the following result:

Proposition 5. *One can decide whether a given regular code X satisfies Cond. (c3) wrt. $\mathcal{P}_k$.*

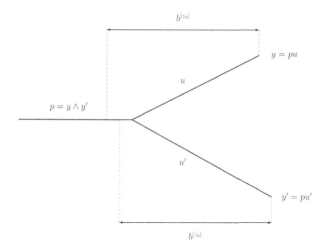

Fig. 3. Proof of Proposition 3: the case where we have $y, y' \in Y$ and $(y, y') \in \mathcal{P}_k$, with $p \notin \{y, y'\}$.

Remark 2. Given a pair of words w, w', their *suffix* distance is $d_S = |w| + |w'| - 2|s|$, where s denotes the longest word in $S(w) \cap S(w')$. Let $\mathcal{S}_k$ be the binary relation defined by $(w, w') \in \mathcal{S}_k$ if, and only if, $d_S(w, w') \le k$. Given a word $w \in A^*$, denote by w^R its *reversal* that is, for $a_1, \cdots, a_n \in A$, we have $w^R = a_n \cdots a_1$ if, and only if, $w = a_1 \cdots a_n$ holds. For every pair $w, w' \in A^*$, we have $d_S(w, w') = d_P(w^R, w'^R)$, hence $(w, w') \in \mathcal{S}_k$ is equivalent to $(w^R, w'^R) \in \mathcal{P}_k$. As a consequence, given a regular code, one can decide whether it satisfies any of Conds. (c1)–(c4) wrt. $\mathcal{S}_k$.

4 Error Detection and the Factor Metric

By definition, given a pair of words $w, w' \in A^*$, at least one tuple of words, say (u, v, u', v'), exists such that $d_F(w, w') = |u| + |v| + |u'| + |v'|$. More precisely, we have $w = ufv$, $w' = u'fv'$, with f being of maximum length. Such a configuration is illustrated by Fig. 4 which, in addition, can provide some support in view of examining the proof of Proposition 8.

Actually the word f, thus the tuple (u, v, u', v'), needs not to be unique (see Example 2(1)). Due to this fact, the construction in the proof of the preceding proposition 1 unfortunately cannot be extended in order to obtain a finite automaton with behaviour $\underline{\mathcal{F}_k}$.

For every positive integer k, the relation $\mathcal{F}_k$ is reflexive and symmetric that is, the equality $\mathcal{F}_k^{-1} = \mathcal{F}_k$ holds. In addition, with the preceding notation, we have $\mathcal{P}_k \cup \mathcal{S}_k \subseteq \mathcal{F}_k \subseteq \mathcal{F}_{k+1}$.

Example 2. (1) Let $w = babababbab$ and $w' = bbabbaababaa$. There are two words of maximum length in $F(w) \cap F(w')$, namely $f_1 = ababa$ and $f_2 = babba$. We have $d_F(w, w') = |u_1| + |v_1| + |u_1'| + |v_1'| = |u_2| + |v_2| + |u_2'| + |v_2'| =$

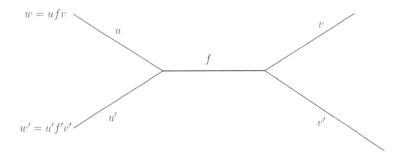

Fig. 4. We have $(w, w') \in \mathcal{F}_k$ iff. $|u| + |v| + |u'| + |v'| \leq k$.

12, where the tuples (u_1, v_1, u'_1, v'_1) and (u_2, v_2, u'_2, v'_2) satisfy the following equations:

$$w = u_1 f_1 v_1, w' = u'_1 f_1 v'_1, \text{ with } u_1 = b, v_1 = bbab, u'_1 = bbabba, v'_1 = a,$$
$$w = u_2 f_2 v_2, w' = u'_2 f_2 v'_2, \text{ with } u_2 = baba, v_2 = b, u'_2 = b, v'_2 = ababaa.$$

(2) Over the alphabet $\{a, b\}$, the code $X = \{a, ba, bb\}$ from Example 1 cannot satisfy Cond. (c1) wrt. $\mathcal{F}_1$. Indeed, we have $ba \in \mathcal{F}_1(a) \cap X$. Since we have $\mathcal{F}_1^{-1} \mathcal{F}_1 (a) \cap X = \{a, ba\}$, X cannot satisfy Cond. (c2). Althought X is complete, since it does not belong to the family of $\mathcal{F}_1$-independent codes, X cannot satisfy Cond. (c3) wrt. $\mathcal{F}_1$. It follow from $\varepsilon \in \mathcal{P}_1(X) \subseteq \mathcal{F}_1(X)$ that X can no more satisfy Cond. (c4).

(3) Take $A = \{a, b\}$ and consider the context-free bifix code $Y = \{a^n b^n : n \geq 1\}$. We have $\mathcal{F}_1(Y) = \bigcup_{n \geq 1} \{a^{n-1} b^n, a^{n+1} b^n, ba^n b^n, a^n b^{n-1}, a^n b^n a, a^n b^{n+1}\}$. This implies $\overline{\mathcal{F}_1(Y)} \cap Y = \emptyset$, thus Y being 1-error-detecting wrt. $\mathcal{F}_1$ (Cond. (c1)). Regarding error correction, we have $a^{n+1} b^{n+1} \in \mathcal{F}_1^2(a^n b^n)$, therefore X cannot satisfy Cond. (c2) wrt. $\mathcal{F}_1$. We have $\mu(X) = \sum_{n \geq 1} \left(\frac{1}{4}\right)^n < 1$, whence Y cannot satisfy Cond. (c3). Finally, since we have $(a^n b^{n-1})(ba^n b^n) = (a^n b^n)(a^n b^n)$, the set $\widehat{\mathcal{F}_1}(Y) = \mathcal{F}_1(Y)$ cannot satisfy Cond. (c4).

The following property allows some noticeable connection between the frameworks of prefix, suffix, and factor metrics:

Lemma 2. *Given a positive integer k we have $\mathcal{F}_k = \mathcal{F}_1^k = (\mathcal{P}_1 \cup \mathcal{S}_1)^k$.*

Proof. – We start by proving that we have $\mathcal{F}_1^k = (\mathcal{P}_1 \cup \mathcal{S}_1)^k$. A indicated above, we have $\mathcal{P}_1 \cup \mathcal{S}_1 \subseteq \mathcal{F}_1$. Conversely, given $(w, w') \in \mathcal{F}_1$, some tuple of words (u, v, u', v') exists such that $w = ufv$, $w' = u'fv'$, with $0 \leq |u| + |v| + |u'| + |v'| \leq 1$, thus $|u| + |v| + |u'| + |v'| \in \{0, 1\}$. More precisely, at most one element of the set $\{|u|, |v|, |u'|, |v'|\}$ is a non-zero integer: this implies $(w, w') \in \mathcal{P}_1 \cup \mathcal{S}_1$. Consequently we have $\mathcal{F}_1 = \mathcal{P}_1 \cup \mathcal{S}_1$, thus $\mathcal{F}_1^k = (\mathcal{P}_1 \cup \mathcal{S}_1)^k$.

– Now, we prove we have $\mathcal{F}_1^k \subseteq \mathcal{F}_k$. Given a pair of words $(w, w') \in \mathcal{F}_1^k$, there is some sequence of words $(w_i)_{0 \leq i \leq k}$ such that $w = w_0$, $w' = w_k$, and $d_F(w_i, w_{i+1}) \leq 1$, for each $i \in [0, k-1]$. We have $d_F(w, w') \leq \sum_{0 \leq i \leq k-1} d_F(w_i, w_{i+1}) \leq k$, thus $(w, w') \in \mathcal{F}_k$.

– For proving that the inclusion $\mathcal{F}_k \subseteq \mathcal{F}_1^k$ holds, we argue by induction over $k \geq 1$. The property trivially holds for $k = 1$. Assume that we have $\mathcal{F}_k \subseteq \mathcal{F}_1^k$, for some $k \geq 1$. Let $(w, w') \in \mathcal{F}_{k+1}$ and let $f \in F(w) \cap F(w')$ be a word with maximum length; set $w = ufv$, $w' = u'fv'$.

  (a) Firstly, assume that at least one of the integers $|w|-|f|$, $|w'|-|f|$ belongs to $[2, k+1]$ that is, without loss of generality $2 \leq |w| - |f| = |u| + |v| \leq k + 1$. With this condition, there are words $s \in S(u)$, $p \in P(v)$ such that $w \in A^h sfpA^{h'}$, with $sp \neq \varepsilon$ and $h + h' = 1$. On a first hand, it follows from $sfp \in F(w)$ that $d_F(w, sfp) = |w| - |sfp| = h + h' = 1$, thus $(w, sfp) \in \mathcal{F}_1$. On the other hand, f remains a word of maximum length in $F(sfp) \cap F(w')$, whence we have $d_F(sfp, w') = |sfp| + |w'| - 2|f|$. Since we have $|w| - |sfp| = 1$, we obtain $d_F(sfp, w') = (|w| - 1) + |w'| - 2|f| = d_F(w, w') - 1 \leq (k + 1) - 1$, thus $(sfp, w') \in \mathcal{F}_k$ that is, by induction, $(sfp, w') \in \mathcal{F}_1^k$. Since we have $(w, sfp) \in \mathcal{F}_1$, this implies $(w, w') \in \mathcal{F}_1^{k+1}$.

  (b) In the case where we have $|w| - |f| \leq 1$ and $|w'| - |f| \leq 1$ that is, $(w, f) \in \mathcal{F}_1$ and $(f, w') \in \mathcal{F}_k$, by induction we obtain $(f, w') \in \mathcal{F}_1^k$, therefore we have $(w, w') \in \mathcal{F}_1^{k+1}$.

In any case the condition $(w, w') \in \mathcal{F}_{k+1}$ implies $(w, w') \in \mathcal{F}_1^{k+1}$, hence we have $\mathcal{F}_{k+1} \subseteq \mathcal{F}_1^{k+1}$. As a consequence, for every $k \geq 1$ the inclusion $\mathcal{F}_k \subseteq \mathcal{F}_1^k$ holds: this completes the proof. □

As a direct consequence of Lemma 2, we obtain the following result:

Proposition 6. *Each of the following properties holds:*

  (i) The relation $\mathcal{F}_k$ is regular.
  (ii) Given a regular code X, it can be decided whether X satisfies Cond. (c4) wrt. $\mathcal{F}_k$.
  (iii) Given a finite code X, one can decide whether X satisfies any of Conds. (c1)–(c4) wrt. $\mathcal{F}_k$.

Proof. In view of Sect. 3, the relations $\mathcal{P}_1$ and $\mathcal{S}_1$ are regular, therefore Property (i) comes from Lemma 2. The proof of Property (ii) is done by merely substituting $\mathcal{F}_k$ to $\mathcal{P}_k$ in the proof of the preceding proposition 2. In the case where X is finite, the same holds for $\mathcal{F}_k(X)$ and $\underline{\mathcal{F}_k}(X)$, furthermore Property (iii) holds. □

For non-finite regular sets, the question of the decidability of Conds. (c1), (c2) remains open. Indeed, presently no finite automaton with behavior $\underline{\mathcal{F}_k} = \mathcal{F}_k \cap \overline{id_{A^*}}$ is known. Actually, the following property holds:

Proposition 7. *For every $k \geq 1$, $\mathcal{F}_k$ is a non-recognizable regular relation.*

Proof. According to Proposition 6, the relation $\mathcal{F}_k$ is regular. By contradiction, assume $\mathcal{F}_k$ recognizable. As indicated in the preliminaries, with this condition a finite set I exists such that $\mathcal{F}_k = \bigcup_{i \in I}(T_i \times U_i)$. For every $n \geq 0$, we have $(a^n b, a^n) \in \mathcal{F}_k$ therefore, since I is finite, there are $i \in I$ and $m, n \geq 1$, with $m - n \geq k$, such that $(a^n b, a^n), (a^m b, a^m) \in T_i \times U_i$. This implies $(a^m b, a^m) \in T_i \times U_i \subseteq \mathcal{F}_k$, thus $d_F(a^m b, a^m) \leq k$: a contradiction with $d_F(a^n b, a^m) = m - n + 1 \geq k + 1$. □

216   J. Néraud

Regarding Cond. (c3), the following result holds:

Proposition 8. *Every non-complete regular $\underline{\mathcal{F}_k}$-independent code can be embedded into some complete one.*

Proof. In the family of $\underline{\mathcal{F}_k}$-independent codes, we consider a non-complete regular set X. In view of Theorem 2, we will construct a convenient word $z_1 \in A^* \setminus F(X^*)$. Take a word $z_0 \notin F(X^*)$, with $|z_0| \geq k$; let a be its initial character, and b be a character different of a. Consider the word z that was constructed in the proof of Proposition 3, that is $z = z_0 a b^{|z_0|}$. Set $z_1 = a^{|z|} b z = a^{2|z_0|+1} b z_0 a b^{|z_0|}$: since by construction, z_1^R, the reversal of z_1, is overlapping-free, the same holds for z_1. Set $U_1 = A^* \setminus (X^* \cup A^* z_1 A^*)$, $Y_1 = z_1 (U_1 z_1)^*$, and $Z_1 = X \cup Y_1$.

According to Theorem 2, the set Z_1 is a regular complete code. In order to prove that it is $\underline{\mathcal{F}_k}$-independent that is, $\underline{\mathcal{F}_k}(Z) \cap Z = \emptyset$, we argue by contradiction. Actually, exactly one of the following conditions occurs:

(a) The first condition sets that $x \in X$, $y \in Y_1$ exist such that $(x,y) \in \underline{\mathcal{F}_k}$. Let f be a word with maximum length in $F(x) \cap F(y)$: we have $y = ufv$, with $|u|+|v| \leq d_F(x,y) \leq k$. It follows from $y \in z_1 (U_1 z_1)^*$ and $|z_1| \geq k+1$ that we have $u \in P(z_1)$ and $v \in S(z_1)$. More precisely, according to the construction of y, we have $u \in P\left(a^{|z_0|}\right)$, $v \in S\left(b^{|z_0|}\right)$, and $u^{-1}y \in a^{|z_0|+1} A^*$, thus $f \in A^* z_0 A^*$. Since we have $f \in F(x)$, we obtain $z_0 \in F(x)$, a contradiction with $z_0 \notin F(X^*)$.

(b) With the second condition, a pair of different words $y, y' \in Y_1$ exist such that $(y,y') \in \mathcal{F}_k$. Let f be a word with maximum length in $F(y) \cap F(y')$. As indicated above, words u, u', v, v' exist such that $w = ufv$, $w' = u'fv'$, with $|u| + |u'| + |v| + |v'| = d_F(w,w') \leq k$.

  (b1) At first, assume that both the words v, v' are different of ε. According to the construction of Y_1, since we have $v, v' \in S(Y_1)$ with $|v|+|v'| \leq k$, a pair of positive integers i, j exist such that $v = b^i$, $v' = b^j$. This implies $fb \in F(y) \cap F(y')$, which contradicts the maximality of $|f|$.

  (b2) As a consequence, at least one of the conditions $v = \varepsilon$, or $v' = \varepsilon$ holds. Without loss of generality, we assume $v' = \varepsilon$, thus $f \in S(y')$. On a first hand, it follows from $z_1 \in F(y) \cap F(y')$ that $|f| \geq |z_1|$: since we have $f, z_1 \in S(y')$, this implies $f \in A^* z_1$. On the other hand, since we have $z_1 \in S(y)$, $fv \in S(y)$, and $|f| \geq |z_1|$, we obtain $z_1 \in S(fv)$. This implies $fv \in A^* z_1 v \cap A^* z_1$, thus $z_1 v \in A^* z_1$. It follows from $|v| \leq |z_1| - 1$ and z_1 being overlapping-free that $v = \varepsilon$. Similar arguments applied to the prefixes of y, y' lead to $u = u' = \varepsilon$: we obtain $y = y'$, a contradiction with $(y,y') \in \underline{\mathcal{F}_k}$.

In any case we obtain a contradiction, therefore Z_1 is $\underline{\mathcal{F}_k}$-independent. $\qquad \square$

As a consequence, by merely substituting $\mathcal{F}_k$ to $\mathcal{P}_k$ in the proof of the propositions 4 and 5, we obtain the following statement:

Proposition 9. *Given a regular code X, each of the following properties holds:*

(i) X is maximal as a $\underline{\mathcal{F}_k}$-independent code if, and only if, we have $\mu(X) = 1$.
(ii) One can decide whether X satisfies Cond. (c3) wrt. $\mathcal{F}_k$.

5 Error Detection in the Topologies Associated to (anti-)automorphisms

Given an (anti-)automorphism θ, we will examine Conds. (c1)–(c4) wrt. the quasi-metric d_θ that is, wrt. the relation $\tau_{d_\theta,1} = \widehat{\theta}$. Regarding error correction, the following noticeable property holds:

Proposition 10. *With respect to $\widehat{\theta}$, a regular code $X \subseteq A^*$ satisfies Cond. (c1) if, and only if, it satisfies Cond. (c2).*

Proof. – Firstly, assume that X is $\underline{\theta}$-independent, and let $x, y \in X$ such that $\tau_{d_\theta,1}(x) \cap \tau_{d_\theta,1}(y) = \underline{\theta}(x) \cap \underline{\theta}(y) \neq \emptyset$. It follows from $\underline{\theta}(x) \cap \underline{\theta}(y) = (\{\theta(x)\} \setminus \{x\}) \cap (\{\theta(y)\} \setminus \{y\})$ that $\theta(x) \neq x$, $\theta(y) \neq y$, and $\theta(x) = \theta(y)$: since θ is one-to-one, this implies $x = y$, therefore X satisfies Cond. (c2).
– Secondly, assume that Cond. (c1) does not hold wrt. $\widehat{\theta}$ that is, $X \cap (\widehat{\theta})(X) = X \cap \underline{\theta}(X) \neq \emptyset$. Necessarily, a pair of different words $x, y \in X$ exist such that $y = \theta(x)$. It follows from $\widehat{\theta}(x) = \{x\} \cup \{\theta(x)\} = \{x, y\}$ and $\widehat{\theta}(y) = \{y\} \cup \{\theta(y)\}$ that $\widehat{\theta}(x) \cap \widehat{\theta}(y) \neq \emptyset$, hence Cond. (c2) cannot hold. $\qquad\square$

Example 3. (1) Let $A = \{a, b\}$ and θ be the automorphism defined by $\theta(a) = b$, and $\theta(b) = a$. The regular prefix code $X = \{a^n b : n \geq 0\}$ satisfies Conds. (c1). Indeed, we have $\underline{\theta}(X) = \{b^n a : n \geq 0\}$, thus $\underline{\theta}(X) \cap X = \emptyset$. According to Proposition 10, it also satisfies Cond. (c2). We have $\mu(X) = \frac{1}{2} \sum_{n \geq 0} \left(\frac{1}{2}\right)^n = 1$, whence X is a maximal prefix code. Consequently X is maximal in the family of $\underline{\theta}$-independent codes (Cond. (c3)). Finally, we have $X \subsetneq \widehat{\theta}(X) = \{a^n b : n \geq 0\} \cup \{b^n a : n \geq 0\}$, hence $\widehat{\theta}(X)$ cannot be a code (we verify that $a, b, ab \in \widehat{\theta}(X)$).
(2) Over the alphabet $A = \{a, b\}$, take for θ the anti-automorphism defined by $\theta(a) = b$, and $\theta(b) = a$. The regular prefix code $X = \{a^n b : n \geq 0\}$ satisfies Conds. (c1), (c2): indeed, we have $\underline{\theta}(X) = \{ab^n : n \neq 1\}$, thus $\underline{\theta}(X) \cap X = \emptyset$. As indicated above, X is a maximal prefix code, thus it satisfies Cond.(c3). At last, it follows from $X \subsetneq \widehat{\theta}(X)$ that X cannot satisfies Cond. (c4).
(3) With the condition above, consider $Y = X \setminus \{b, ab\} = \{a^n b : n \geq 2\}$. By construction, we have $\underline{\theta}(Y) \cap Y = \emptyset$, hence Y satisfies Conds. (c1), (c2). However, by construction, Y cannot satisfy Cond. (c3). Finally, we have $\widehat{\theta}(Y) = \bigcup_{n \geq 2}\{a^n b, ab^n\}$, which remains a prefix code that is, Y satisfies Cond. (c4).

(4) Over the alphabet $\{A, C, G, T\}$, let θ denotes the Watson-Crick anti-automorphism (see e.g. [11,13]), which is defined by $\theta(A) = T$, $\theta(T) = A$, $\theta(C) = G$, and $\theta(G) = C$.
Consider the prefix code $Z = \{A, C, GA, G^2, GT, GCA, GC^2, GCG, GCT\}$. We have $\theta(Z) = \{T, G, TC, C^2, AC, TGC, G^2C, CGC, AGC\}$, hence Z satisfies Conds. (c1), (c2). By making use of the uniform distribution, we have $\mu(Z) = 1/2 + 3/16 + 1/16 = 3/4$, hence Z cannot satisfy Cond. (c3). At last, it follows from $G, G^2 \in \widehat{\theta}(Z) = \theta(Z) \cup Z$ that Cond. (c4) is not satisfied.

(5) Notice that, in each of the preceding examples, since the (anti-)automorphism θ satisfies $\theta^2 = id_{A^*}$, the quasimetric d_θ is actually a metric. Of course, (anti-)automorphisms exist in such a way that d_θ is only a quasimetric. For instance over $A = \{a, b, c\}$, taking for θ the automorphism generated by the cycle (a, b, c), we obtain $d_\theta(a, b) = 1$ and $d_\theta(b, a) = 2$ (we have $b = \theta(a)$ and $a \neq \theta(b)$).

(a, a), with $a \notin B$
$(a, \theta(a))$, with $a \in B$

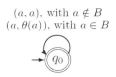

Fig. 5. An automaton with behavior $\tau_{d_\theta, 1} = \hat{\theta}$, in the case where θ is an automorphism: a represents every character in A and we set $B = \{a \in A : \theta(a) \neq a\}$.

Regarding regulary and recognizability of relations we state:

Proposition 11. *With the preceding notation, the following result holds:*

(i) *If θ is an automorphism, then $\tau_{d_\theta, 1} = \hat{\theta}$ and $\tau_{d_\theta, 1} = \underline{\theta}$ are non-recognizable regular relations.*

(ii) *If θ is an anti-automorphism, then it cannot be a regular relation.*

(iii) *Given an (anti-)automorphism θ, if X is a regular subset of A^*, then the same holds for $\hat{\theta}(X)$.*

Proof. Let θ be an (anti-)automorphism onto A^*.

(i) In the case where θ is an automorphism of A^*, it is regular: indeed, trivially θ is the behavior of a one-state automaton with transitions $(0, (a, \theta(a)), 0)$, for all $a \in A$. Starting with this automaton we obtain a finite automaton with behaviour $\hat{\theta}$ by merely adding the transitions $(0, (a, a), 0)$, for all $a \in A$ (see Fig. 5): consequently the relation $\hat{\theta}$ is regular.
By contradiction, assume θ recognizable. As in the proof of Proposition 7, a finite set I exists such that $\hat{\theta} = \bigcup_{i \in I}(T_i \times U_i)$. Since I is finite, there are $i \in I$, $a \in A$, and $m, n \geq 1$ such that $\left(a^n, \left(\hat{\theta}(a)\right)^n\right), \left(a^m, \left(\hat{\theta}(a)\right)^m\right) \in T_i \times U_i$,

with $m \neq n$. This implies $\left(a^n, \left(\hat{\theta}(a) \right)^m \right) \in T_i \times U_i$ that is, $\hat{\theta}(a^n) = (\hat{\theta}(a))^m$.
If we have $\hat{\theta}(a) = a$, we obtain $a^n = a^m$. Otherwise, we have $\hat{\theta}(a) = \theta(a)$, thus $\theta(a^n) = (\theta(a))^m$. In each case, this contradicts $m \neq n$.

Finally, the binary relation $\overline{(\hat{\theta})} = \underline{\theta}$ is the set of all the pairs (uas, ubs') that satisfy both the three following conditions:

(1) $u \in A^*$;
(2) $b = \theta(a)$, with $a, b \in A$ and $a \neq b$;
(3) $s' = \theta(s)$, with $s \in A^*$.

Consequently, $\underline{\theta}$ is the behaviour of the automaton in Fig. 6, hence it is a regular relation. In addition, by merely substituting the relation $\underline{\theta}$ to $\hat{\theta}$ in the argument above, it can be easily prove that $\underline{\theta}$ recognizable implies $\theta(a^n) = (\theta(a))^m$, for some $a \in A$ and $m \neq n$: a contradiction with θ being a free monoid automorphism.

(ii) For every anti-automorphism θ, the relation $\hat{\theta}$ is the result of the composition of the so-called *transposition*, namely $t : w \to w^R$, by some automorphism of A^*, say h. As shown in [24, Example IV.1.10], the transposition is not a regular relation. Actually, the same argument can be applied for proving that the resulting relation $\hat{\theta}$ is non-regular.

(iii) Let X be a regular subset of A^*. If θ is an automorphism, the relation $\hat{\theta} = \theta \cup id_{A^*}$ is a regular relation, hence $\hat{\theta}(X)$ is regular. In the case where θ is an anti-automorphism, with the preceding notation, although the transposition is not a regular relation, the set $t(X)$ itself is regular (see e.g. [24, Proposition I.1.1]). Consequently $\hat{\theta}(X) = h(t(X))$ is also regular. □

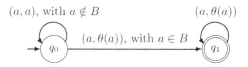

Fig. 6. With the notation in Example 6, an automaton with behavior $\underline{\theta}$, in the case where θ is an automorphism.

As a consequence of Proposition 11, we obtain the following result:

Proposition 12. *Given an (anti-)automorphism θ onto A^* and a regular code $X \subseteq A^*$, each of the two following properties holds:*

(i) In any case, X satisfies Cond. (c1), (c2).
(ii) It can be decided whether X satisfies Cond. (c4).

Proof. (i) Firstly, since θ is an (anti-)automorphism of A^*, it is a one-to-one mapping. For every $x \in X$, we have $\theta(x) \neq \emptyset$, moreover we have $\theta^{-1}(\theta(x)) \cap X = \{x\}$, consequently, in any case X satisfies Cond (c2). Secondly, according to Proposition 10, necessarily X also satisfies Cond. (c1).

(ii) According to Proposition 11, in any case the set $\theta(X)$ is regular: by applying Sardinas and Patterson algorithm, one can decide whether X satisfies Cond. (c4). $\square$

It remains to study the behavior of regular codes with regard to Cond. (c3):

Proposition 13. *Every non-complete regular $\underline{\theta}$-independent code can be embedded into some complete one.*

Proof. According to Theorem 2, the result holds if θ is an automorphism: indeed the action of such a transformation merely consists of rewriting words by applying some permutation of A.

Now, we assume that θ is an anti-automorphism. Classically, some positive integer n, the *order* of the permutation θ, exists such $\theta^n = id_{A^*}$. As in the propositions 3 and 8, in view of Theorem 2, we construct a convenient word in $A^* \setminus F(X^*)$.

Let $z_0 \notin F(X^*)$, a be its initial character, and b be a character different of a. Without loss of generality, we assume $|z_0| \geq 2$ and $z_0 \notin aa^*$, for every $a \in A$ (otherwise, substitute $z_0 b$ to z_0). By definition, for every integer i, we have $|\theta^i(z_0)| = |\theta(z_0)|$, therefore it follows from $z_0\theta(z_0) \cdots \theta^{n-1}(z_0) \in A^* \setminus F(X^*)$ that $z_2 = z_0\theta(z_0) \cdots \theta^{n-1}(z_0)ab^{n|z_0|}$ is an overlapping-free word in $A^* \setminus F(X^*)$. Set $U_2 = A^* \setminus (X^* \cup A^* z_2 A^*)$, $Y_2 = (z_2 U_2)^* z_2$, and $Z_2 = X \cup Y_2$.

According to Theorem 2, the set $Z_2 = X \cup Y_2$ is a complete regular code. In order to prove that it is $\underline{\theta}$-independent, we argue by contradiction. Actually, assuming that $\underline{\theta}(Z) \cap Z \neq \emptyset$, exactly one of the three following cases occurs:

(a) The first condition sets that $x \in X$ exists such that $\underline{\theta}(x) \in Y_2$. By construction, the sets X and Y_2 are disjoint (we have $z_2 \notin F(X^*)$), therefore we have $\theta(x) \neq x$: this implies $\underline{\theta}(x) = \theta(x)$. According to the definition of Y_2, we have $z_2 \in F(\theta(x))$, thus $\theta(z_0) \in F(\theta(x))$. It follows from $x = \theta^{n-1}(\theta(x))$, that the word $z_0 = \theta^{n-1}(\theta(z_0))$ is a factor of x, a contradiction with $z_0 \notin F(X^*)$.
(b) With the second condition some pair of words $x \in X$, $y \in Y_2$ exist such that $x = \theta(y)$. It follows from $\theta^{n-1}(z_0) \in F(z_2) \subseteq F(y)$ that $z_0 = \theta(\theta^{n-1}(z_0)) \in F(\theta(y)) = F(x)$: once more this contradicts $z_0 \notin F(X^*)$.
(c) The third condition sets that there are different words $y, y' \in Y_2$ such that $y' = \theta(y)$. Since θ is an anti-automorphism, $ab^{n|z_0|} \in S(y)$ implies $\theta(b^{n|z_0|}) \in P(y')$. Since we have $z_0 \in P(Y_2)$, this implies $(\theta(b))^{|z_0|} = z_0$. But we have $|z_0| \geq 2$, and $\theta(b) \in A$: this is incompatible with the construction of z_0.

In any case we obtain a contradiction, therefore Z_1 is $\underline{\theta}$-independent. $\square$

As a consequence, given a regular code $X \subseteq A^*$, X is maximal in the family of $\underline{\theta}$-independent codes of A^* if, and only if, the equation $\mu(X) = 1$ holds. In other words, one can decide whether X satisfies Cond. (c3).

Finally, the following statement synthesizes the results of the whole study we relate in our paper:

Theorem 3. *With the preceding notation, given a regular code X one can decide whether it satisfies each of the following conditions:*

(i) Conds. (c1)–(c4) wrt. $\mathcal{P}_k$, $\mathcal{S}_k$, or $\widehat{\theta}$, for any (anti-)automorphism θ of A^*.
(ii) Conds. (c3), (c4) wrt. $\mathcal{F}_k$.
(iii) In the case where X is finite, Conds. (c1), (c2) wrt. $\mathcal{F}_k$.

6 Concluding Remark

With each pair of words (w, w'), the so-called *subsequence metric* associates the integer $\delta(w, w') = |w| + |w'| - 2s(w, w')$, where $s(w, w')$ stands for a maximum length common subsequence of w and w'. Equivalently, $\delta(w, w')$ is the minimum number of one character insertions and deletions that have to be applied for computing w' by starting from w. We observe that, wrt. relation $\tau_{\delta,k}$, results very similar to the ones of the Propositions 6 and 9 hold [20]. Moreover, in that framework, we still do not know whether Conds. (c1), (c2) can be decided, given a regular code X.

It is noticeable that, although the inclusion $\mathcal{F}_k \subseteq \tau_{\delta,k}$ holds (indeed factors are very special subsequences of words), we do not know any more whether the relation $\mathcal{F}_k$ is regular or not. In the case where the answer is no, can after all Conds. (c1), (c2) be decidable?

From another point of view, wrt. each of the relations we mentionned, presenting families of codes satisfying all the best Conds. (c1)–(c2) would be desirable.

References

1. Berstel, J., Perrin, D., Reutenauer, C.: Codes and Automata. Cambridge University Press, Cambridge (2010). https://doi.org/10.1017/CBO9781139195768
2. Bruyère, V., Wang, L., Zhang, L.: On completion of codes with finite deciphering delay. Eur. J. Comb. **11**, 513–521 (1990). https://doi.org/10.1016/S0195-6698(13)80036-4
3. Choffrut, C., Pighizzini, G.: Distances between languages and reflexivity of relations. Theor. Comp. Sci. **286**, 117–138 (2002). https://doi.org/10.1016/S0304-3975(01)00238-9
4. Cohn, P.M.: Universal algebra. In: Mathematics and Its Applications, vol. 6. Springer, Dordrecht (1981). https://doi.org/10.1007/978-94-009-8399-1
5. Ehrenfeucht, A., Rozenberg, S.: Each regular code is included in a regular maximal one. RAIRO - Theor. Inf. Appl. **20**, 89–96 (1986). https://doi.org/10.1051/ita/1986200100891
6. Eilenberg, S.: Automata, Languages and Machines. Academic Press (1974). eBook ISBN: 9780080873749
7. Elgot, C., Meizei, J.: On relations defined by generalized finite automata. IBM J. Res. Develop. **9**, 47–68 (1965). https://doi.org/10.1147/rd.91.0047
8. Hamming, R.: Error detecting and error correcting codes. Bell Tech. J. **29**, 147–160 (1950). https://doi.org/10.1002/j.1538-7305.1950.tb00463.x

9. Jürgensen, H., Katsura, M., Konstantinidis, S.: Maximal solid codes. J. Autom. Lang. Combinatorics **6**, 25–50 (2001). https://doi.org/10.25596/jalc-2001-025

10. Jürgensen, H., Konstantinidis, S.: Codes. In: Rozenberg, G., Salomaa, A. (eds.) Handbook of Formal Languages, vol. 1, no. 8, pp. 511–607. Springer, Berlin, Heidelberg (1997). https://doi.org/10.1007/978-3-642-59136-5

11. Kari, L., Konstantinidis, S., Kopecki, S.: On the maximality of languages with combined types of code properties. Theor. Comp Sci. **550**, 79–89 (2014). https://doi.org/10.1016/j.tcs.2014.07.015

12. Konstantinidis, S., Mastnak, M.: Embedding rationally independent languages into maximal ones. J. Aut. Lang. and Comb. **21**, 311–338 (2016). https://doi.org/10.25596/jalc-2016-311

13. Kruskal, J.B.: An overview of sequence comparison: time warps, string edits, and macromolecules. SIAM Rev. **25**, 201–237 (1983). https://www.jstor.org/stable/2030214

14. Lam, N.: Finite maximal infix codes. Semigroup Forum **61**, 346–356 (2000). https://doi.org/10.1007/PL00006033

15. Lam, N.: Finite maximal solid codes. Theot. Comput. Sci. **262**, 333–347 (2001). https://doi.org/10.1016/S0304-3975(00)00277-2

16. Lam, N.: Completing comma-free codes. Theot. Comput. Sci. **301**, 399–415 (2003). https://doi.org/10.1016/S0304-3975(02)00595-9

17. Levenshtein, V.I.: Binary codes capable of correcting deletions, insertions and reversals. In: Cybernetics And Control Theory, vol. 10, pp. 707–710 (1966). Doklady. Academii. Nauk. SSSR, vol. 163, no. 4, pp. 845–848 (1965)

18. Néraud, J.: On the completion of codes in submonoids with finite rank. Fundam. Informaticae **74**, 549–562 (2006). ISSN 0169-2968

19. Néraud, J.: Completing circular codes in regular submonoids. Theoret. Comput. Sci. **391**, 90–98 (2008). https://doi.org/10.1016/j.tcs.2007.10.033

20. Néraud, J.: Variable-length codes independent or closed with respect to edit relations. Inf. Comput. (2021). (in press). arXiv:2104.14185

21. Néraud, J., Selmi, C.: Embedding a θ-invariant code into a complete one. Theoret. Comput. Sci. **806**, 28–41 (2020). https://doi.org/10.1016/j.tcs.2018.08.022

22. Ng, T.: Prefix distance between regular languages. In: Han, Y.-S., Salomaa, K. (eds.) CIAA 2016. LNCS, vol. 9705, pp. 224–235. Springer, Cham (2016). https://doi.org/10.1007/978-3-319-40946-7_19

23. Restivo, A.: On codes having no finite completion. Discrete Math. **17**, 309–316 (1977). https://doi.org/10.1016/0012-365X(77)90164-9

24. Sakarovitch, J.: Elements of Automata Theory. Cambridge University Press, Cambridge (2009). https://doi.org/10.1017/CBO9781139195218

25. Sardinas, A., Patterson, G.W.: A necessary and sufficient condition for the unique decomposition of coded messages. IRE Internat Con. Rec. **8**, 104–108 (1953)

26. Van, D.L., Van Hung, K., Huy, P.T.: Codes and length-increasing transitive binary relations. In: Van Hung, D., Wirsing, M. (eds.) Theoretical Aspects of Computing - ICTAC 2005. LNCS, vol. 3722, pp. 29–48. Springer, Berlin, Heidelberg, New York (2005). https://doi.org/10.1007/11560647_2

27. Wilson, W.A.: On quasi-metric spaces. Am. J. Math. **53**, 675–684 (1931). https://doi.org/10.2307/2371174

28. Zhang, L., Shen, Z.: Completion of recognizable bifix codes. Theoret. Comput. Sci. **145**, 345–355 (1995). https://doi.org/10.1016/0304-3975(94)00300-8

Author Index

Printed in the United States
by Baker & Taylor Publisher Services